Regulation of Gene Expression

Regulation of Gene Expression

Molecular Mechanisms

By

Gary H. Perdew, PhD

John P. Vanden Heuvel, PhD

Jeffrey M. Peters, PhD

Center for Molecular Toxicology and Carcinogenesis
Department of Veterinary and Biomedical Sciences
The Pennsylvania State University
University Park, PA

999 Riverview Drive, Suite 208
Totowa, New Jersey 07512

www.humanapress.com

This publication is printed on acid-free paper. ∞
ANSI Z39.48-1984 (American Standards Institute) Permanence of Paper for Printed Library Materials.

Cover design by Patricia F. Cleary

For additional copies, pricing for bulk purchases, and/or information about other Humana titles, contact Humana at the above address or at any of the following numbers: Tel.: 973-256-1699; Fax: 973-256-8341; E-mail: orders@humanapr.com; or visit our Website: www.humanapress.com

Printed in the United States of America. 10 9 8 7 6 5 4 3 2 1

eISBN 1-59745-228-9

Library of Congress Cataloging-in-Publication Data

Perdew, Gary H.
Regulation of gene expression : molecular mechanisms / by Gary H. Perdew, John P. Vanden Heuvel, Jeffrey M. Peters.
p. ; cm.
Includes bibliographical references and index.
ISBN 1-58829-265-7 (alk. paper)
1. Genetic regulation. I. Vanden Heuvel, John P. II. Peters, Jeffrey M. (Jeffrey Maurice) III. Title.
[DNLM: 1. Gene Expression Regulation--physiology. 2. Molecular Biology. QU 475 P433r 2006]
QH450.P45 2006
572.8'65--dc22

2006020090

Preface

The use of molecular biology and biochemistry to study the regulation of gene expression has become a major feature of research in the biological sciences. Many excellent books and reviews exist that examine the experimental methodology employed in specific areas of molecular biology and regulation of gene expression. However, we have noticed a lack of books, especially textbooks, that provide an overview of the rationale and general experimental approaches used to examine chemically or disease-mediated alterations in gene expression in mammalian systems. For example, it has been difficult to find appropriate texts that examine specific experimental goals, such as proving that an increased level of mRNA for a given gene is attributable to an increase in transcription rates. *Regulation of Gene Expression: Molecular Mechanisms* is intended to serve as either a textbook for graduate students or as a basic reference for laboratory personnel. Indeed, we are using this book to teach a graduate-level class at The Pennsylvania State University. For more details about this class, please visit http://moltox.cas.psu.edu and select "Courses." The goal for our work is to provide an overview of the various methods and approaches to characterize possible mechanisms of gene regulation. Further, we have attempted to provide a framework for students to develop an understanding of how to determine the various mechanisms that lead to altered activity of a specific protein within a cell. We expect the reader will have a good working knowledge of basic biochemistry and cell biology, although detailed understanding of molecular biology techniques is not required.

Each of the three parts of *Regulation of Gene Expression: Molecular Mechanisms* is self-contained. Thus the order of reading does not need to follow the order of presentation, although the parts have been arranged in the way that investigators often approach the study of gene regulation. We have thoughtfully selected key references only and included their details in the page margins for ready reference, as this work is intended as a textbook, not a review of the literature. Key points as well have been placed in the margins in order to emphasize important issues.

Part I, written by John P. Vanden Heuvel, presents the experimental approaches that can be utilized to study control of mRNA expression and the determination of target genes for a given transcription factor. Part II, written by Gary H. Perdew, examines the experimental approaches utilized to determine how proteins can regulate each other by mediating synthesis, degradation, protein–protein interactions, and posttranslational modification. Finally, Part III, written by Jeffrey M. Peters, explores how gene targeting techniques in mice can provide insight into protein function. The point of view is that of a molecular toxicologist, but we have kept in mind a wider range of graduate students and professionals in the biological sciences. As toxicologists, however, we are primarily concerned with mammalian systems and with determining how chemicals can modify gene expression. This has clearly influenced the biological systems utilized in the experimental approaches suggested throughout this text.

We thank those who contributed to the completion of this book, in particular, Marcia H. Perdew and Cheryl Brown for their excellent editorial assistance. Also, thanks go to the many students who have directly or indirectly contributed to the overall concept of this book, and who also read many segments of this book. We are indebted to Dr. C. Channa Reddy for his support and vision in providing an excellent research environment and establishing a molecular toxicology group at The Pennsylvania State University. Finally, we would like to acknowledge all of our mentors who have contributed to our careers and have inspired us to be the best scientists and mentors possible.

Gary H. Perdew, **PhD**
John P. Vanden Heuvel, **PhD**
Jeffrey M. Peters, **PhD**

Contents

Companion CD

Illustrations appearing in the book may be found on the Companion CD attached to the inside back cover. The image files are organized into folders by chapter number and are viewable in most Web browsers. The number following "f" at the end of the file name corresponds to the figure in the text. The CD is compatible with both Mac and PC operating systems.

PART I

Gene Expression Control at the mRNA Level

John P. Vanden Heuvel, PhD

Contents

1

Overview

1. Concepts

In the Part I, the basic approaches to understanding how a treatment or condition results in mRNA accumulation will be described. Although the focus and many of the examples will center on gene regulation by xenobiotics, the approaches are applicable to any treatment/condition that alters gene expression. The subsequent chapter will contain an overview of the molecular biology involved in each step of the process; however, details can be found elsewhere. For a good basic overview of transcriptional control of gene expression, the reader is directed to Molecular Biology of the Cell [1]. In addition, there have been several excellent review articles on eukaryotic transcriptional control [2–8]. Posttranscriptional gene regulation is, in general, more difficult to examine experimentally but is a very important determinant of cellular events. We will briefly discuss mRNA processing and stability, with emphasis on events altered by xenobiotics and the methods used to examine these events. Detailed laboratory procedures are available from several sources [9–13] and will not be emphasized here. Instead, we focus on the approaches to be used and the rationale behind these decisions. Translational and posttranslational regulation of gene control will not be examined in detail in this section, but they are discussed subsequently in Part II.

A xenobiotic or disease can affect the accumulation of mRNA for a particular gene in many ways, as shown in Fig. 1-1. The first level of control is at the level of chromatin packaging of DNA (chromatin control). It has been well established that

The term xenobiotic refers to any chemical that is foreign to that organism and includes drugs, pollutants, and so forth

1. Alberts, B. et al. In: Molecular Biology of the Cell. Garland Publishing, New York, NY, 1994, pp. 401–476.

2. Pugh, B.F. Gene 25 (2000) 1–14.

3. Dillon, N., Sabbattini, P. Bioessays 22 (2000) 657–665.

4. Huang, L. et al., Crit. Rev. Eukaryot. Gene Expr. 9 (1999) 175–182.

5. Kornberg, R.D. Trends Cell. Biol. 9 (1999) M46–9.

6. Chin, J.W. et al. Curr. Biol. 9, (1999) R929–932.

7. Roeder, R.G. Cold Spring Harb. Symp. Quant. Biol. 63 (1998) 201–218.

8. Franklin, G.C. Results Probl. Cell Differ. 25 (1999) 171–187.

9. Ausubel, F.M. et al. Current Protocols in Molecular Biology (John Wiley, New York, NY, 1994).

10. Sambrook, J. et al. (eds.), Molecular Cloning: A Laboratory Manual. Cold Spring Harbor Press, Cold Spring Harbor, NY, 1989.

11. Davis, L. G. et al. Basic Methods in Molecular Biology. Appleton & Lange, Norwalk, CT, 2nd ed. 1994.

12. Kaufman, P.B. Handbook of Molecular and Cellular Methods in Biology and Medicine. CRC Press, Boca Raton, FL. 1995.

13. Vanden Heuvel, J.P. et al. In: PCR Protocols in Molecular Toxicology (J.P. Vanden Heuvel, ed). CRC Press, Boca Raton, FL. 1998.

Gene Expression Control at the mRNA Level by J. P. Vanden Heuvel
From: *Regulation of Gene Expression*
By: G. H. Perdew et al. © Humana Press Inc., Totowa, NJ

Messenger RNA accummulation may be caused by effects at the level of chromatin, transcription, or RNA processing, transport, or stability

Receptors are macromolecules that contain functions for recognition (i.e., binding to a xenobiotic) and transduction (ability to regulate gene expression).

remodeling of DNA structure is required for gene transcription. In the fully packaged state ("closed"), the DNA is inaccessible to the transcriptional machinery, whereas once remodeled ("open"), transcription may proceed. Second, and perhaps most importantly from a xenobiotic standpoint, is controlling how and when a given gene is transcribed (transcriptional control). This is the predominant, but not the sole, manner in which ligands for nuclear receptors (NRs) and other soluble receptors affect gene expression. Many important signal transduction molecules including kinases and phosphatases will ultimately result in regulation of transcription factor activity and control mRNA levels in this manner. Also, certain oncogenes (including c-myc, fos and jun) encode for proteins that are transcriptionally active. Third, a chemical may control how the primary RNA transcript is spliced or otherwise processed (RNA processing control). Alternatively spliced transcripts may result in different protein products or may result in mRNA with different rates of turnover or translation. Although not studied in great detail, there are several examples of chemically-induced or disease-specific splice variants. Fourth, a chemical may select which completed mRNAs in the nucleus are exported to the cytoplasm (RNA transport control). To date, there are no examples of this means of regulation by xenobiotics, although the potential exists. Fifth, chemicals may affect gene expression by selecting which mRNAs in the cytoplasm are translated by ribosomes (translational control). Similar to RNA processing control, there are few concrete examples of chemicals specifically regulating the expression of a gene via this pathway. However, many chemicals affect the expression of ribosomal proteins that may in turn affect translation efficiency. It is also of note that the control of the rate of translation is coupled to RNA processing (i.e., capping) and transport. Sixth, certain mRNA molecules may be stabilized or destabilized in the cytoplasm (mRNA degradation control). This may be the most under-

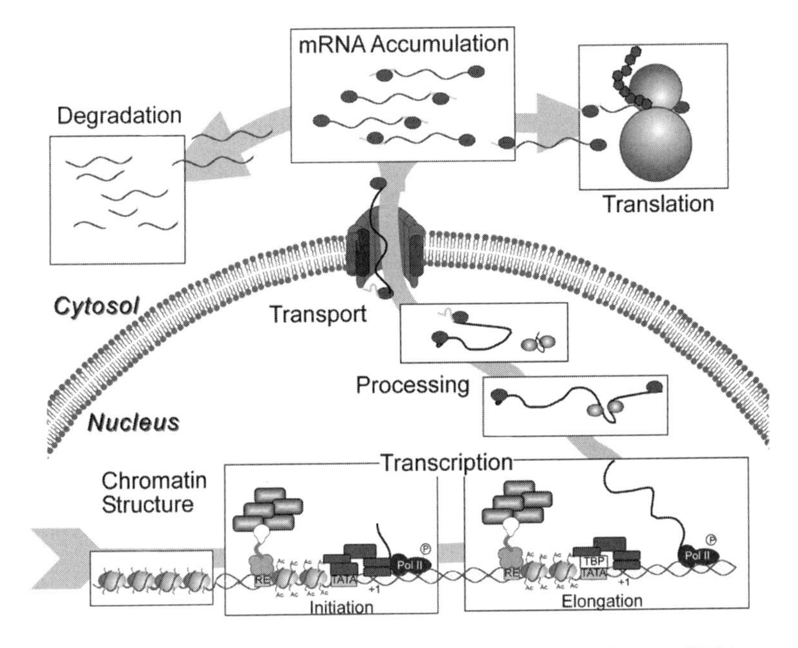

Fig. 1-1. Control of gene expression. In the Part I, the focus is on mRNA accumulation and hence transcription. mRNA processing and stability will be described. (Adapted from ref. [1]).

appreciated area in chemically-induced and disease-mediated alteration in mRNA levels. There are many examples of soluble receptor ligands that affect the stability of mRNA thereby resulting in alteration in corresponding protein levels.

Certainly, affecting protein activity is an important mechanism by which chemicals result in their biological and toxicological effects and how diseases are manifested. However, often it is the mRNA levels that are being examined and it is assumed (often falsely) that altered protein levels and activity will result. This is partially attributable to the fact that the examination of mRNA levels is at a much higher level of sophistication than are equivalent protein measurements. For example, microarray techniques make the concurrent examination of thousands of mRNA transcript levels feasible. Other high-throughput quantitation, such as real-time and competitive reverse transcriptase-polymerase chain reaction (RT-PCR), allow for detailed examination of when and how a transcript is affected by a xenobiotic. Equivalent methods for protein and enzyme activity are being developed but are not readily available at this time.

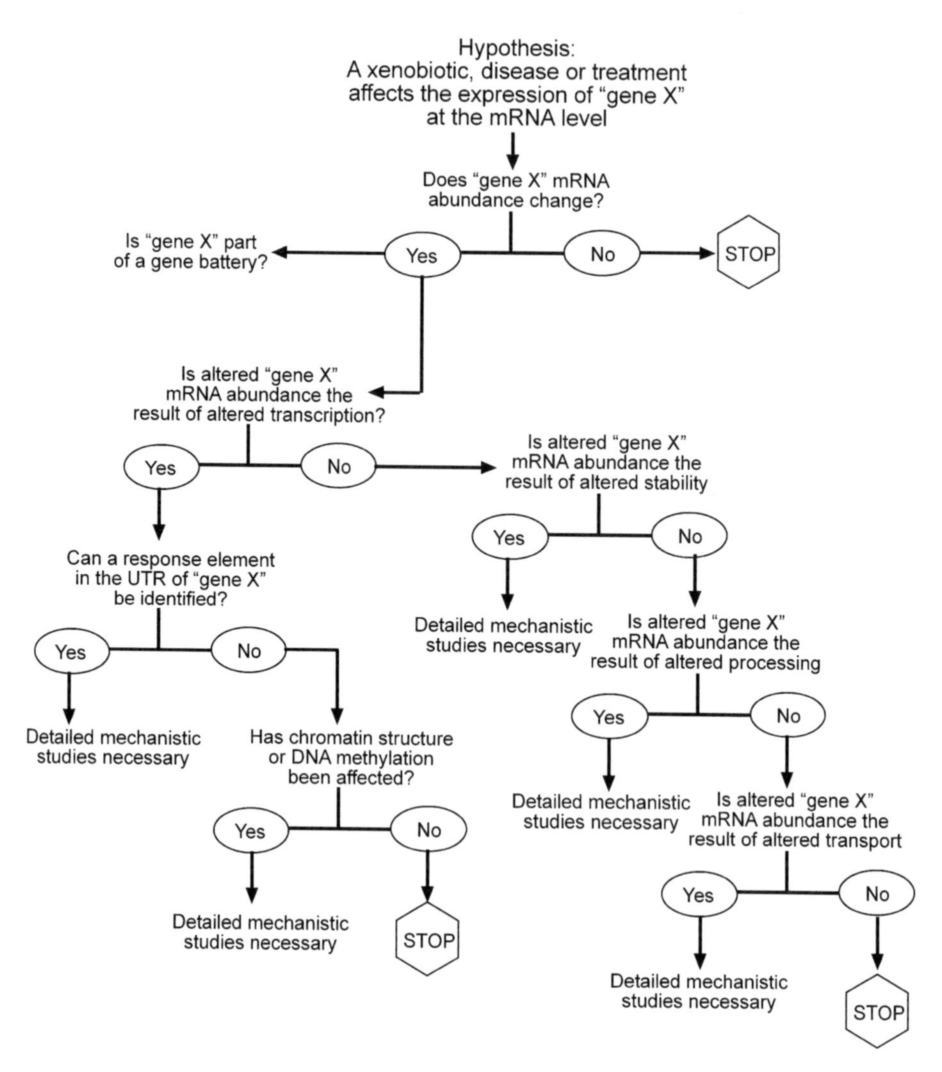

Fig. 1-2. Decision tree to determine how a xenobiotic treatment or disease affects gene expression at the mRNA level. *See* text for details.

Although the linear scheme depicted in Fig. 1-1 conveys the mechanism of mRNA accumulation, it is rarely utilized as an outline to determine the mechanism by which a gene is being regulated. More often, the scheme of events is more of a decision tree as shown in Fig. 1-2. Briefly, the hypothesis to be tested is generally "Does the particular treatment or condition affect mRNA accumulation?" If accumulation occurs, the goals of the subsequent experiments are to determine the mechanism of this regulation. The approaches to answering each of the hypotheses listed in this figure will be discussed in detail in the subsequent chapters.

2. Example

The decision tree (Fig. 1-2) depicts a situation in which a xenobiotic results in altered gene expression via one mechanism only. By answering a series of yes/no questions (utilizing the approaches in the subsequent chapters), the mechanism by which a gene is being regulated can be found. However, this is rarely, if ever, the case and multiple mechanisms are likely for any particular gene and xenobiotic. A good example of this phenomena is altered expression of the proto-oncogene c-myc mRNA via the tumor promoter tetradecanoyl phorbol acetate (TPA).

The c-myc proto-oncogene is an important regulator of cellular differentiation, proliferation, growth, and apoptosis. Because of its central role in determining cell fate, the concentration of c-myc is under extremely tight and integrated control at both the mRNA and the protein level. For c-myc transcript production, control is exerted at the levels of transcription initiation and elongation, RNA processing, RNA stability, and translation. TPA and other phorbol esters are activators of protein kinase C (PKC) and regulate gene expression through AP-1 as well as other transcription factors. The effects of TPA on cell fate depend on the type of cells being examined and can include increased (hepatocytes) or decreased (hematopoietic cells) proliferation. Although not an exhaustive analysis of all the studies performed on TPA regulation of c-myc mRNA, Table 1-1 illustrates the complexity that is inherent in studies performed on mRNA levels for a growth regulatory gene.

Table 1-1
Multifactorial Regulation of c-myc mRNA by Phorbol Esters

Question	Answer	Comment	References
Does c-myc mRNA abundance change?	Yes (either up or down)	Induction of c-myc mRNA by phorbol esters can occur through a protein kinase C (PKC)-dependent pathway in human astrocytoma cells.	Blackshear, P.J., et al. J. Biol. Chem. 262 (1987) 7774–7781.
		In proliferating JURKAT cells, high levels of c-myc mRNA were found, which diminished rapidly following TPA-induced cessation of growth.	Makover D., et al. Oncogene 6 (1991) 455–460.
Is c-myc part of a gene battery?	Yes	Subtractive hybridization identified 16 sequences, including c-myc, that are down regulated by TPA-induced monocytic differentiation of HL60 cells.	Herblot, S., et al. FEBS Lett. 414 (1997) 146–152.
Is altered c-myc mRNA the result of altered transcription?	Yes	PMA decreases c-myc mRNA by blocking transcription elongation at sites near the first exon/ intron border.	Chen, L., et al. J. Biol. Chem. 275 (2000) 32,227–32,233.
		In THP-1 cells PMA altered c-myc expression dependent on the AP-1 enhancer activity.	Matikainen, S., Hurme, M. Int. J. Cancer 57 (1994) 98–103.
		PMA stimulated transcriptional activity dependent on the phorbol 12-myristate 13-acetate-responsive element (TRE) of c-myc in MCF-7 cells.	Wosikowski, K., et al. Biochem. Biophys. Res. Commun. 188 (1992) 1067—1076.
Is altered mRNA abundance the result of altered stability?	Yes/No	Treatment of WEHI 231 cells, derived from a murine B-cell lymphoma, with PMA caused dramatic increase in the rate of c-myc gene transcription, as well as from partial stabilization of the mRNA in the cytoplasm.	Levine, R. A., et al. Mol. Cell. Biol. 6 (1986) 4112-4116.

	Yes	In human leukemia cell lines c-myc mRNA was regulated primarily by posttranscriptional mechanisms with altered mRNA half-life after TPA treatment.	Meinhardt, G., Hass, R. Leuk. Res. 19 (1995) 699-705.
		TPA-induced destabilization of c-myc mRNA requires post-ribosomal and polysome-associated factors. In dividing cells c-myc mRNA decays via a sequential pathway involving removal of the poly(A) tract followed by degradation of the mRNA body, TPA activates a deadenylation-independent pathway.	Brewer, G. J. Biol. Chem. 275 (2000) 33,336-33,345.
	No	The treatment of transfected cells with phorbol ester (TPA) revealed that polyadenylated mRNAs from some genes were stabilized, while mRNAs bearing 3'-end structure of c-myc genes were not.	Sato, K. et al. Nucleic Acids Symp. Ser. (1989) 23–24.

2

Messenger RNA Accumulation

1. Concepts

The detection of mRNA levels of a particular gene is one of the cornerstones of molecular biology. There are many ways that mRNA can be detected, each with its strengths and weaknesses. Surprisingly, few investigators give much thought as to whether their methodologies and approaches are appropriate. Yet, the quantification and interpretation of results depends on understanding some key points. Three key factors contribute to the complexity and difficulty in examining differentially expressed genes. First, genes are not present at the same abundance and can vary from less than one up to thousands of copies per cell. This has implications for methods that must be used to accurately detect and quantify the expression of an mRNA. Second, the intensity of response varies greatly from gene to gene; that is, when comparing two treatments or conditions, an mRNA can be twofold or several orders of magnitude different between samples. Certain methods have a robust linear range and can handle both levels of response, while others are biased toward either the low or high responder. Last, there are many ways to alter gene expression, and some or all of the particular mechanisms may be at play. Approaches must be used that are capable of isolating, or at least accounting for, the competing possibilities so that hypotheses can be tested confidently. We will briefly discuss these parameters as they pertain to examining altered gene expression, and how these factors impinge on developing an optimal model system. Much of the following has been described in more detail in reference [14] as it pertained to cloning xenobiotically-induced genes.

1.1. mRNA Abundance

The mammalian genome of 3×10^9 base pairs (bp) has sufficient DNA to code for approx 300,000 genes, assuming a length of 10,000 bp per gene [15]. Obviously, not every gene is expressed by every cell; also, not every segment of DNA may be associated with a gene product. In fact, hybridization experiments in the mammalian cell have shown that approx 1 to 2% of the total sequences of nonrepetitive DNA are represented in mRNA [15]. Thus, if 70%

Gene Expression Control at the mRNA Level by J.P. Vanden Heuvel
From: *Regulation of Gene Expression*
By: G. H. Perdew et al. © Humana Press Inc., Totowa, NJ

14. Vanden Heuvel, J.P. In: Toxicant-Receptor Interactions, Denison, M.S., Helferich, W.G., eds. Taylor and Francis, Philadelphia, PA. 1998, pp. 217–235

15. Lewin, B. Genes IV, Oxford University Press, Oxford UK.1990

of the total genome is nonrepetitive, 10,000–15,000 genes are expressed at a given time.

The average number of molecules of each mRNA per cell is called its representation or abundance. Of the 10,000 to 15,000 genes being expressed, the mass of RNA being produced per gene is highly variable. Usually, only a few sequences are providing a large proportion of the total mass of mRNA. Hybridization and kinetic experiments between excess mRNA and cDNA in solution identifies several components of mRNA complexity. Most of the mass of RNA (50%) is accounted for by a component with few mRNA species. In fact, approx 65% of the total mRNA may be accounted for in as few as ten mRNA species. The remaining 35% of the total RNA represents other genes being expressed in that tissue. Of course, the genes present in each category may be present in very different amounts and represent a continuum of expression levels. For means of this discussion, we will divide the three major components [16] into abundant, moderate, and scarce, representing approx 100,000 copies, 5000 copies and <10 copies per cell, respectively.

16. Bishop, J. O. et al. Nature 270 (1974) 199–204.

There are several reasons for discussing the components of mRNA. First, in a differential screen procedure (i.e., subtractive hybridization, differential display, microarray) what is actually being compared is two populations of mRNA, and you are examining the genes that overlap or form the intersection between these groups. When comparing two extremely divergent populations, such as liver and oviduct, as much as 75% of the sequences are the same [15] amounting to 10,000 genes that are identical and approx 3000 genes that are specific to the oviduct or liver. This suggests that there may be a common set of genes, representing required functions, that are expressed in all cell types. These are often referred to as housekeeping or constitutive genes. Second, there are overlaps between all com-

15. Lewin, B. Genes IV, Oxford University Press, Oxford UK.1990

ponents of mRNA, regardless of the number of copies per cell. That is, differentially expressed genes may be abundant, moderate, or scarce. In fact, the scarce mRNA may overlap extensively from cell to cell, on the order of 90% for the liver-to-oviduct comparison. However, it worth noting that a small number of differentially expressed genes are required to denote a specialized function to that cell, and the level of expression does not always correlate with importance of the gene product.

As discussed in sections following, the key to developing an effective model for the study of differential gene expression may be to keep the differences in the abundant genes to a minimum. This is due to the fact that a small difference in expression of a housekeeping gene, say twofold, will result in a huge difference in the number of copies of that message from cell type to cell type (i.e., an increase of 10,000 copies per cell). Also, it is important to have a screening method that can detect differences in the scarce component. If the two populations to be compared have little difference in the abundant genes and you have optimized your screening technique to detect differences in the scarce population, the odds of cloning genes that are truly required for a specialized cellular function have increased dramatically.

1.2. Intensity of Response

A basic pharmacologic principle is that drugs and chemicals have different affinities for a receptor, and drug-receptor complexes will have different efficacies for producing a biological response, i.e., altering gene expression. A corollary of this principle states that not every gene being effected by the same drug-receptor complex will have identical dose-response curves. That is, when comparing two responsive genes, the affinity of the drug-receptor complex for the DNA response elements found in the two genes and the efficacy of the drug-receptor—DNA complex at effecting transcription could be quite different. In fact, similar DNA response elements may cause a repression or an induction of gene expression, depending on the context of the surrounding gene. Therefore, when comparing two populations of mRNAs (i.e., control versus treated), there may be dramatic differences in the levels of induction and repression regardless of the fact that all the genes are affected by the same drug-receptor complex.

Needless-to-say, the extent of change is important in the detection of these differences but not the importance of that deviation. For technical reasons, it is often difficult to detect small changes in gene expression (less than twofold). However, a twofold change in a gene product may have dramatic effects on the affected cell, especially if it encodes a protein with a very specialized or nonredundant function. Also, the detection of a difference between two cell populations is easier if the majority of the differences are in scarce mRNAs. Once again, this is owing to technical aspects of analyzing gene expression whereby the change from 500 to 1000 copies per cell is a dramatic effect compared to a change from 1×10^5 to 2×10^5, an effect that may be virtually unnoticed.

1.3. Specificity of Response

The last factor we will discuss regarding the complexity of mRNA species is that regulation of gene expression is multifaceted. The analysis of differential gene expression is most often performed by comparing steady-state levels of mRNA; that is, the amount of mRNA that accumulates in the cell is a function of the rate of formation (transcription) and removal (processing, stability, degradation). If differences in protein products are being compared, add translation efficiency and posttranslational processing to the scenario. With all the possible causes for altered gene expression, the specificity of response must be questioned. Is the difference in mRNA or protein observed an important effect on expression or is it secondary to a parameter in your model system you have not controlled or accounted for?

In the best-case situation, the key mechanism of gene regulation that results in the end point of interest should be known. At least, one should have criteria in mind for the type of response that is truly important. With most receptor systems, early transcriptional regulation may predominate as this key event. However, the other modes must be acknowledged, at least when novel genes are being characterized. By assuming that the key event is mRNA accumulation, the true initiating response such as protein phosphorylation or processing may be overlooked. Also, the extent and diversity of secondary events, i.e., those that require the initial changes in gene expression, may far exceed the primary events. The amplification of an initial signal (i.e., initial response [gene A] causes regulation of secondary response [gene B]) can confuse the interpretation of altered mRNA accumulation. Once again, one must have a clear understanding of whether a primary or secondary event is the key response and design your model accordingly.

2. Basic Methods and Approaches

For specific procedures on the analysis of gene expression the reader is directed to laboratory manuals such as Current Protocols in Molecular Biology [9] and Molecular Cloning [10]. Popular methods for examining mRNA accumulation are outlined in Fig. 2-1. As with any laboratory procedure, you must keep in mind the limitations of the procedure and your own technical expertise. In this section, the methods will be outlined broadly but with emphasis on the positive and negative aspects of each approach.

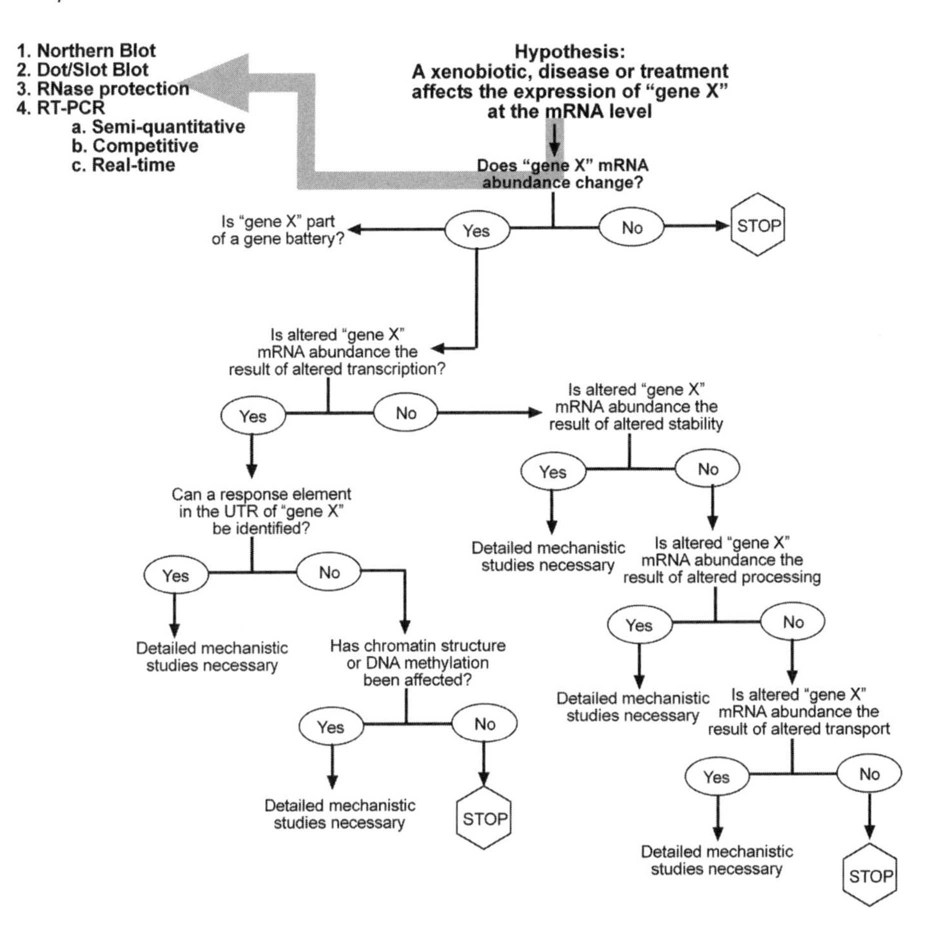

Fig. 2-1. Outline of approaches to examine mRNA accumulation.

Many are aware of the need for defining the model system when examining genome-wide mRNA accumulation, as when microarrays or serial analyses of gene expression (SAGE) are performed (Chapter 3). However, similar considerations should be made when using Northern blots or reverse transcriptase-polymerase chain reaction (RT-PCR) to examine a specific gene. Proper controls are necessary to minimize variability, and the studies must be designed in a way that eliminates extraneous factors (e.g., stress, contamination) that are not under investigation. In this section, approaches used to quantify the differences between two populations of mRNA for a particular gene will be examined. An assumption will be made that the model system is adequately designed to specifically test the hypothesis.

2.1. Northern Blots

An excellent resource for RNA analysis methods is www.ambion.com.

Despite the advent of more powerful techniques (RT-PCR, RNase protection), Northern blot analysis remains a standard method for detection and quantitation of mRNA levels. Northern blot analysis provides a relative comparison of message abundance between samples, is the preferred method for determining transcript size, and can be used for detecting alternatively spliced transcripts. The Northern blot procedure is straightforward and inexpensive, using common equipment and supplies present in most molecular biology laboratories.

The Northern blot measures mRNA levels. Some other "directional" assays include Southern (DNA), Western (protein) and Southwestern (protein–DNA) blots.

There are significant limitations associated with Northern blot analysis. First, despite the relative simplicity of the methodology, great pains must be taken to avoid RNase contamination of solutions and plasticware. If RNA samples are even slightly degraded, the quality of the data and the ability to quantitate expression are severely compromised. Second, Northern blotting is much less sensitive than nuclease protection assays and RT-PCR. Rare genes require highly specific probes and large amounts of RNA. Enriching the sample for polyadenylated mRNA and optimizing hybridization conditions can improve sensitivity to some extent. Third, Northern blots are not amenable to high-throughput analysis. For example, although it is possible to detect more than one gene per blot, often this requires stripping the nylon membrane and reprobing,which is time consuming and problematic, since harsh treatment is required to strip conventional probes from blots. Last, this procedure is most often used as a relative quantitation method, since the gene in question is being examined as a function of the expression of a housekeeping gene. Although internal standards can be synthesized and spiked into the sample (e.g., a synthetic gene with the hybridization site of the probe) in order to generate a standard curve and to obtain absolute quantification, this is rarely performed. Thus Northern blotting requires a large difference between samples (five- to tenfold) to be significant.

The general approach of Northern blot analysis is as follows (*see* Fig. 2-2). Extraction of high-quality intact RNA is a critical step in performing Northern analysis. This is generally performed by cell lysis with detergents or solvent, inhibition of ribonuclease, and ultimately separation of proteins and DNA from the RNA. This later event can be performed by liquid phase separation or via oligo(dT) chromatography. Once RNA samples are isolated, denaturing agarose gel electrophoresis is performed. Formaldehyde has traditionally been used as the denaturant, although the glyoxal system has several advantages over formaldehyde. All buffers and apparatus must be painstakingly treated with RNase inhibitors. Following separation by denaturing agarose gel electrophoresis, the RNA is transferred to a positively charged nylon membrane and then immobilized for subsequent hybridization. The transfer may be performed using a passive, slightly alkaline elution or via commercially available active transfer methods (electroblotter, pressure blotter). The membrane is crosslinked by ultraviolet light or by baking. Northern blots can be probed with radioactively or nonisotopically labeled RNA, DNA, or oligodeoxynucleotide probes. DNA probes may originate from plasmids or PCR. RNA probes can be produced by in vitro transcription reactions. Radioactivity may be incorporated during the PCR or in vitro transcription reactions or may be performed via end-labeling or random priming. Prehybridization, or blocking, is required prior to probe hybridization to prevent the probe from coating the membrane. Good blocking is necessary to minimize background problems. Although double-stranded DNA probes must be denatured prior to use, RNA probes and single-stranded DNA probes can be diluted and then added to the prehybridized blot. After hybridization, unhybridized probe is removed by washing in several changes of buffer. Low-stringency washes (e.g., with 2X SSC) remove the hybridization solution and unhybridized probe. High-stringency washes (e.g., with 0.1X SSC)

There are many commercially available kits for the extraction of total RNA (mRNA, tRNA, rRNA) or poly(A) RNA (mRNA). Most methods for total RNA utilize chloroform:phenol extraction, whereas mRNA can be enriched via oligo(dT) chromatography.

SSC buffers are commonly employed in nucleic acid hybridizations. 20X SSC consists of 3 M NaCl, 0.3 M Na3 citrate.

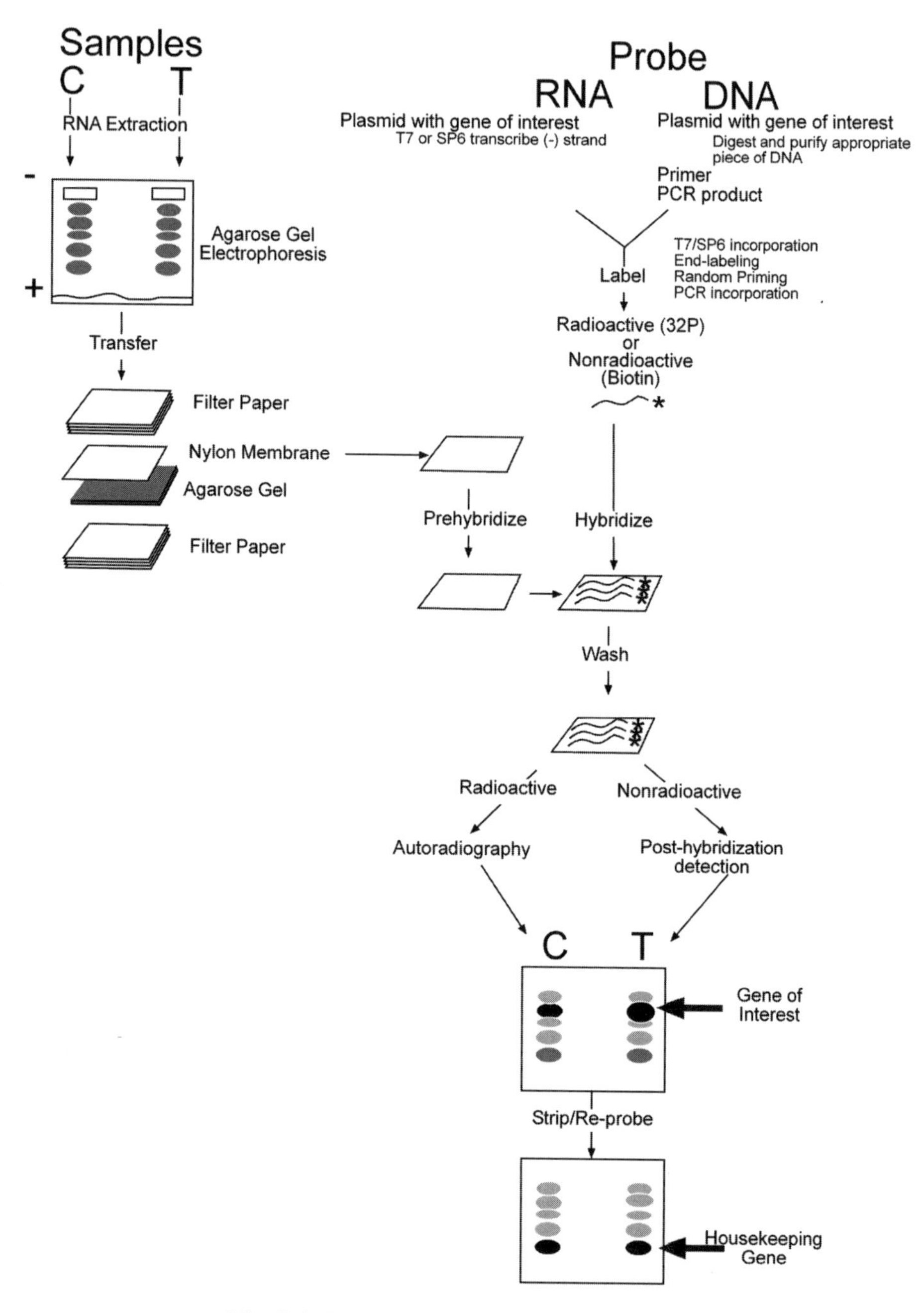

Fig. 2-2. Basic procedures in Northern blotting.

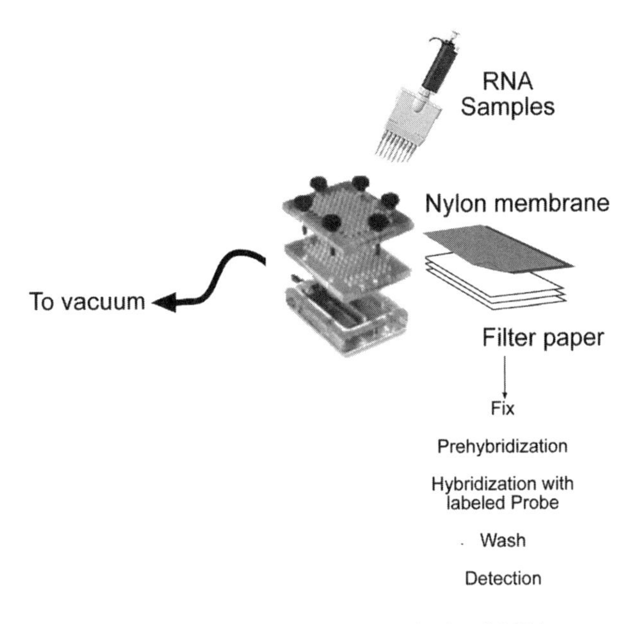

Fig. 2-3. Dot- or slot-bot analysis of RNA.

remove partially hybridized molecules. If a radiolabeled probe was used, the blot can be wrapped in plastic wrap to keep it from drying and then immediately exposed to film for autoradiography. If a nonisotopic probe was used, the blot must be treated with nonisotopic detection reagents prior to film exposure. Standard methods for removing probes from blots to allow subsequent hybridization with a different probe often include harsh treatments with boiling 0.1% SDS or autoclaving.

2.2. Dot/Slot Blots

Dot or slot blot analysis takes its name from the apparatus used to apply the sample to the nylon membrane, as shown in Fig. 2-3. This method is analogous to the Northern blot in most ways but has a much higher throughput, as dozens of samples can be examined simultaneously. RNA samples are applied to the membrane and fixed by UV or heat crosslinking. The subsequent application of probe, washing and detection is as described for Northern blots. The main advantage of the dot/slot blot is the number of samples that can be run simultaneously and the relative ease of quantitation resulting from uniformity in bands to be analyzed. There may be some increased sensitivity in the assay because more sample can be loaded onto the matrix, as long as the nitrocellulose or nylon does not become saturated. In addition, the RNA can be of slightly lower quality and still give a detectable signal. However, sensitivity is still signifi-

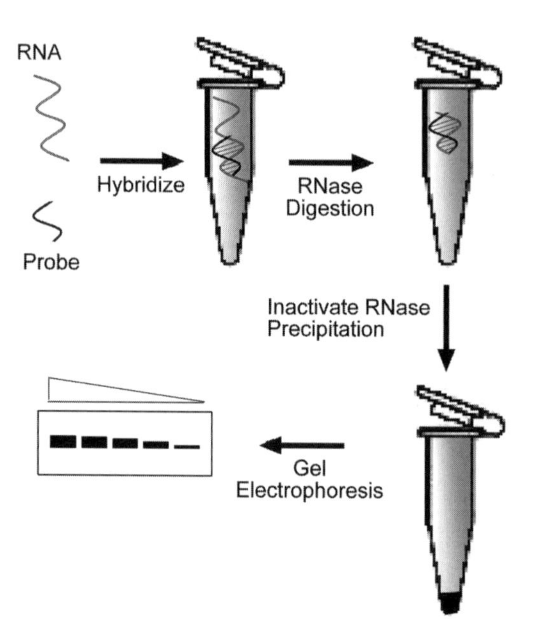

Fig. 2-4. Basics of RNase protection assays (RPAs).

cantly less than that for ribonuclease protection assay or RT-PCR. The dot/slot blot does not result in any information about transcript size, and conditions must be optimized to ensure that the probes are specific for the target gene.

It is also possible to perform a "reverse Northern" using the dot/slot blot format. In this procedure, the probe is spotted onto the nylon and the RNA sample is labeled (isotopic or nonisotopic) and added to the hybrizidation solution. This allows for the examination of multiple genes in one sample. The reverse Northern is predecessor to the microarray assay that will be discussed in detail in Chapter 3.

2.3. RNase Protection Assays (RPA)

Ribonuclease (or RNase) protection assays (RPAs) are an extremely sensitive method for the detection and quantitation of specific RNAs in a complex mixture of total cellular RNA. RPA is a solution hybridization of a single-stranded antisense probe to an RNA sample (*see* Fig. 2-4). The small volume solution hybridization is far more efficient than more common membrane-based hybridization and can accommodate much more RNA (hence increasing sensitivity over Northern blotting). After hybridization, any remaining unhybridized probe and sample RNA are removed by digestion with a mixture of nucleases (usually RNase A and T1; *see* Table 2-1). The nucleases are inactivated, usually by extraction with chloroform/phenol, and the remaining RNA-probe hybrids are precipitated. These products are separated on a denaturing polyacrylamide gel and are visualized by autoradiography or secondary detection. To quantitate mRNA levels using RPAs, the intensities of probe fragments protected by the sample RNA are compared to the intensities generated from either an endogenous internal control (relative quantitation) or known amounts of sense strand RNA (absolute quantitation).

As just mentioned, one major advantage of RPAs over membrane-constrained hybridization is sensitivity. Low-abundance genes are detectable in RPAs, or a small sample size may be used for moderate or high expressing species. In addition, RPAs are the method of choice for the simultaneous detection of several RNA species. During solution hybridization and subsequent analysis, individual probe-target interactions are for all intents and purposes independent of one another. Thus, several RNA targets and housekeeping genes can be assayed simultaneously if the protected fragments are of different lengths. In addition to their use in quantitation of RNA, various RNases can be used to map the structure of transcripts, as will be discussed in more detail in Chapter 4 and 5.

The basic steps of the typical RNase protection assay are relatively straightforward and utilize common materials found in molecular biology laboratories (Fig. 2-4). RNA is extracted as described for the Northern blot analysis. RPA requires RNA probes, most often prepared using in vitro transcription assays and either radiolabeled or nonisotopically labeled. Oligonucleotides and other single-stranded DNA probes can only be used in assays containing S1 nuclease. Optimization of an RPA assay is very probe- and gene-specific and depends mostly on transcript concentration. Probe concentration must be in molar excess over the target mRNA to ensure rapid hybridization. Thus, highly abundant targets

Table 2-1
RNase Properties

RNase	Specificity
RPA	
A	3' of ss C and U
I	3' of ssNTP
T1	3' of ss G
S1 nuclease	ssRNA
RNA structure	
H1	RNA of RNA/DNA
V1	dsRNA

ss, single stranded; ds, double stranded

Ambion (www.ambion.com) suggests the following dilutions of labeled to unlabeled probe:

- *Low abundance, 1:0*
- *Moderate abundance, 1:50*
- *Abundant, 1:10,000*

Comparison of S1 nuclease and RNase A/T1 RPA can be found at: http:/www.promega.com/pnotes/38/38_01/38_01.htm. Accessed 03/24/06.

require the use of low specific-activity probes for abundant targets (dilute labeled probe with unlabeled) whereas probes for scarce RNA targets require no dilution. Generally, 10 μg total RNA is hybridized overnight with approx 50,000 cpm for each probe.

Although S1 nuclease is able to cleave after every residue in both RNA and single-stranded DNA, some problems are associated with its use in RPA assays. Lower incubation temperatures and high salt concentrations (>200 m*M*) must be used to favor single-stranded over double-stranded cleavage, and S1 nuclease is prone to nonspecific cleavage in AU-rich regions. Thus, mixtures of RNase A and T1 typically have been used for digestion of unhybridized RNA in solution hybridization experiments. Nuclease P1 has been substituted for RNase A when using AU-rich probes which require lower reaction temperatures. The amount of enzyme(s), buffer constituents and incubation conditions vary widely from protocol to protocol and may require some optimization. Following digestion, the RNases need to be inactivated. Phenol/ chloroform extraction is required to inactivate Nuclease P1 and RNases A and T1. This extraction is not required for some of the commercially available RNase mixtures and addition of ethylenediaminetetraacetic acid (EDTA) and ethanol precipitation is sufficient.

The percentage of polyacrylamide in gel electrophoresis depends on the size of the products to be examined. Typically a 6% denaturing acrylamide gel is used to resolve fragments of 300–1000 nucleotides. The detection and analysis of the band intensities depends on whether radioactivity or nonisotopic methods were used.

2.4. RT-PCR

Polymerase chain reaction is an enzymatic assay which is capable of producing large amounts of a specific DNA fragment from a small amount of a complex mixture (Fig. 2-5; reviewed in [17]). In RT-PCR, the mRNA must first be converted to a

17. Mattes, W. B. In: PCR Protocols in Molecular Toxicology, Vanden Heuvel, J.P., ed. CRC Press, Boca Raton, FL. 1997, pp. 1–40.

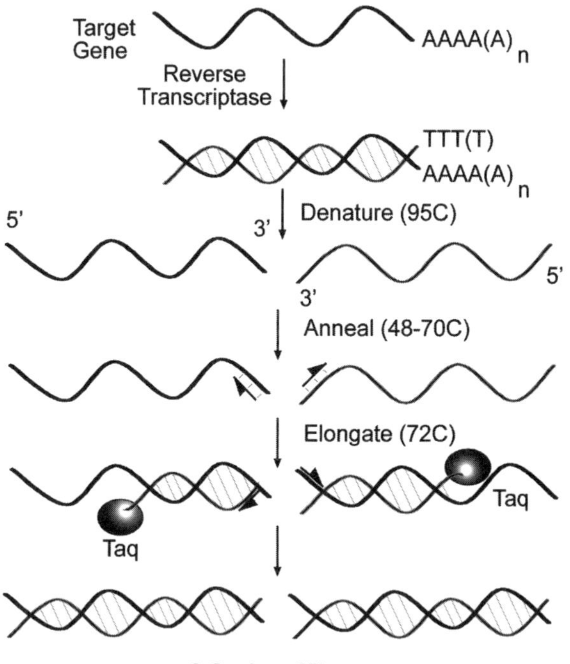

Fig. 2.5. Reverse transcription and one cycle of PCR.

double-stranded molecule by using the enzyme reverse transcriptase (RT). The thermostable DNA polymerase (i.e., Taq) and the use of specific "primers" are the key features of any PCR reaction. All known DNA polymerases require deoxyribonucleotide triphosphates (dNTPs), a divalent cation (Mg^{2+} or Mn^{2+}), a DNA or cDNA template, and a region of that template that is double stranded adjacent to a single-stranded nick or gap. The double-stranded region is provided by the primer annealing to its complementary region of the DNA template. If the starting mixture includes not only a single-stranded polynucleotide template, but also (1) its complementary strand and (2) two oligonucleotide primers that hybridize to both strands, copies of both of these strands will be produced each cycle, and these copies can be used as templates for subsequent cycles. Short DNA fragments whose ends are defined by the position of the two oligonucleotide primers will accumulate in an exponential fashion, i.e., like a chain reaction. If 30 cycles of PCR are performed, theoretically one will achieve a 2^{30} amplification of the target gene's cDNA. The product which is formed is specific for a particular transcript as dictated by the design of the oligonucleotide primers.

Common RT-PCR Definitions

Internal standard (IS): A type of control molecule that can be used to minimize tube-to-tube variability in amplification efficiency. Normally an IS is a synthetic molecule that contains the same recognition sequences as the gene of interest. This type of amplification control is spiked into the PCR reaction.

External standard (ES): A type of control that can be used to minimize differences in template (mRNA, cDNA, or DNA) concentration from sample to sample. Most often an ES is a housekeeping gene that is used in a "coamplification" type of quantitation. This control is not added to each sample, as it is present in a finite amount in each tube. Care must be taken to ensure that the housekeeping gene is not affected by the treatment condition. Typical external standards include actin, tubulin, 18S rRNA, or glyceraldehyde 3-phosphate dehydrogenase (GAPDH)

Template: Any cDNA or DNA that contains primer recognition sites and can be PCR amplified.

Target gene: The gene of interest; to be differentiated from the internal standard, external standard, or artifact templates.

Linker gene: Specifically used to describe a template used to create a synthetic molecule in the synthesis of an internal standard

Forward primer (FP): Analogous to the 5' or "upstream" primer (usp).

Reverse primer (RP): Same as the 3' or "downstream" primer (dsp).

Cross-over point: In competitive RT-PCR, the concentration of internal standard at which the PCR products for the target and the internal standard are equivalent.

Therefore, RT-PCR is a tool to examine the messenger RNA expression of a target gene in that the amount of product formed is a function of the amount of starting template. Of course, the examination of mRNA accumulation can be determined in many cases by hybridization procedures such as Northern blots, dot or slot blots, and RNase protection assays. Nonetheless, in terms of amount of sample required, detection of small differences in expression and ability to examine many genes in a large number of samples, RT-PCR stands above the more conventional procedures.

2.4.1. Competitive RT-PCR

The predominant negative of RT-PCR is related to its ability to amplify the products. That is, an internal standard (IS) is necessitated in these assays owing to the fact that there is a large amount of tube-to-tube variability in amplification efficiency. For example, if ten tubes of seemingly identical reagents are PCR amplified, there could be as much as a threefold difference in the amount of product formed. If an IS was coamplified with the target, the efficiency of amplification in each tube could be corrected and this threefold difference could easily be negated.

A good internal standard for quantifying mRNA levels depends on many factors. The reader is directed to other sources [18] for a description of IS construction methods and considerations. In addition, once an IS is produced or obtained, the investigator must select from several methods for the actual RT-PCR assay. However, in general, the "competitive" RT-PCR approach is the easiest and most adaptable procedure. The standard competitive RT-PCR approach is shown in Fig. 2-6. The basis for this method is that the greater the quantity of competitor (i.e., IS) that is present, the less likely it is that the primers and Taq DNA polymerase will bind to the target cDNA and amplify it. Therefore, despite the fact that all reagents are present in excess (i.e., primers, enzyme, $MgCl_2$, nucleotides) the reaction appears to be competitive. As the amount of competitor is increased, less and less target PCR product is formed until eventually only the IS product is observed. When a dilution series of IS is spiked into a constant amount of RNA, it is possible to estimate the amount of a specific product present in the sample. As depicted in Fig. 2-6, an increase or decrease in target mRNA is easy to visualize using competitive RT-PCR. In addition, calculation of the amount of transcript mRNA in each sample is quite straightforward if one has access to standard laboratory photographic and densitometric equipment [18]. In general terms, the amount of IS required to result in a 1:1 ratio of IS to target PCR product is representative of the amount of transcript present in the original sample; for example, in Fig. 2-6, 10^6 molecules target mRNA in total RNA derived from control versus 10^7 from treated cells. As small as a twofold difference between exposure groups can be routinely identified using a very small amount of RNA (i.e., 10–100 ng total RNA, the equivalent of approximately 1000–10,000 cells).

18. Vanden Heuvel, J.P. In: PCR Protocols in Molecular Toxicology, Vanden Heuvel, J.P., ed. CRC Press, Boca Raton, FL. 1997, pp. 41–98.

18. Vanden Heuvel, J.P. In: PCR Protocols in Molecular Toxicology, Vanden Heuvel, J.P., ed. CRC Press, Boca Raton, FL. 1997, pp. 41–98.

2.4.2. Relative RT-PCR

Relative RT-PCR uses primers for an external control that are coamplified ("multiplexed") in the same RT-PCR reaction with the gene-specific

An optimal internal standard for competitive RT-PCR should amplify with equal efficiency as the target transcript. This can be achieved by designing the competitor with the same primer recognition sequence as the target but with slightly different product size.

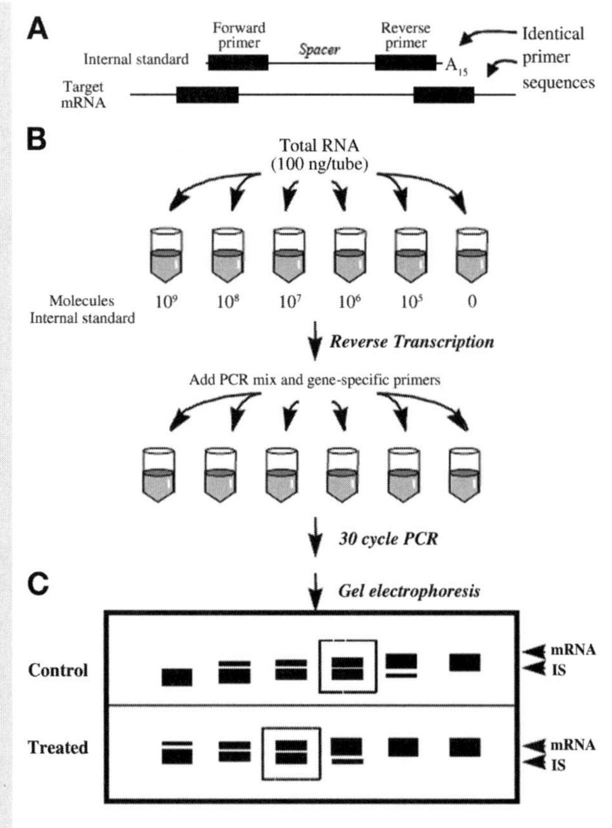

Fig. 2-6. Basic competitive RT-PCR. An internal standard (IS) is constructed, as depicted in **(A)**, that contains the same primers as the target gene but whose resultant product is of different length. In basic competitive RT-PCR, a constant amount of RNA is used and a dilution series of IS is spiked into each tube (10^9–10^5 molecules/tube) as shown in **(B)**. A representative ethidium bromide-stained agarose gel is shown in **(C)**. Note that as the amount of IS is increased the intensity of the target product decreases. Also, the equivalency, or crossover, point is higher in the treated samples than control, indicative of induction of target mRNA. (Adapted from [19].)

19. Gilliland, G.S. et al. In: PCR Protocols: A Guide to Methods and Applications, Innis, M.A., ed. Academic, San Diego, CA. 1990, pp. 60–69.

primers. Common internal controls include β-actin, tubulin, and glyceraldehyde phosphate dehydrogenase (GAPDH) mRNAs and 18S rRNA. This method is quite straightforward but requires more optimization than one would anticipate. External control and target gene primers must be compatible and not produce additional bands or hybridize to each other. Also, salt concentration, annealing temperature, and pH must be controlled to assure that both products are amplifying efficiently. Quantitation must be performed during the nonplateau phase of the amplification for both the target and external control; often, this affects the sensitivity of the assay since samples are examined at a relatively low number of cycles (20–25 cycles). If the efficiency of the primer pairs are quite dissimilar, it may be necessary to dilute the primers for the more robust reaction. The expression of the external control should be constant across all samples being analyzed and this signal can be used to normalize sample data to account for tube-to-tube differences caused by variable RNA quality or RT efficiency, inaccurate quantitation, or pipetting.

Real-time RT-PCR has become the method of choice for quantifying mRNA accumulation due to its high-throughput capabilities

2.4.3. Real-Time RT-PCR

Real-time RT-PCR combines the best attributes of both relative and competitive RT-PCR in that it is accurate, precise, high throughput, and relatively easy to perform. Real-time PCR gets its name from the fact that reaction products are quantitated for each sample in every cycle. The result is a large (10^7-fold) dynamic range, with high sensitivity and speed. Preoptimized kits for thousands of genes are available and they greatly reduce the time required to generate data. However, the procedure requires expensive equipment and reagents, which may be prohibitive to some investigators.

Preoptimized kits for real-time PCR quantitation can be obtained from Applied Biosystems Inc. (ABI) at http://appliedbiosystems.com. Accessed 03/24/06.

Real-time PCR systems rely on the detection and quantitation of a fluorescent reporter. Since the products are analyzed in real time, there is no need for post-PCR manipulation. This decreases the chance for experiment-to-experiment contamination and eliminates the variability inherent in agarose gel electrophoresis and quantitation. Currently, there are two major methods that are used in real-time

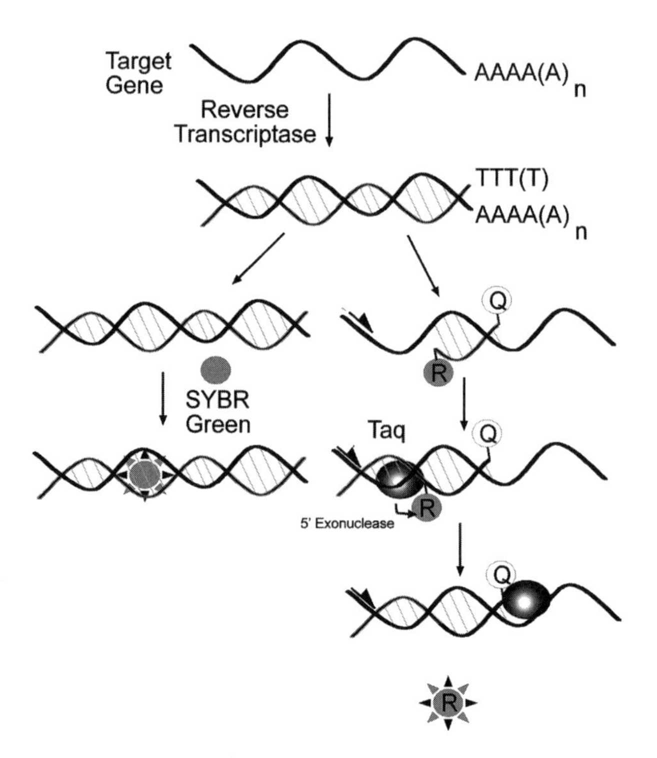

Fig. 2-7. Real-time PCR.

PCR: DNA fluorescent dyes (SYBR green) and fluorescent resonance energy transfer (FRET; i.e., TaqMan(r), Molecular Beacons). The amount of fluorescence in each case increases as the amount of PCR product increases. *See* Fig. 2-7 for a summary of the two methods.

Of the DNA fluorescent dyes, SYBR green is the most common. SYBR green is an intercalating agent that binds double-stranded DNA, and in this bound form upon excitation emits light. Thus, as a PCR product accumulates, fluorescence increases. The advantages of SYBR green are that it's inexpensive, easy to use, and sensitive. The disadvantage is that SYBR green will bind to any double-stranded DNA in the reaction, including primer-dimers and other nonspecific reaction products, which results in an overestimation of the target concentration.

The TaqMan and Molecular Beacons approaches both utilize fluorescent probes and FRET for quantitation. This assay utilizes the 5' nuclease activity of Taq polymerase. TaqMan probes are oligonucleotides that contain a fluorescent dye on the 5' end (i.e., carboxyfluorescein [FAM]), and a quenching dye on the 3' base (i.e., tetramethylrhodamine [TAMRA]). These probes are designed to hybridize between the standard PCR primers. Prior to enzymatic activity, the excited fluorescent dye (FAM) transfers energy to the nearby quenching dye molecule (TAMRA), which has a much weaker emission, resulting in a nonfluorescent probe. During PCR, when the polymerase replicates a template on which a TaqMan probe is bound, the 5' exonuclease activity of the polymerase cleaves the probe. This separates the fluorescent and quenching dyes and FRET no longer occurs. Fluorescence increases in each cycle, proportional to the rate of probe cleavage.

Probes used as molecular beacons are utilized in a similar manner to TaqMan, but exonuclease activity is not required. Molecular beacons also contain fluorescent and quenching dyes and utilize FRET, but the two dyes are incorporated within the probe and are not accessible to Taq's exonuclease function. Molecular beacons are designed to adopt a hairpin structure while free in solution, bringing the fluorescent dye and quencher in close proximity. Upon hybrization to the target gene, the fluorescent dye and quencher are separated, interrupting the FRET and enhancing the signal from the probe. Unlike TaqMan probes, molecular beacons are designed to remain intact during the amplification reaction, and must rebind to target in every cycle for signal measurement.

3. Summary

The examination of transcript concentrations is one of the fundamental experiments in modern biology. The methods currently in use for the quantification of mRNA accumulation are summarized in Table 2-2, with the identifiable strengths and weaknesses of each. Perhaps the two most common methods are Northern blots and RT-PCR. Northern blots are insensitive and relatively nonquantitative. However, they allow for visualization of transcript size, utilize well-established, straightforward techniques, and are inexpensive to perform. In contrast, RT-PCR (quantitative, real-time) are exquisitely sensitive and robust, with the capability for high-throughput gene expression analysis. RT-PCR requires expensive equipment and must be performed with care to minimize contamination and tube-to-tube variability. Realtime PCR has become the method of choice for high-throughput gene expression analysis.

Table 2-2
Summary of Pros and Cons of Methods to Examine mRNA Accumulation

Method	Pros	Cons	Quantitation
Northern blot	• Technically simple and inexpensive • No amplification or RNA manipulation • Can gain information on transcript size or splicing	• Insensitive, requiring large sample size • Low throughput • Sensitive to RNA quality	• Generally requires large differences in expression between two samples • Concentration is often expressed relative to a housekeeping gene • An exogenous standard may be spiked for absolute quantitation
Dot/Slot blot	• Technically simple and inexpensive • No amplification or RNA manipulation • High throughput	• Relatively insensitive • No information on transcript size or quality	• Concentration is often expressed relative to a housekeeping gene • Standard curves may be generated for absolute quantitation
RNase protection	• Increased sensitivity • Less sensitive to partial RNA degradation • Several genes can be examined concurrently	• No information on transcript size or quality • Difficult to optimize	• Concentration is often expressed relative to a housekeeping gene • Standard curves may be generated for absolute quantitation
RT-PCR	• Most sensitive and quantitative method • High throughput and rapid	• Technically challenging • Potential for contamination (DNA in sample; experiment to experiment; tube to tube) • Competitive PCR requires synthesis of internal standard • Requires expensive equipment (in particular, real-time PCR)	• Relative PCR requires amplification in exponential phase; coamplification performed with housekeeping gene. Generally requires large differences between samples • Competitive PCR uses a synthetic molecule with primer sites. Standard curves are prepared. Small differences in expression can be quantified. • Real-time PCR detects the number of transcripts produced during amplifica--tion. Sensitive and high throughput with small differences between samples detected.

3

Transcript Profiling

1. Concepts

1.1. Differential Gene Expression

An important concept about gene expression in the disease state, and the effects of xenobiotics or any external stimuli, is that often the events observed are cell-, species-, sex-, and development-specific. This is owing to the fact that combinatory gene expression is the norm in eukaryotes. In general, a combination of multiple gene-regulatory proteins, rather than a single protein, determines where and when a gene is transcribed. A single gene may be necessary but not sufficient to alter the phenotype of a cell. A good example of combinatorial gene control comes from muscle cell differentiation. Myogenic proteins (MyoD, myogenin; both are helix-loop-helix proteins) are key to causing certain fibroblasts to differentiate into muscle cells. It appears that MyoD regulates myogenin. If myogenin is removed, the cells will not differentiate, and if this gene is overexpressed in fibroblasts, they will convert to muscle. However, other cell types are not converted to muscle by myogenic proteins. This suggests that some cells have not accumulated the other regulatory proteins required.

As will be discussed in this chapter, multiple gene-regulatory proteins can act in combination to regulate the expression of an individual gene. But, combinatorial gene expression means much more; not only does each gene have multiple "inputs," each regulatory protein contributes to the control of multiple genes ("outputs"). Although some regulatory proteins, such as MyoD and myogenin, are specific to a cell type, more typically production of a given regulatory protein is switched "on" in a variety of cell types. With combinatorial control, a given gene regulatory protein does not have a single and simply definable function. It is the combination of gene products that conveys this information. The consequence of adding a new gene is dependent on the past history of the cell. The metaphor often used is that proteins are the words and the phenotype of a cell is the language. Words require context to convey their meaning.

Differential gene expression is a key component of many complex phenomena including cellular development, differentiation, maintenance, and injury or death. In fact, the subset of genes that are being expressed determines to a large

Gene Expression Control at the mRNA Level by J. P. Vanden Heuvel
From: *Regulation of Gene Expression*
By: G. H. Perdew et al. © Humana Press Inc., Totowa, NJ

extent the phenotype of that cell. Also, a loss of control of differential gene expression underlies many disease states, not the least of which is cancer. The identification of genes that are being expressed in one cell type versus another (i.e., control versus treated; tumor versus normal) can help in explaining the function of those genes as well as lend insight into the system being examined. For this reason, the identification of differentially expressed genes has been pursued for diverse stimuli such as responses to biological programs (developmental and circadian cues), physical agents (i.e., UV irradiation, X-rays), and chemical agents (hormones and xenobiotics). In fact, hundreds of examples of these pursuits can be found [20] and have resulted in a much greater understanding of cellular biology and our responses to physical and chemical insult.

20. Wan, J.S. et al. Nat. Biotechnol. 14 (1996) 1685–1691.

1.2. Model Development

The comparison of two cell types, i.e., control versus treated tissue, represents a complex analysis of two different populations of mRNA species. Any study of receptor-mediated effects on gene expression is most efficient and effective when it *utilizes the simplest, most well-defined model possible.* Figure 3-1 shows a basic model system that starts with drug-receptor binding and ultimately leads to a biological response. If a segment of the response can be isolated, i.e., primary responses, the comparisons between populations will be more facile. A comparison of normal versus tumor cells requires the analysis of thousands of different mRNA species, most of which have little or no connection to the development of the disease (cancer). However, comparing two identical populations of cells, one that had been treated with a hormone or xenobiotic, the other with the appropriate vehicle for a short period of time, may result in a very small subset of genes that represent an important initial response. The guidelines provided next should aid in the design of a good system for examining genes as

Great care must be taken to develop an appropriate biological model system to study differential gene expression. These studies can take years to complete and are expensive endeavors. The adage, "garbage-in, garbage-out" applies to these analyses.

they are affected by a particular treatment; they will also aid in interpreting data found in the literature from similarly designed studies.

- **Utilize a model system that has a reproducible, biologically pertinent response.**
  After genes are identified, an attempt must be made to equate this change in mRNA levels with some biological response. For example, if the interest is in identifying genes that are causally linked to altered patterns of growth and differentiation, choose cells or tissues that have the capacity to respond to the treatment with the prototypical changes associated with this process, such as increased DNA synthesis. Remember, the pertinence of the novel genes identified is only as good as the model system will allow.
- **Compare two populations of mRNA where the abundant component is unaffected.**
  This can be accomplished by comparing similar types or populations of cells (not tumor versus normal) and by looking at earlier events. As mentioned previously, a small difference in expression of an abundant gene, say a twofold induction in a 10,000 copy-per-cell gene, will result in a huge difference in the mass of the message. The increase in the abundant gene may dilute the possibility of detecting more pertinent changes.
- **Identify the subset of genes you desire to pursue.**
  Are you interested in genes involved in primary or secondary events? Are genes regulated by transcription more important than those affected by mRNA stabilization? Many subsets of genes can be selected in your model system. For example, if you are interested in primary events, early time points after treatment are warranted. If genes affected by mRNA stability are your forte, treat the two-cell populations with a transcription inhibitor such as actinomycin D or α-amanitin prior to the experimental treatment.
- **Utilize a method that can detect differences in the scarce component of mRNA.**
  Quite often the scarce mRNA component represents the types of genes that infer tissue-specific or chemical-specific responses. In the following sections, we will discuss the differences, strengths, and weaknesses of the various technical procedures to identify differentially expressed genes. The need to identify genes of a particular expression level may determine which method is ultimately chosen.
- **Identify a positive control gene for verification of response.**
  Prior to initiating a screen, make sure that the cells have responded as expected. A gene that is known to be regulated by your treatment can be checked by Northern blots, RNase protection, or RT-PCR. In addition, this gene should be one that is subsequently identified from your differential screening.

2. Methods and Approaches

This is not intended to discuss the details of the different screening methods but to serve as a guide to the basic theory behind the more popular methods.

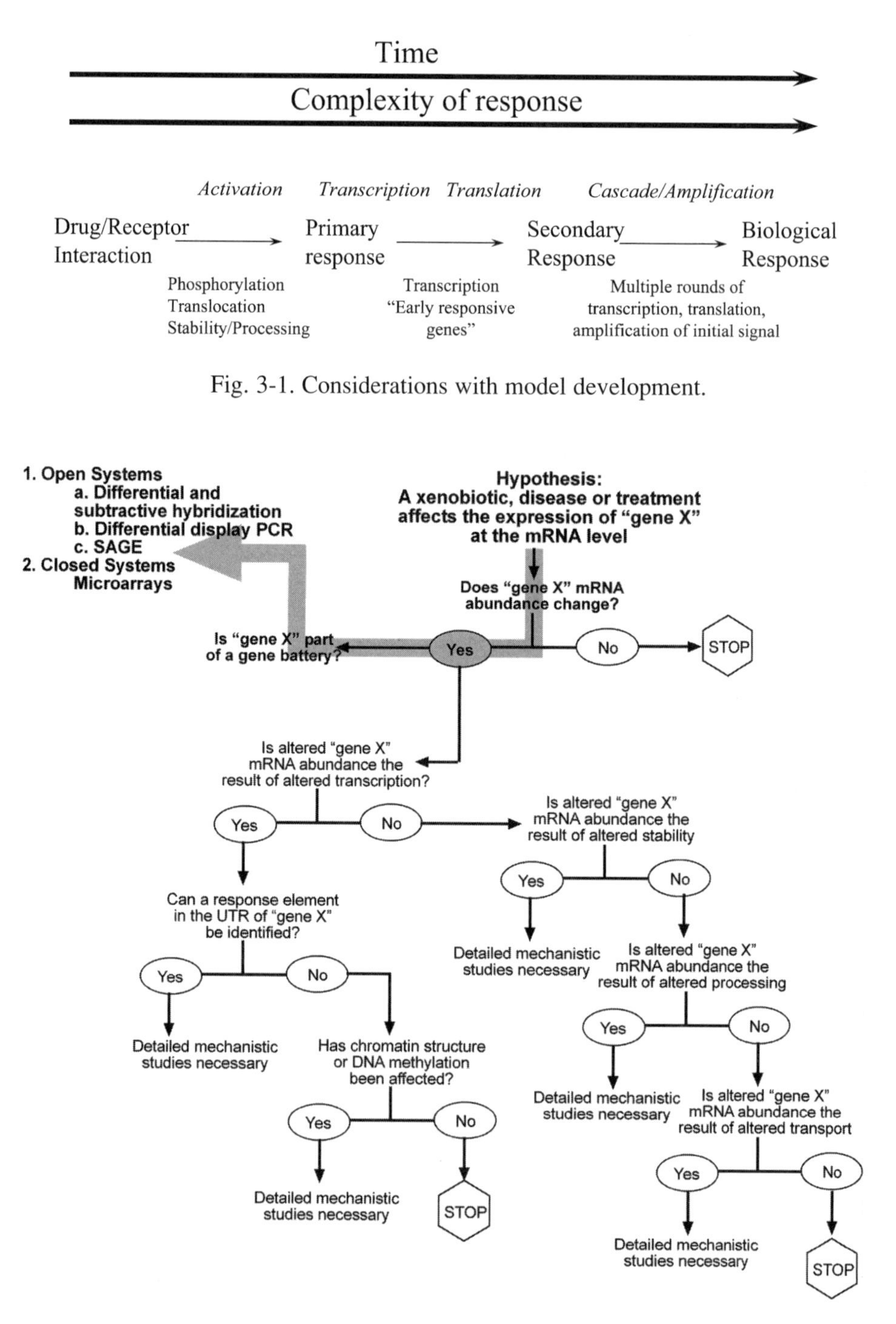

Fig. 3-1. Considerations with model development.

Fig. 3-2. Outline of approaches to determine if a gene network exists.

For specific procedures the reader is directed toward laboratory manuals such as Current Protocols in Molecular Biology [9] and Molecular Cloning [10]. Each project and model system is different and personal preference and experience may dictate the choice of method to identify novel responsive genes. We will discuss some of the more common procedures for identifying the differences between two cell populations, often referred to as "transcript profiling" or "transcriptonomics" (Fig 3-2). Key to an understanding of transcript profiling is the concept of open and closed systems. An "open" system is a means of examining differentially expressed transcripts that does not require prior sequence information. In contrast, "closed" systems use probes of known sequence and thus are biased. We will describe both types of systems.

9. Ausubel, F.M. et al. Current Protocols in Molecular Biology. Wiley, New York, NY. 1994.

10. Sambrook, J. et al., eds. Molecular Cloning: A Laboratory Manual. Cold Spring Harbor Press, Cold Spring Harbor, NY. 1989.

Gene expression profiling, transcript profiling, *and transcriptonomics are interchangeable terms to describe the examination of mRNA being expressed at a given time under a certain condition or treatment.*

2.1. "Open" Systems

2.1.1. Differential Hybridization

For many techniques used to examine gene expression, one must be familiar with the construction of cDNA libraries. These libraries are essential for differential and subtractive hybridization and electronic subtraction, and also aid in the identification of full length mRNAs in differential display PCR. The premise behind library construction is to convert mRNA from a given cell type into double stranded cDNA using reverse transcriptase. Linkers or adapters are added to the cDNA and subsequently ligated into a bacteriophage vector. Following packaging of the virus and infection of bacteria, the library is amplified, titered and characterized.

The experimental procedures for differential hybridization (DH) is shown in Fig. 3-3A. A cDNA library of the "treated" cells is established using the basic procedures described above. Transformed bacteria at low density are replica plated onto nylon membranes, fixed by heat or uv irradiation, and probed with ^{32}P-cDNA made to either the "treated" or "control" mRNA. The autoradiograms of these membranes are examined and compared.

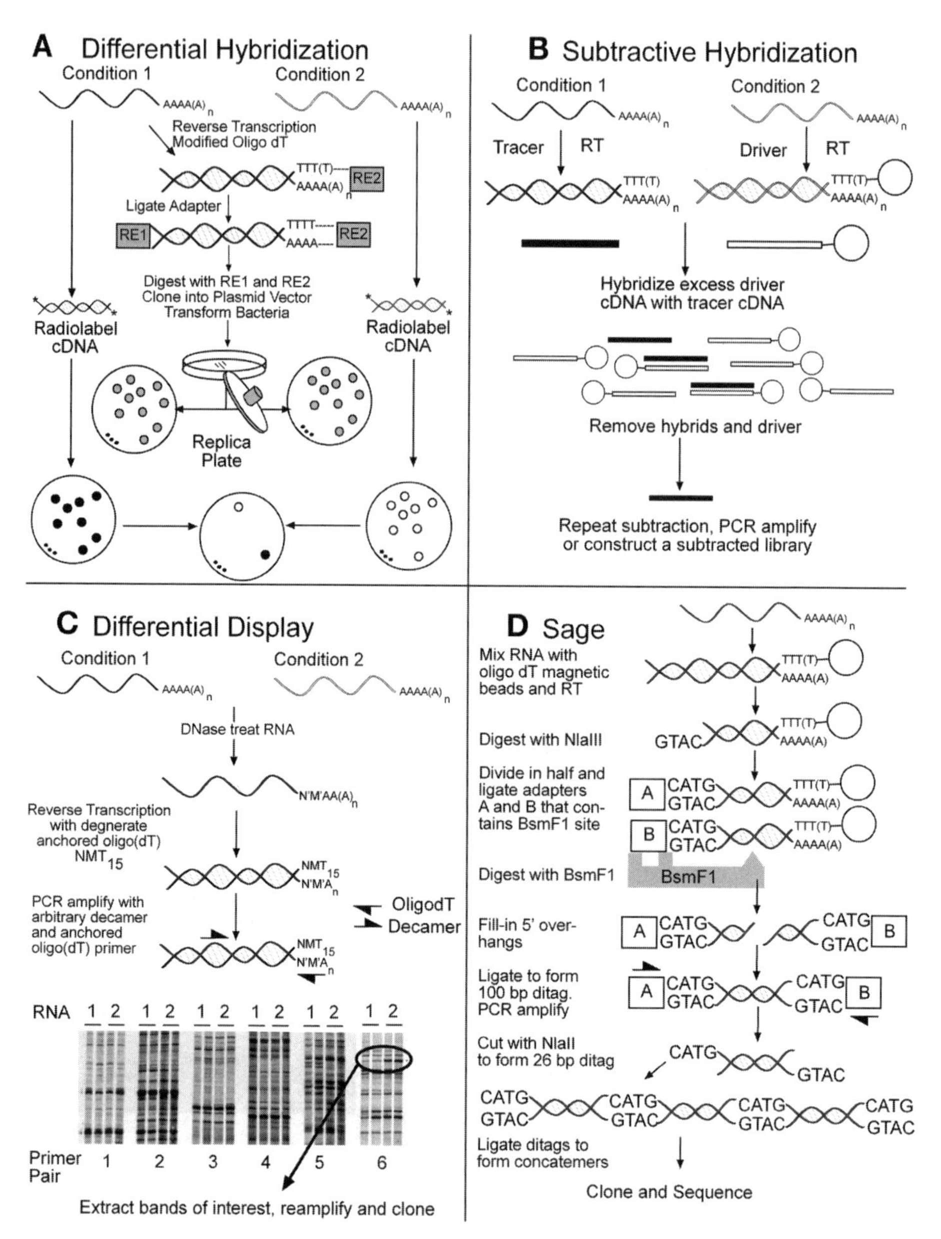

Fig. 3-3. Open systems for gene expression batteries.

Alternatively, a color coding method may be used to aid in the comparisons. The positions of the differences between the two hybridization conditions are marked and the clones removed from the original plate. Probes are generated from these clones for slot-blot, Northern blot, or RNase protection assays.

DH was the first method routinely used to clone differentially expressed genes. This procedure has an advantage over subtractive hybridization in being able to examine either repressed or induced messages. However, the procedures are technically challenging and may be biased toward more abundant sequences or for mRNAs that have a large difference in expression.

2.1.2. Subtractive Hybridization

Subtractive hybridization (SH) has proven to be one of the most useful methods to clone differentially expressed genes, although it too has fallen out of favor as more rapid procedures have developed. Although many adaptations of SH have been described, the basic procedures remain the same (*see* Fig. 3-3B). Double stranded cDNA from treated cells ("tracer") is produced and hybridized to a molar excess of mRNA from control ("driver") cells. The conditions and efficiency of hybridization is critical and should be optimized for the desired abundance level of differentially expressed cDNAs. The mRNAs that are common to the two populations (the double stranded hybrids) are separated from the differentially expressed sequences, usually using a streptavidin-biotin interaction (i.e., the driver mRNA is biotinylated). The resulting single stranded ("subtracted") cDNA is either re-selected, amplified or used to construct a subtracted cDNA library.

SH may represent the best chance to identify and clone differentially expressed genes in the scarce component. This is due to the fact that the subtraction is an enrichment of the differences between two populations of mRNAs. However, the differences between the two populations must be dramatic in order for the differentially expressed gene to be selected. Also, the comparison is "one way" in that studying genes that are induced or repressed would require two separate experiments.

2.1.3. Differential Display Polymerase Chain Reaction (DD-PCR)

Differential display (DD)-PCR is a relatively new addition to the battery of common techniques used to study differential gene expression. Since the initial description [21] of DD, there have been numerous modifications and improvements. The basic principle (as shown in Fig. 3-3C) is to reverse transcribe mRNA from control and treated cells and to systematically amplify the 3' termini. This is accomplished by using a set of four anchored oligo(dT) primers (T_{12}MN primers, where M is G, A, or C and N is any base) and an arbitrary decamer. The region between these primers is amplified by PCR in the presence of a radioactive nucleotide. Subsequently, the PCR products from the control and treated cDNA are resolved on a denaturing SDS-polyacrylamide gel and the differences in "fingerprints" determined. The differentially expressed products are eluted from the gel, re-amplified, and used to make probes or are cloned for sequencing.

Differential display is a very versatile, technically facile procedure that can examine either repression or induction in the same set of samples. In addition, multiple comparison can be made (i.e., a time-course or triplicate samples) on the same gel. Also, since PCR is a sensitive procedure, it may be possible to examine the scarce component of mRNA. However, DD has been associated with a high rate of false positives and its ability to identify low-expression genes has been questioned. Also, DD identifies short fragments of cDNA at the 3' end of the message, making identification of gene products and verification more difficult. This particular concern may be circumvented using an adaptation of DD which utilizes two arbitrary decamers instead of one anchored primer and a decamer and does not require radioactivity, resulting in a larger fragment of cDNA from internal sequences at higher concentrations [22].

22. Sokolov, B.P., Prokop, D.J. Nucl. Acids Res. 22 (1994), 4009–4015.

2.1.4. Electronic Subtraction and SAGE

Serial analysis of gene expression (SAGE) [23,24], a form of electronic subtraction (ES), is a brute-force method to examine the differences between two populations of mRNA. With ES, randomly selected cDNAs from libraries prepared from the two treatment conditions are sequenced. Usually 1000–3000 cDNAs are sequenced from each library. A particular mRNA is described as differentially expressed if its frequency in this random sampling is different between the two conditions. This type of approach has led to the establishment of dbEST, the database for "expressed sequence tags," which contains thousands of short sequences from several species. SAGE is a modification of ES in which short tags are prepared for high-throughput sequencing (Fig. 3-3D) [23,24]. SAGE is based on the premise that a sequence as short as nine nucleotides is enough to uniquely identify the transcript (i.e., 4^9 or >260,000 transcripts can be distinguished). The key innovation in SAGE is the utilization of the the novel restriction enzyme ("tag-

23. Velculescu, V.E. et al. Science 270 (1995) 484–487.

24. Velculescu, V.E. et al. Trends Genet. 16 (2000) 423–425.

ging enzyme"), which cuts downstream from its recognition site. Originally BsmF1 was utilized, although other enzymes are also available that leave tags of 9–21 bp in length. Using a combination of restriction enzymes and the ligation of PCR primers, a concatenation of these tag segments and intervening 4-bp "punctuations" is produced. In this manner a continuous string of data with multiple sequences per clone (i.e., 40 tags/clone) can be obtained. Public domain and private software is available that automatically retrieves these tags and searches for known sequences in the databases. Comparison between the tag frequency is made between the two (or more) mRNA populations under study. Gene expression profiles of various normal and cancer tissues is readily available at the National Center for Biotechnology Information (http://www.ncbi.nlm.nih.gov/).

Detailed protocols for performing SAGE can be found at SAGENET: http://www.sagenet.org/protocol/MANUAL1e.pdf as well as at http:// www. genzyme moleculironcology.com/sage/process.pdf.

Serial analysis of gene expression has several advantages over other methods, not the least of which being its ease of use. Automation of plasmid purification and sequencing makes ES/SAGE the least technically challenging method for examining differential gene expression. In addition, the data obtained is digital and reusable. The major disadvantage of ES is the preferential identification of abundant mRNA species. When sequencing 1000 templates for each treatment group, there is a very low likelihood of observing differences in the scarce component of mRNA. In order to identify an mRNA of scarce abundance (i.e., 1 out of 20,000 mRNAs), it would require sequencing 126,000 cDNAs from each condition, and to identify all mRNAs (i.e., 1 out of 70,000 mRNAs) would result in sequencing >400,000 templates for each population. Even if utilizing SAGE technology, this represents a monumental undertaking. Another major drawback of SAGE is the effect of sequencing or PCR amplification error. With such a small amount of information being incorporated by the tags, a sequencing error will affect the frequency of occurrence of a transcript. (For example, an error will

The National Center for Biotechnology Information (NCBI) has SAGE and other molecular abundance data at the gene expression omnibus (GEO); http://www.ncbi.nlm.nih.gov/GEO.

cause the appearance of a "new" transcript and decrease the occurrence of the "true" tag.) With more sequencing, the effects of sequencing and PCR errors can be minimized. One also must accept the fact that many of the tags are going to match ESTs or other uncharacterized sequences in the database.

2.2. "Closed" Systems

2.2.1. Nylon-Membrane Arrays

A "closed" system for transcript profiling is one that monitors a predetermined set of sequences. The first closed system for examination of mRNA levels was Northern blotting and the closest predecessor to the microarray (described next) was the "reverse dot blot" (*see* Fig. 2-3 for summary). In the latter approach, cDNA fragments corresponding to different transcripts are bound to a nylon membrane in a grid pattern. RNA is converted to labeled cDNA and hybridized to the membrane, and the intensity of label at any given spot or grid location gives the abundance of that transcript in the original RNA pool. Such membrane-based arrays of cDNAs are commercially available and remain a flexible and cost-effective way to monitor a number of selected genes simultaneously. However, the number of genes to be examined is relatively small (less than 1000) and the dynamic range and convenience of radioactive detection systems are generally less than those utilizing fluorescent probes. Also, when comparing two RNA samples, many control probes must be included to achieve a direct comparison between the two separate blots.

25. Mattes, W.B. In: Cellular and Molecular Toxicology, Vanden Heuvel, J.P. et al., eds. Elsevier, Amsterdam, The Netherlands. 2002, vol. 14, pp. 463–492.

2.2.2. High-Density cDNA and Oligonucleotide Arrays

The development of the high-density microarray heralded the expanded use of gene expression profiling and the ability to routinely examine gene expression "on a genomic scale" [25]. The basic approach to microarrays is similar to that of the

Although many websites are devoted to microarrays, one of the most comprehensive is http://www.gene-chips.com.

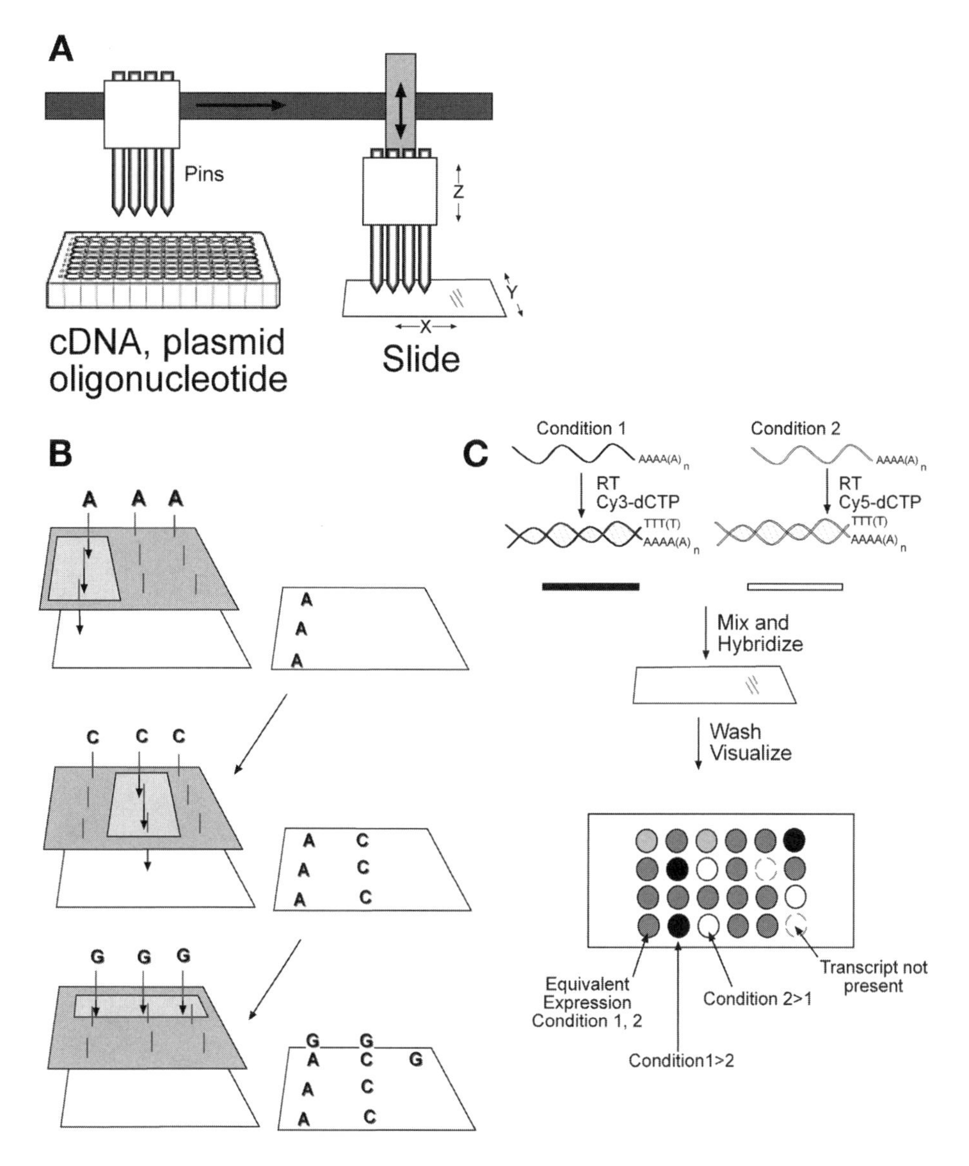

Fig. 3-4. Overview of microarray printing and gene expression profiling.

"reverse dot blot" but incorporates several technological innovations. The production of the microarray generally requires sophisticated equipment. In the spotted array, the probes (prepared from PCR amplification of cloned genes or synthetic oligonucleotides) are arrayed onto coated slides using a robotic device capable of printing thousands of spots in an 18 mm^2 area (Fig. 3-4A). The second approach synthesizes probes *in situ* on nylon or glass surfaces using photochemical synthesis or an ink-jet printing process. With photolithographic masks, oligonucleotides of different sequence may be synthesized in 25 μm^2 areas, yielding a chip with thousands of probes (Fig. 3-4 B). As designed by Affymetrix these GeneChip® arrays make use of several (16–20) 20-mer probe pairs for each RNA being monitored, a pair consisting of an oligonucleotide complementary to the RNA and another with one mismatched base in the sequence. The mismatch probe for each pair serves as an internal control for hybridization specificity. A particular strength of the ink-jet approach is the ability to create different array configurations rather simply, allowing rapid optimization of probe oligonucleotide sequence as well as creation of many custom arrays.

The generation of labeled probe material from RNA utilizes similar approaches regardless of the type of high-density microarray (Fig. 3-4C). A reverse transcriptase reaction is performed with fluorescently labeled nucleotides (or may be post-labeled); RNA from one condition is labeled with one dye (Cy3), while the RNA from the second sample is labeled with a second dye (Cy5). The two probe samples are then competitively hybridized to the cDNA microarray. Finally, the fluorescence at each spot on the slide is determined with a confocal laser scanning microscope; the ratio of the intensities for the two dyes at each spot gives the ratio of abundance for that transcript in the two conditions. The only significant variation with the photolithographic systems (Affymetrix) is that RNA samples are converted to a cDNA using an oligo(dT) primer that also incorporates a bacteriophage T7 RNA polymerase promoter sequence. This allows subsequent production of "complementary RNA" in the presence of fluorescently labeled nucleotides. Also, unlike the approach with the spotted microarrays, only one labeled sample is hybridized at a time. The analysis of spot intensities over all the probe pairs for an RNA allows quantification of the abundance of the transcript in the total sample.

There are many advantages of high-density microarrays versus their "open-system" counterparts. The most obvious is the speed at which differentially regulated genes can be determined. If the slides are readily available, a fairly comprehensive examination of gene expression can be performed in a matter of days. Currently configurations allow for examination in excess of 30,000 genes in a single experiment. The predominant down-side of this technology is cost. From printing to scanning to analysis the equipment is specialized and expensive. If high-density microarrays are purchased from commercial sources the cost can be hundreds of dollars for one slide. (However, this approach does save on the expense of probes and the printing equipment.) This latter point is in direct conflict with the second major drawback of high-density microarrays, the need for many replicates. The variability in the experimental portions of the procedures, as well as the inherent statistical issues in dealing with thousands of data points, requires that each condition be examined as many times as possible. Great pains must be taken to assure that proper controls (internal and external) have been included in the array design and that proper normalization of data has been performed [26–28].

26. Bilban, M. et al. Curr. Issues Mol. Biol. 4 (2002) 57–64.

27. Hegde, P. et al. Biotechniques 29 (2000) 548–556.

28. Quackenbush, J. Nat. Genet. 32 suppl. (2002) 496–501

In comparing the two basic formats of arrays (spotted versus *in situ*-printed), there are more similarities than differences. The spotted arrays have the advantage of being easily customized and are less expensive. The array printer can be custom built and plans for a robot are available on the web. Free software is available for data analysis. The *in situ*-printed arrays, in particular those available from Affymetrix, have the advantage of multiple probes per gene and built-in controls. The hybridization methods are automated and hence are less variable. Density is another factor with piezoelectric printing affording many more genes per unit area. The major negative to these model systems is the cost of the arrays themselves (hundred to thousands of dollars per slide) and the equipment for hybridization and scanning.

2.2.3. Chromatin Immunoprecipitation (ChIP) Cloning

Chromatin immunoprecipitation (ChIP) following by screening a cytidine-phosphate-guanosine (CpG)-containing DNA microarray is described in detail in Chapter 4. Briefly, an antibody directed against a particular transcription factor is used to precipitate DNA associated with this protein. The immunoprecipitated DNA is fluorescently labeled (Cy5) and is hybridized to a microarray of CpG-containing DNA. These CpG-enriched DNA libraries are available from several individual and commercial laboratories and contain the regulatory regions of potential target genes. Also included in the hybridization is DNA (Cy3-labeled) from DNA–protein complexes immunoprecipitated with a control antibody or from cells that lack the particular transcription factor. This method allows for the identification of genes that are under direct control of a transcription factor. It is also possible to perform this assay as an open system, by cloning the co-precipitated DNA and identifying the potential target genes by sequencing.

2.3. Data Analysis

As mentioned above, the statistical examination of microarray data is complicated by the shear numbers of genes being examined, as well as by the variability from the experimental conditions. Perhaps the foremost challenge in transcript profiling data analysis is the handling of large data sets [25]. Ideally, these data sets have not only the thousands of differential expression values, but also include functional and/or pathway information for the genes examined to allow later functional grouping. Currently, this step is complicated by the limited (but growing) amount of functional annotation for database entries and the need for standard gene and gene function nomenclature (i.e., gene ontology). The next challenge is that of determining patterns of gene expression relevant to the experimental hypotheses. Even if the question is only "what genes are truly regulated under a single experimental condition," data analysis is in order. Thus, microarray experiments, despite their expense, must be replicated, with statistical methods applied to determine random variability. In terms of identifying trends in groups of transcripts, a variety of clustering methods has been described, including the original hierarchical clustering, interactive clustering, k-means clustering, and self-organizing maps. Many of these methods have been incorporated into commercially available software packages, and all remain valuable approaches to data analysis. Several reviews of statistical methods and clustering of microarray data have been published (for example [28,29]).

25. Mattes, W. B. In: Molecular Toxicology, Vanden Heuvel, J. P. et al., eds. Elsevier, Amsterdam, The Netherlands, 2002, vol. 14, pp. 463–492.

28. Quakenbush, J. Nat. Genet. 32 suppl. (2002) 496–501.

29. Pennie, W.D. Toxicology 181,182 (2002) 551–554.

The following is a recommendation of how to proceed with data analysis aimed at finding biologically pertinent changes in gene expression and gene batteries under coordinate control. Of course, one must start with the simplest, best-designed model system possible. With the relative simplicity and speed with which data can be generated by microarray, researchers have become flippant with model development. This was not the case with approaches such as differential hybridization that could take months to years to generate transcript profiles. Hence, one was very cognizant of the dictum "garbage in, garbage out". Thus, assuming that the model is appropriate and the experiment was replicated numerous times (n=3–6, for example), here is one analysis set that can be performed with relative ease and generate some meaningful, biologically relevant information.

1. Determine genes that are statistically different between your two conditions.

   This requires the appropriate normalization of data and inclusion of the proper controls to minimize slide-to-slide variability. In the past, many have used the "twofold rule". If the gene is either twofold higher or lower than the mean for the slide, in the majority of the slides, then it is considered to be different between the samples. However, this approach does not take into consideration the variability of the gene's expression. A more thorough statistical analysis would add confidence to the generation of lists of differentially regulated genes. Several software packages are available to determine statistically different levels of gene expression. For example, GeneSpring (Silicon Genetics, Redwood City, CA) performs a t-test to determine if a gene's expression is different from the mean, taking into account slide-to-slide and gene-to-gene variability. A software program available from the Stanford University microarray website called SAM (Statistical Analysis of Microarray, http://www-stat.stanford.edu/~tibs/SAM/) uses a more sophisticated algorithm to decrease the rate of false positives. Specifics of statistical analysis are beyond the scope of this book , and often the approach used depends on the investigator's comfort level with different programs. At some point one must go from normalized, replicated data (Fig. 3-5A) to lists of genes you think may be differentially regulated (Fig. 3-5B).

2. Classify the regulated genes based on function.

   This is the beginning of functional genomics, where the lists generated in step 1 are scrutinized to find an insight into the biological hypothesis being examined. For example, is there a particular signal transduction pathway that is key to the condition under question? Are the responsive genes coordinately affecting cell cycle or apoptosis? Some people also equate this with "hypothesis generation" versus "hypothesis testing." Most of the genes present on commercial arrays, or those produced from purchased oligonucleotide or clone sets, are associated with a wide range of biological information. The GenBank accession number can be used to identify a gene's molecular function, cellular localization, or biological process. Information on homologs, chromosome localization, and polymorphisms can also be easily retrieved. Commonly used programs for performing this func-

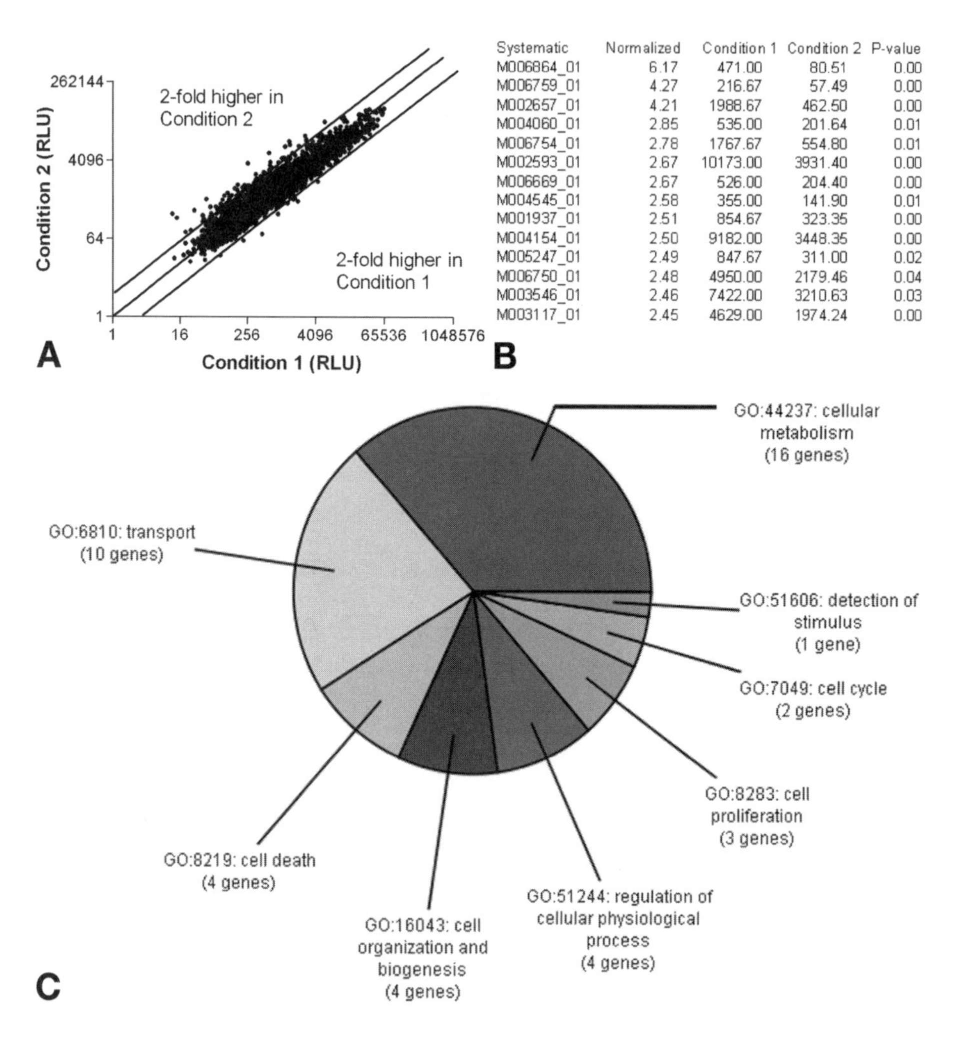

Systematic	Normalized	Condition 1	Condition 2	P-value
M006864_01	6.17	471.00	80.51	0.00
M006759_01	4.27	216.67	57.49	0.00
M002657_01	4.21	1988.67	462.50	0.00
M004060_01	2.85	535.00	201.64	0.01
M006754_01	2.78	1767.67	554.80	0.01
M002593_01	2.67	10173.00	3931.40	0.00
M006669_01	2.67	526.00	204.40	0.00
M004545_01	2.58	355.00	141.90	0.01
M001937_01	2.51	854.67	323.35	0.00
M004154_01	2.50	9182.00	3448.35	0.00
M005247_01	2.49	847.67	311.00	0.02
M006750_01	2.48	4950.00	2179.46	0.04
M003546_01	2.46	7422.00	3210.63	0.03
M003117_01	2.45	4629.00	1974.24	0.00

Fig. 3-5. Basic data analysis for microarray experiments. (**A**) Normalized data showing fluorescence obtained from Cy5 and Cy3. Data found outside the solid lines are two-fold different between the two conditions. (**B**) List of genes determined to be differentially-regulated showing pertinent information. (**C**) Functional classification of genes from table in (B). (Produced using GeneSpring, Redwood City, CA.)

tional classification simply retrieve this information and create a database that can be updated as new information is obtained and can be used to sort (Fig. 3-5C).

3. Verify a subset of potential target genes.

   It is not likely that all the genes added to the list of potential target genes can be systematically verified. Thus a subset of genes can be chosen for further scrutiny. This is where the functional classification can have an impact. A subset(s) of genes

28. Quakenbus, J. Nat. Genet. 32 suppl. (2002) 496–501.

Table 3-1
Methods to Identify the Differences Between Populations of RNA

	"Open systems"				"Closed systems"
	Differential hybridization	Subtractive hybridization	Differential display	SAGE	High-density microarray
RNA required	1-5 (µg poly(A)	1–5 (µg poly(A)	5 (µg total	1-5 (µg poly(A)	10-20 µg
Time required until genes identified	Months	Months	Weeks	Weeks	Days
Expertise required	Much	Much	Little	Moderate	Little
Prevalence of mRNA surveyed	Abundant	Abundant and rare	Abundant and rare	Abundant	Abundant
Types of differences identified	>Twofold	All or none	>Twofold	>Twofold	>Twofold
Considerations	Technically difficult	One-way comparisons, technically difficult	Reported to have high rate of false positives. Technically simple	High cost of operation. Simple to perform if experienced	Relatively insensitive, high cost. Simple and fast to perform

that share a common functional classification can be chosen for verification. Other factors may influence how the subset is chosen, such as strength of response, importance of the product, and whether or not it has been shown to be affected by this condition in other experiments. Although any method described in Chapter 2 can be utilized, more and more microarray labs are utilizing real-time PCR owing to its quantifiability and throughput.

3. Summary

The ability to identify the differences between two (or more) populations of mRNA is an important biological pursuit. There are several approaches that can be used to perform this analysis (Table 3-1). "Open systems" are unbiased and can be used to identify transcripts that have little or no previous information. Of the "open systems," differential display and SAGE are much more common than differential and subtractive hybridization. The advent of high-density microarrays has resulted in "transcript profiling" where the expression of thousands of genes can be performed simultaneously.

4

Transcriptional Regulation of Gene Expression

1. Concepts

The previous chapters have discussed how to determine whether a gene's mRNA accumulates under a particular treatment or condition. In the present chapter, the exploration of the reason for this accumulation commences. The two major ways in which the steady-state levels of a transcript are altered are via transcriptional or posttranscription processes. For simplicity, effects at the chromatin level will be included in this chapter. A summary of the events leading to transcriptional regulation are given in Fig. 4-1 and will be discussed in more detail. Further description of the molecular events associated with transcriptional regulation can be found elsewhere [1,30].

1. Alberts, B. et al. In: Molecular Biology of the Cell. Garland Publishing, New York, NY. 1994, pp. 401–476.

30. Vanden Heuvel, J. P. In: Cellular and Molecular Toxicology, J. P. Vanden Heuvel et al., eds. Elsevier, Amsterdam, The Netherlands. 2002, vol. 14, pp. 57–79.

1.1. Binding of Activator Proteins to Enhancer Sequences

The general configuration of the upstream regulatory region of a gene comprises enhancer or operator sequences as well as promoter elements. Together, enhancers and promoters are defined as *cis*-elements. *Cis*-elements are the site where *trans*-acting proteins (proteins encoded from a separate gene) bind. There are many nuances to this configuration of the gene's regulatory region including, but not limited to: multiple, overlapping, and competing enhancer sequences; enhancers within an intron, 3' untranslated region (UTR) or far upstream elements; alternative promoter sites; and TATA-less promoters. Enhancer elements that associate with

Gene Expression Control at the mRNA Level by J. P. Vanden Heuvel
From: *Regulation of Gene Expression*
By: G. H. Perdew et al. © Humana Press Inc., Totowa, NJ

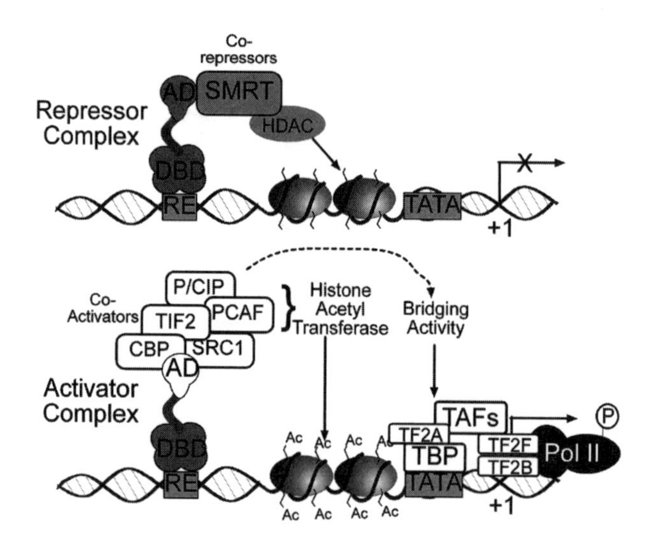

Fig. 4-1. General scheme of gene transcription in eukaryotic cells.

transcription factors (also called activator proteins) are often defined as response elements (RE). The specificity of this DNA–protein interaction is driven by physiochemical properties of both constituents (*see* Table 4-1) and is an important factor in transcriptional control of gene expression. These *cis*-acting regulatory sequences are predominantly located upstream (5') of the transcription initiation site, although some elements occur downstream (3') or within the genes. The activator proteins themselves contain DNA-binding domains (DBD) that specifically recognize certain enhancer sequences. Also contained in many proteins of this sort are activation domains (AD) that form a scaffold for numerous protein–protein interactions. The number and type of regulatory elements to be found varies with each mRNA gene. Different combinations of transcription factors also can exert differential regulatory effects upon transcriptional initiation. Various cell types each express characteristic combinations of transcription factors; this is the major mechanism for cell-type specificity in the regulation of mRNA gene expression.

1.2. Promoter Elements, TBP, and TAFs

Almost all eukaryotic mRNA genes contain basal promoters of two types and any number of different transcriptional regulatory domains. The basal promoter elements are termed CCAAT boxes and TATA boxes because of their sequence motifs. The TATA box resides 20–30 bases upstream of the transcriptional start site and is similar in sequence to the prokaryotic Pribnow box (consensus TATA($^T/_A$) A($^T/_A$), where ($^T/_A$) indicates that either base may be found at

Table 4-1
Representative Transcription Factor Response Elements

Factor	Sequence motif
c-Myc and Max	CACGTG
c-Fos and c-Jun	TGA($^{C}/_{G}$)T($^{C}/_{A}$)A
CREB	TGACG($^{C}/_{T}$)($^{C}/_{A}$)($^{G}/_{A}$)
SP1	GGG GCG GGG
c-ErbA; also TR	GTGTCAAAGGTCA
c-Ets	($^{G}/_{C}$)($^{A}/_{C}$)GGA($^{A}/_{T}$)G($^{T}/_{C}$)
GATA	(T/A)GATA
c-Myb	($^{T}/_{C}$)AAC($^{G}/_{T}$)G
MyoD	CAACTGAC
NFκB and c-Rel	GGGA($^{A}/_{C}$)TN($^{T}/_{C}$)CC[(1)]
ARE (antioxidant RE)	RTGACNNNGC
SRF (serum response factor)	GGATGTCCATATTAGGACATCT
HSE (heat shock RE)	CNNGAANNTCCNNG
GRE (glucocorticoid RE)	TGGTACAAATGTTCT
RARE (retinoic acid RE)	ACGTCATGACCT
MRE (mineralocortoid RE)	CGNCCCGGNCNC
PPRE (PPAR RE)	AGGCCAAGGTCA
TRE (thyroid receptor RE)	TGACTCA
DRE or XRE (dioxin or xenobiotic RE)	TNGCGTG

[1]N signifies that any base can occupy that position.
CREB, c-AMP response-element binding protein.
(Adapted from ref. [30].)

that position). Numerous proteins identified as TFIIA, B, C, and so on (for transcription factors regulating RNA polymerase II) interact with the TATA box. The CCAAT box (consensus GG($^{T}/_{C}$) CAATCT) resides 50–130 bases upstream of the transcriptional start site. The protein identified as C/EBP (for CCAAT box/enhancer binding protein) binds to the CCAAT box element.

The general transcription machinery assembles over the core promoter and drives transcription at the initiator site (+1). The presence of the TATA-binding protein (TBP) at the TATA box appears to be a pivotal step in transcriptional regulation. (For a review of TBPs and its function, *see* [2]). TBP becomes associated with the TATA box, partially the result of activator proteins associating with enhancer

30. Vanden Heuvel, J. P. In: Cellular and Molecular Toxicology, Vanden Heuvel, J. P. et al., eds. Elsevier, Amsterdam, The Netherlands. 2002, vol. 14, pp. 57–79.

2. Pugh, B. F. Gene 255 (2000) 1–14

elements. In fact, the preinitiation complex is not assembled in the cell unless facilitated by activator proteins. Activator proteins recruit a variety of chromatin remodeling factors that may be required for TBP binding. Therefore, it appears that the core promoter sequence is hidden within nucleosomes, thus preventing assembly of the general transcription machinery. Once TBP is bound to DNA, it forms the scaffolding by which other factors assemble. Another important aspect of TBP's action is that it is able to bend DNA, which may bring upstream factors in contact with the promoter or with downstream elements.

1.3. RNA Polymerases and Transcription Initiation

Transcription is the mechanism by which a template strand of DNA is utilized by specific RNA polymerases to generate one of the three different classifications of RNA. All RNA polymerases are dependent upon a DNA template in order to synthesize RNA. Transcription of the different classes of RNAs in eukaryotes is carried out by three different polymerases (RNA polymerase [pol] I, II, III; Table 4-2) *RNA pol I* synthesizes the rRNAs, except for the 5S species. These classes of RNAs are assembled, together with numerous ribosomal proteins, to form the ribosomes. *RNA pol II* synthesizes the mRNAs and some small nuclear RNAs (snRNAs) involved in RNA splicing. *RNA pol III* synthesizes the 5S rRNA and the tRNAs. The vast majority of eukaryotic RNAs are subjected to posttranscriptional processing. The most complex regulation observed in eukaryotic genes are those transcribed by RNA pol II; It is the regulation of mRNA genes that will be the focus of this chapter.

Transcription proceeds in the following general scheme. After activators and TBP/TAFs bind to the 5' regulatory region of the gene, the RNA pol II haloenzyme and general transcription machinery is recruited. Bending of the TATA-box DNA around TBP confers a context for interaction with TFIID. This large protein complex becomes the scaffolding upon which the rest of the transcription machinery can assemble. Similarly, TFIIH includes an ATP-dependent helicase activity that can unwind the promoter around the transcription start site. Working together, they open the DNA double helix and pol II proceeds down one strand working in the 3' to 5' direction. As it moves down this strand, pol II assembles ribonucleotides into the strand of RNA following the rules of base pairing. Synthesis of the RNA proceeds in the 5' to 3' direction. The primary transcript produced is called heterogeneous nuclear RNA (hnRNA) and it will undergo processing as described in Chapter 5, including adding a 5' cap, poly(A) tail, and splicing. The mammalian pol II large subunit contains an essential, multifunctional carboxy terminal domain (CTD) that is a target for many kinases and phosphatases and is comprised of 52 seven-amino acid

Table 4-2
Eukaryotic RNA Polymerases

Polymerase	RNA	Inhibitor
Pol I	rRNA	Actinomycin D
Pol II	mRNA	α-Amanitin
Pol III	tRNA	Species specific

Inhibitors are not absolutely specific. At higher concentration actinomycin D will also inhibit RNA pol II. Pol, RNA polymerase; rRNA, ribosomal RNA; mRNA, messenger RNA; tRNA, transfer RNA.

repeats [31,32]. This particular region of pol II is an important point of convergence of mRNA transcription initiation, elongation, and mRNA processing.

31. Conaway, J. W. et al. Trends Biochem. Sci. 25(8) (2000), 117–172.

32. Bregman, D. B. et al. Front. Biosci. 5, (2000), D244–57.

1.4. Transcription Elongation

The transition from transcription initiation to elongation is loosely defined in eukaryotes. One key step must be breaking the initial ties to the promoter and the accessory initiation factors, but it may also include conversion of RNA pol II to an elongation-competent form [33]. This conversion involves phosphorylation of the CTD followed by association with other accessory proteins. The CTD is phosphorylated by TFIIH, which itself is composed of cyclin dependent kinase 7 and cyclin H. Other cyclin- and cell cycle-dependent kinases may also phosphorylate RNA pol II. During the preinitiation and initiation stage of transcription, several key proteins bind to the hypophosphorylated CTD. Subsequently, this region of pol II becomes hyperphosphorylated, thereby allowing transcription elongation to occur.

33. Uptain, S. M. et al., Annu. Rev. Biochem. 66 (1997) 117–72.

Transcription elongation is discontinuous and may encounter a pause, an arrest, or may reach termination. During a pause state, RNA pol II temporarily stops RNA synthesis for a finite period of time. An arrested elongation complex is unable to resume transcript elongation without the aid of accessory factors. Both transcription pause and arrest may be intrinsic (due to RNA sequence or chromatin structure) or extrinsic (DNA binding proteins, nucleotide concentration, CTD phosphorylation) [33]. Transcription termination is the result of RNA pol II becoming catalytically inactive. Once the elongation apparatus is terminated, or falls off the template, it cannot simply resume its function; transcription initiation must be restarted at the promoter. Pausing or arresting the elongation process allows for more control over the amount of transcript being produced. Termination may be important in negating the expression of mutant mRNAs or may be closely tied to DNA damage.

33. Uptain, S. M. et al., Annu. Rev. Biochem. 66 (1997) 117–172.

1.5. Structure of the Nucleosome

The nucleosome is the fundamental structural unit of eukaryotic chromosomes. It consists of pairs of the core histones, a single linker histone, and 160 bp of DNA.

Chromatin is a term designating the structure in which DNA exists within the nucleus of cells. The structure of chromatin is determined and stabilized through the interaction of the DNA with DNA-binding proteins. There are two classes of DNA-binding proteins. The histones are the major class involved in maintaining the compacted structure of chromatin. There are five different histone proteins, identified as H1, H2A, H2B, H3, and H4. The other class of DNA-binding proteins is a diverse group of proteins called simply, nonhistone proteins. This class of proteins includes the various transcription factors, polymerases, soluble receptors, and other nuclear enzymes. The binding of DNA by the histones generates a structure called the nucleosome. The nucleosome core contains an octamer protein structure consisting of two subunits each of H2A, H2B, H3, and H4. The nucleosome core contains approx 160 bp of DNA in two helical loops of 80 bp each. The initial histone–DNA association begins when the $(H3/H4)_2$ tetramer forms a stable complex with approx 120 bp of DNA. Domains within the histone contain α-helix structures (histone-fold domains) that are key sites of DNA contact. The core nucleosome conformation is similar for most genes, regardless of the DNA sequence itself. In addition to histone-fold domains, a critical region of these DNA binding proteins is the tail domain. This tail, if fully extended, can project into the helices of DNA in the nucleosome. In addition, the histone tails are involved in internucleosomal contact and formation of higher order chromatin structure (coiled-coil). Posttranslational modification of the histone tails may evoke a change in the chromatin fiber, affecting the accessibility to other modifying proteins (*see* Section 1.7.).

The linker DNA between each nucleosome can vary from 20 to more than 200 bp. These nucleosomal core structures would appear as beads on a string if the DNA were pulled into a linear structure. The nucleosome cores themselves coil into a solenoid shape, which coils to further compact the DNA. These final coils are compacted further into the characteristic chromatin seen in a karyotyping spread. The protein–DNA structure of chromatin is stabilized by attachment to a nonhistone-protein scaffold, the nuclear matrix.

1.6. Transcription States

The nucleosome is a very stable physical structure and under physiological conditions folds into higher order structures of high concentration [34]. In spite of the nucleosome's stability and coiled-coil structure, many metabolic processes are capable of modifying this structure in vivo. In fact, coiling of DNA around histones to form the nucleosomal structure is an important regulator of transcriptional control of gene expression [5], as outlined in Fig. 4-1. The physical state of a gene can be categorized as repressed, basal, or induced [2], each of which is associated with a different structure of the nucleosome. A repressed gene is likely to be encased in chromatin to such an extent that the transcription machinery cannot access the promoter DNA. A basally expressed gene might have a more permissive structure of the chromatin that allows for

34. Wolffe, A. P., Hayes, J. J. Nucl. Acids Res. 27 (1999) 711–720.

5. Kornberg, R. D. Trends Cell. Biol. 9 (1999) M46–49.

2. Pugh, B. F. Gene 255 (2000) 1–14.

a low level of gene expression. Last, an induced gene is likely to have an open chromatin structure and be bound by transcriptional activators and efficient recruitment of the transcriptional machinery.

5. Kornberg, R. D. Trends Cell Biol. 9 (1999) M46-49.
35. Struhl, K., Cell 98 (1999)1-4.

Unlike prokaryotes, which utilize repressor proteins to maintain an "off state" of genes, eukaryotic genes are repressed by the nucleosome complex [5,35]. This repression occurs in one of three ways [5]. In one, nucleosomes may block the DNA binding sites for activator proteins (REs) such as receptors, transcription factors, or DNA modifying enzymes. In another, higher-ordered folding into the solenoid configuration may repress entire chromosomal domains. In the third, interactions of nucleosomes with additional proteins to form heterochromatin may result in a hereditably-repressed gene. The significance of nucleosomal structure is that the binding of the TBP to the promoter (Pro) is prevented, and hence the polymerase machinery (i.e., pol II) is absent from the gene context [35].

35. Struhl, K., Cell 98 (1999)1-4.

36. Mannervik, M., et al. Science 284 (1999) 606–609.

An important regulatory mechanism that involves chromatin structure is the recruitment of chromatin modifying enzymes by activator and repressor proteins. For example, a DNA-binding repressor may inhibit the transcription machinery by recruiting histone deacetylase (HDAC) activity to the promoter [36] resulting in a repressed state. Similarly, activator proteins may recruit proteins that acetylate histone tails (HAT), thereby resulting in an activated state [35].

1.7. Regulation of Chromatin Structure

Coactivators and corepressors are collectively called "coregulators." The coregulator complex can contain dozens of proteins and as a unit can supply chromatin remodeling activity as well as pol II recruitment ("bridging").

That chromatin structure is important in gene regulation is underscored by its importance in development [36]. The activation of developmentally regulated genes may occur as a result of sequential changes in chromatin structure [35]. Recent studies have shown that coactivators and corepressors mediate communication between upstream regulatory proteins and RNA pol II. These transcription coregulators carry DNA modifying activities, in particular histone acetylation and deacetylation capabilities [36]. Ultimately, the end result of this complex, tiered approach to transcription regulation is that a particular gene is expressed in a proper temporal and spatially localized manner and is sensitive to small differences in levels of any extracellular signaling molecule. Three means by which chromatin structure may be modified are briefly described in the following subsections and include acetylation of histone tails, other posttranslational modification (phosphorylation, methylation, and ubiquitinization) of the core histones [34], and DNA methylation.

34. Wolffe, A. P., Hayes, J. J. Nucl. Acids Res. 27 (1999) 711–720.

1.7.1. Consequences of Acetylation of Core Histones

Histones are modified on specific lysine residues in the tail region and are the targets of histone acetyltransferase (HAT)- and histone deacetylase (HDAC)-containing coregulators. The acetylation state of histones affects chromatin structure on several levels [34]. First, acetylation of histone tails may reduce the stability of interaction with nucleosomal DNA. Acetylated histones wrap DNA less tightly and are more mobile than are hypoacetylated histone tails. Second, acetylation may disrupt the protein–protein interactions between histones, in particular the H3–H4 junction. Third, acetylated histones are less able to interact with adjacent nucleosomal arrays, thereby decreasing the stability of the compacted fiber. Last, acetylation of the core histones decreases the association of the linker histone H1, further destabilizing the higher order structure. Ultimately the end result of histone acetylation is an increase in transcription-factor access to nucleosomal DNA. This fact may be observed by DNase sensitivity assays, in which regions of sensitivity are shown to correlate with histone acetylation.

A number of coactivators that are recruited to activator proteins have intrinsic HAT activity. These include the p160 family, important in nuclear receptor (NR) function, and the more general coactivators CREB-binding protein (CBP)/p300 and p300/CBP-

37. Bevan, C., Parker, M. Exp. Cell Res. 253 (1999) 349–356.

38. Workman, J. L., Kingston, R. E. Annu. Rev. Biochem. 67 (1998) 545–579.

associated protein (PCAF) [37]. The p160 family includes steroid receptor coactivator-1 (SRC-1), transcription intermediary factor-2 (TIF2), glucocorticoid receptor interacting protein (GRIP), SRC-3 and many others. These proteins associate with the NR's activator function-2 (AF2) domain, a latent domain that is revealed upon ligand binding. Thus, in the presence of hormone, the activated receptor binding to DNA and the HAT activity is recruited to the histone core.

HDAC activity is equally important in the regulation of gene expression [38]. Deacetylation results in an increase in histone/DNA interaction, internucleosomal interactions, and higher order DNA structure (reversal of HAT activity). Similar to HAT-containing proteins, HDAC proteins are often parts of large complexes of proteins. Several co-repressors are associated with histone deacetylase 1 (HDAC1), including Sin3, N-CoR, and SMRT. Transcriptional regulation by co-repressor complexes can be blocked by chemical HDAC inhibitors (tricostatin A, trapoxin), showing the role of this enzymatic activity in gene control. The recruitment HDAC to DNA by NRs is often the reverse to that of HAT proteins. That is, co-repressors bind to NRs in the absense of ligand and are released upon addition of hormone.

1.7.2. Consequences of Phosphorylation, Ubiquitination, and Methylation of Core Histones

In contrast to the many studies on the structural and functional consequences of histone acetylation, the impact of other posttranslational modifications of the core histones is much less characterized. The phosphorylated residue (serine 10) is located within the basic N-terminal domain of histone H3 and may interact with the ends of DNA in the nucleosomal core. Phosphorylation of histones might be expected to have structural consequences comparable to acetylation, presumably the result of change in protein charge and decreased histone-DNA binding.

Ubiquitination of histones, in particular H2A, is associated with transcriptional activity. Ubiquitin is a 76-amino acid peptide that is attached to the C-terminal tail of approximately one histone H2A in every 25 nucleosomes in an inactive gene. This increases to one nucleosome in two for the transcriptionally active gene. Enrichment in ubiquitinated H2A is especially prevalent at the 5'-end of transcriptionally active genes. Since the C-terminus of histone H2A contacts nucleosomal DNA at the dyad axis of the nucleosome, ubiquitination of this tail domain might be anticipated to disrupt higher order chromatin structures.

Lysine residues are also targeted for modification by methylation. Most methylation in vertebrates occurs on histone H3 at Lys9 and Lys27 and histone H4 at Lys20, not known to be sites of acetylation. The consequence of this modification has not been elucidated.

1.7.3. Consequences of DNA Methylation

Mammalian cells possess the capacity to modify their genomes via the covalent addition of a methyl group to the 5-position of the cytosine ring within the context of the CpG dinucleotide [39]. Approximately 70% of the CpG residues in the mammalian genome are methylated, although this does not mean that the event is nonspecific. The distribution of CpG is mainly in the 5' UTR region of genes whereas the majority of the genome is CpG-poor. Certain regions of the genome that possess the high CpG frequency are termed CpG islands. DNA methylation has been shown to be essential for normal development, X-chromosome inactivation, and imprinting.

39. Robertson, K. D. Jones, P. A. Carcinogenesis 21 (2000) 461–467.

A strong correlation between DNA hypermethylation, transcriptional silence, and tightly compacted chromatin has been established in many different systems. As described previously, at least a portion of chromatin remodeling appears to be accomplished through acetylation and deacetylation of the histone tails. Inactive regions of DNA that were demonstrated to be heavily methylated were enriched in hypoacetylated histones. Only one DNA methyl transferase (DNMT) protein has been examined in detail, DNMT1. DNMT1 is a large enzyme (~200 kDa) composed of a C-terminal catalytic domain with homology to bacterial cytosine-5 methylases and a large N-terminal regulatory domain with several functions, including targeting to replication foci. Disruption of DNMT1 in mice results in abnormal imprinting, embryonic lethality, and greatly reduced levels of DNA methylation. Targeting of DNMT1 to replication foci via the N-terminal domain is believed to allow for copying of methylation patterns from the parental to the newly synthesized daughter DNA strand. The enzymatic removal of 5-methylcytosine from DNA has also been described but has been far less extensively studied. One mechanism identified involves a 5-methylcytosine DNA glycosylase activity.

2. Methods and Approaches

2.1. Transcription Analysis

The determination of whether a treatment or condition affects mRNA accumulation at the transcriptional level is usually examined in one of two ways (Fig. 4-2). The first uses RNA polymerase inhibitors and is easier but less definite. Nuclear run-on (or run-off) assays are the preferred method to determine transcriptional activation. These two methods are described in the following subsections. Other potential approaches are to perform reporter assays (described in Section 2.2.3.) or to exclude mRNA stabilization (described in Chapter 5). Another assay that may be considered early in the characterization of a particular gene is the determination of the transcriptional start site. Often genes will contain different promoters that have tissue-, sex- and disease-specific regulation, and assessment of a transcription start site will assist in identifying where to search for response elements and *trans*-acting proteins.

2.1.1. RNA Polymerase Inhibitors

As shown in Table 4-2, there are two commonly employed eukaryotic RNA polymerase inhibitors. Actinomycin D preferentially inhibits RNA pol I but at higher concentrations it also affects RNA pol II. A better choice for inhibiting mRNA production is to use α-amanitin because it is more specific for RNA pol II. (Nonetheless, actinomycin D is still used more frequently.) The approach to examine transcriptional activation using RNA polymerase inhibitors is quite straightforward, especially if cell culture systems are used. Cells are pretreated with actinomycin D (1–5 μg/mL) or α-amanitin (1–5 μg/mL) or vehicle control prior to the treatment (or establishment of the condition or state). At various times after treatment, RNA is extracted and the target gene examined by any means discussed in Chapter 2.

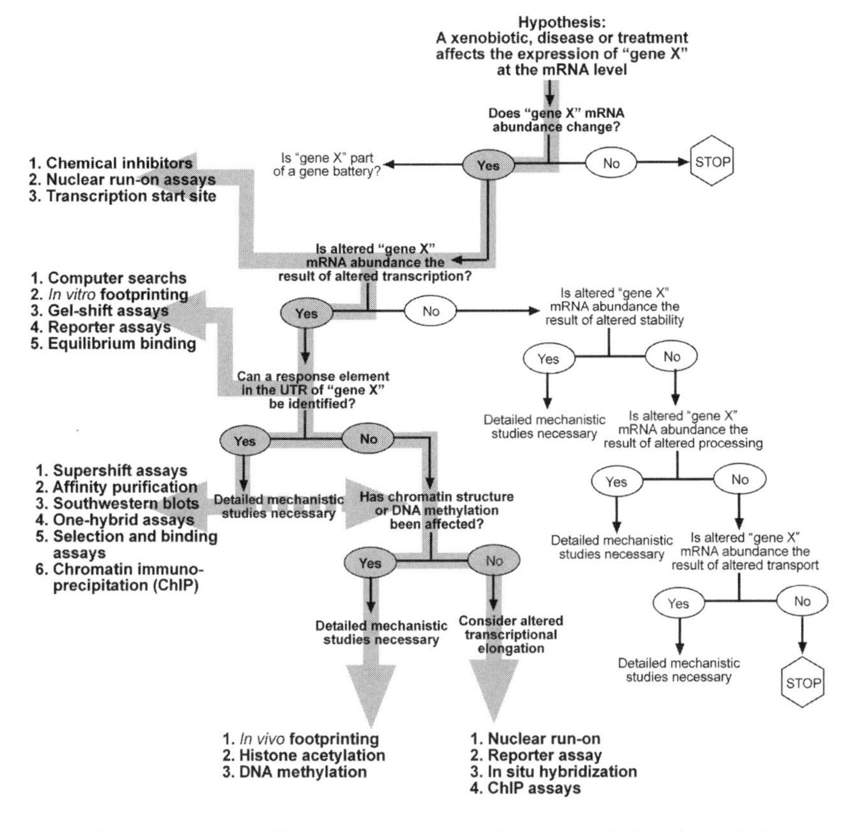

Fig. 4-2. Outline of approaches to examine transcriptional regulation.

2.1.2. Nuclear Run-On Assays

Nuclear run-on (or run-off) assays are the most sensitive and specific procedure to measure gene transcription as a function of cell state. As mentioned previously, this method is also more difficult as it requires the isolation of nuclei that are free of cell membranes and have intact enzymatic activity. The basic outline is shown in Fig. 4-3 and specifics can be found elsewhere [9]. The basic premise of this method is that preinitiated transcription complexes will continue to elongate the transcript in vitro and the amount of these complexes formed is affected by the cell state/condition. This method works equally well for in vivo samples or culture

9. Ausubel, F. M. et al. Current Protocols in Molecular Biology. John Wiley, New York, NY. 1994.

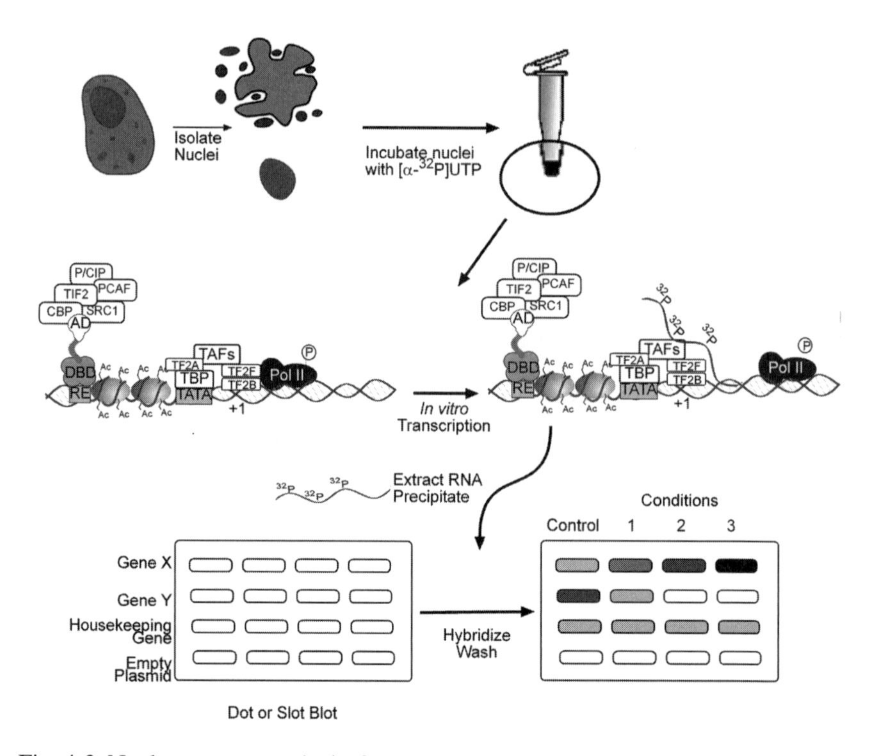

Fig. 4-3. Nuclear run-on analysis. Shown is the rate of transcription of three genes (and control plasmid); Gene X is transcriptionally up-regulated, whereas Gene Y is repressed. The housekeeping gene shows no regulation under these treatment conditions.

treatments, in particular for adherent cells. Several possibilities exist for isolating nuclei from lysed cells, and the resultant nuclei can be stored in glycerol under liquid nitrogen for long periods of time. The most common isolation method is a simple detergent lysis (NP-40) on ice followed by centrifugation. The extent of lysis and purity of the nuclei can be checked by phase contrast microscopy. Other intact nuclei preparation procedures include Dounce homogenization or cell lysis with isoosmotic solution and centrifugation through a sucrose cushion. Thawed nuclei are incubated in the presence of [α-^{32}P] UTP (or CTP) and unlabeled nucleotides, and the reaction is allowed to proceed for a relatively short period of time (30 min). The RNA goes through two rounds of RNase-free DNase and proteinase K treatment, extraction, and precipitation.

The subsequent analysis steps have been outlined in Chapter 2 and most often include the dot or slot blot. Linearized plasmids or PCR products, along with control plasmids and housekeeping genes, are spotted onto nitrocellulose. The labeled RNA is incubated with the blots and washed and extent of transcription assessed by autoradiography.

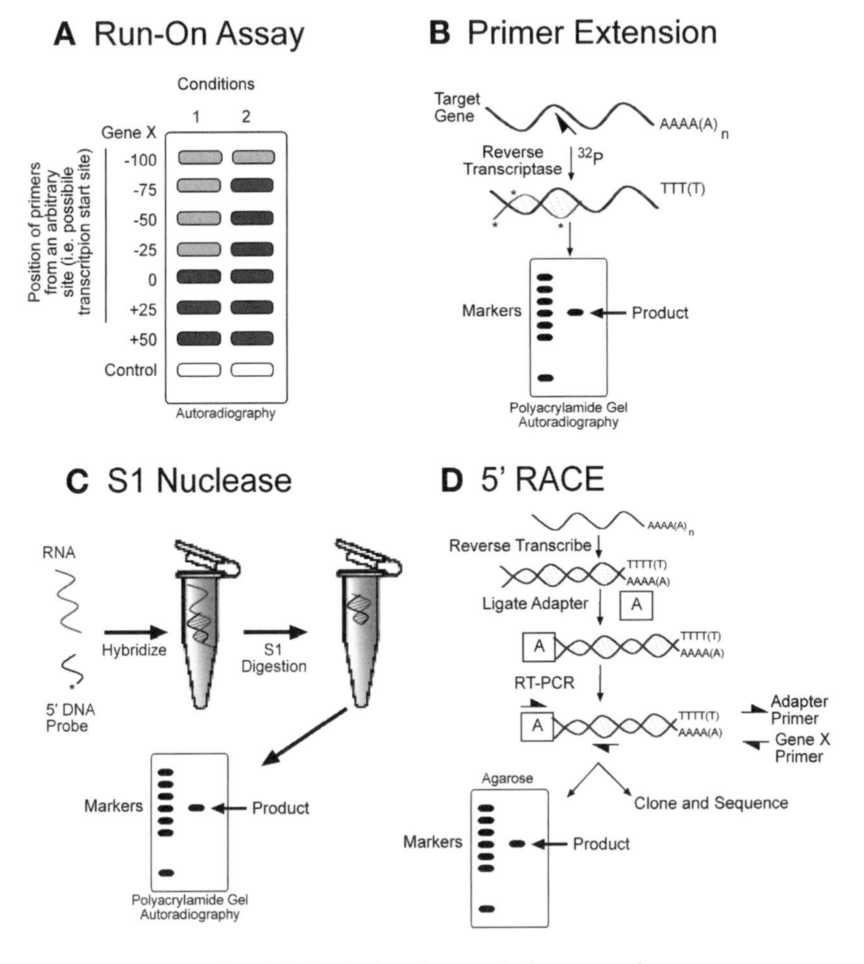

Fig. 4-4. Analysis of transcription start sties.

2.1.3. Mapping the Transcription Start Site

Many genes are more complicated than depicted in Fig. 4-1 with the possibility of alternative transcription start sites. Once transcription is determined to be the means by which a gene is regulated, examination of the predominant promoter sequence may be of great use in the determination of *cis*-sequences (*see* Section 2.2.). Commonly employed methods include the Nuclear run-on assay, primer extension, S1 nuclease mapping and 5' Rapid Amplification of cDNA ends (RACE; Fig. 4-4).

The use of the nuclear run-on assay to determine 5' ends is a logical choice if this was the method used to determine transcriptional activation. This approach simply utilizes a sequential set of probes designed to the 5' end of the cDNA. Generally, synthetic oligonucleotide probes are designed to regions around a suspected transcription start site.

Primer extension analysis is used to determine the location and to quantitate the amount of the 5' end of specific RNAs. A ^{32}P end-labeled oligonucleotide is hybridized to RNA and is utilized as a primer for reverse transcriptase. In the presence of deoxynucleotides, the RNA is reverse transcribed into cDNA and is analyzed on a denaturing polyacrylamide gel. The length of the cDNA reflects the number of bases between the labeled nucleotide of the primer and the 5' end of the RNA; the quantity of cDNA product is proportional to the amount of targeted RNA.

Detailed primer extension protocols can be found at www.promega.com, technical bulletin 113.

The use of S1 nuclease is very common in the analysis of RNA, including mapping of potential transcription start sites. Denatured 5'-labeled DNA (PCR product, plasmid insert, or oligonucleotide probe containing the target gene's promoter region) is hybridized to mRNA. S1 nuclease digests the 3' single-stranded end of DNA and does not digest double stranded DNA/RNA hybrid. The size of the product is determined by autoradiography of a polyacrylamide sequencing gel and can be used to determine the RNA transcription start site.

RACE analysis is a very simple method to find the 5' end of a gene that has not been extensively examined. That is, this method is the only one that does not require prior knowledge of the 5' sequence of the gene of interest. Following reverse transcription, an adapter oligonucleotide is ligated to the cDNA. This adapter contains a primer recognition site and can also include a restriction enzyme or nested primer site as well. Following ligation, the DNA is amplified using the adapter primer and a gene-specific reverse (downstream, 3') primer. The product can be analyzed for size and may be cloned and sequenced to determine the sequence of the gene. Detailed protocols can be found elsewhere [13].

13. Vanden Heuvel, J. P. et al. In: PCR Protocols in Molecular Toxicology, J. P. Vanden Heuvel, ed. CRC Press, Boca Raton, FL. 1998.

2.2. Analyzing *cis*-Sequences

The next step in the examination of a particular target gene is to determine how it is being transcriptionally regulated by a particular treatment or condition. This can take a variety of forms and may be dependent on the amount of information known about your gene of interest. For example, if you are interested in an oncogene such as c-myc, there is a wealth of information about potential transcription start sites, promoters, enhancers, and other *cis*-acting elements. In this case you could probably use information already in the literature or in public databases to predict what *cis*-elements are being affected by your treatment and what *trans*-acting factors are required. However, in the vast majority of cases there will not be sufficient data to hypothesize, or you will find your initial predictions were incorrect. In the following discussion, it will be assumed that the genomic sequence of the gene of interest is known and that the appropriate constructs are readily available.

Cis-elements *are nucleotide sequences found within the gene of interest.* Trans-elements *or* trans-acting *factors often refer to nucleotide binding proteins that are the products of other genes.*

2.1.1. Computer Searches

The analysis of *cis*-acting elements can be initiated without ever setting foot in the lab. However, the result of the computer-based analysis of a regulatory region on a gene will probably result in more questions than answers and at best will generate hypothesis that must be tested. A very simple outline of identifying potential transcription-factor binding sites in the human *c-myc* gene is shown in Fig. 4-5. Although this scheme shows the results obtained with a commercial software package, there are numerous websites that can serve the same purpose (*see* URLs in the marginal note, as well as Appendix)

NCBI website: http://www.ncbi.nlm.nih.gov/
Motif discovery and search: http://meme.sdsc.edu/meme/
Transcription-factor database: http://www.gene-regulation.com

The first step in this analysis is to obtain the sequence of interest in a form that can be used by other programs. This can be accomplished most easily through the National Center for Biotechnology

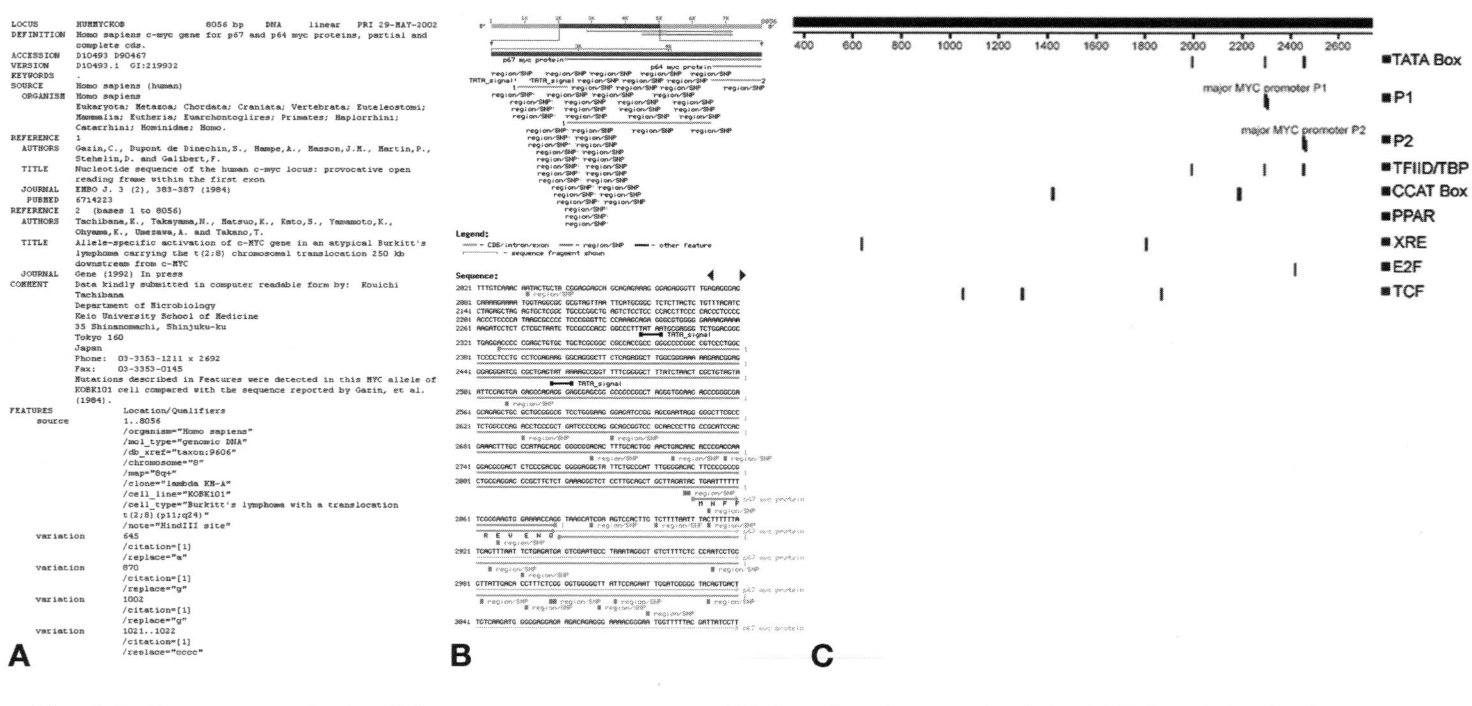

Fig. 4-5. Computer analysis of the *c-myc* protooncogene. (**A**) Results of a search of the NCBI website for human *c-myc*. (**B**) The sequence was visualized using the "Graph" output at the NCBI site. (**C**) This sequence was imported into Lasergene and the GeneQuest program (DNAStar, Madison, WI). Known transcription binding sites were examined in the first 2700 bp of the sequence and a small sampling of potential response elements are depicted.

Information (NCBI) website. Many web-based applications will require that you simply cut-and-paste the sequence from NCBI. Commercial programs may contain a search routine that will directly access a database of sequences, or you will be able to save the sequence file as a text file for import into the program you are using. Once the sequence is obtained, some of the most pertinent information may already be contained within the features section. For example, in the case of human *c-myc*, the identification of the two predominant promoters (P1 and P2) as well as exon–intron borders are listed.

The analysis of potential enhancer elements within the sequence can be "directed" or "undirected." In the former situation, a prediction is made regarding what transcription factor may be affecting the expression of the gene of interest. Some known response elements are listed in Table 4-1. It is not essential to use a fancy software package to find these sites as a word processor, or simply visual inspection, may be utilized. However, in many instances it may be advantageous to use a program that allows for ambiguous bases within the site. Current nomenclature for common ambiguous bases is given in Table 4-3.

An undirected search for potential REs requires the use of a search algorithm and a database of consensus sequences. There are several excellent web-based search programs to perform this search. The region that is believed to contain the regulatory sequence of interest is compared to the database of consensus sequences; a generated list (often quite long) will show the location of the matches. Most programs allow for flexibility in the percentage of match required and perhaps how many hits are acceptable. From the resulting list, and based on what is known about the biology of your model system and the gene of interest, a subset of likely players in the transcriptional initiation response may be tested.

Table 4-3
Ambiguous Bases

Code	Nucleotide
R	A or G
Y	C or T
S	G or C
W	A or T
K	G or T
M	A or C
B	C or G or T
D	A or G or T
H	A or C or T
V	A or C or G
N (or X)	Any base

2.1.2. In Vitro Footprinting

The method of nuclease footprinting is used to study the sequence-specific binding of proteins to DNA. There are many variations of the assay with different nucleases (DNase I, MNase, exonuclease III) as well as time of application of nuclease (in vitro vs in vivo). In the present case we will focus on in vitro footprinting as a means to identify regions of DNA that bind to purified transcription factors or crude cellular extracts. When used in this manner, in vitro footprinting is similar to the "undirected" computer search as well as the gel-shift assay (*see* the next section). As will be discussed in Section 5, in vivo footprinting is useful for examination of chromatin structure and is technically more challenging.

The most common nuclease for in vitro footprinting is DNase I. This enzyme binds in the minor groove of DNA and cuts the phosphodiester backbone of both strands. It is approx 40 Å in diameter and its bulk prevents it from cutting the DNA that is under and around a bound protein. A bound protein may have some other effects on nuclease cleavage, resulting in hypersensitive sites that are usually immediately adjacent to a protected region. DNase I does not cleave the DNA indiscriminately, some regions of naked DNA being more sensitive to cleavage than others; this is the reason behind using uncomplexed DNA as a comparison for detecting footprints.

This technique requires a DNA probe that has been end-labeled on one strand (details of the protocols can be found elsewhere [40]; *see* Fig. 4-6). To obtain such a probe, an isolated DNA fragment is labeled with ^{32}P and digested with a restriction enzyme that releases one of the labeled ends. This procedure may be facilitated by subcloning the fragment into a polylinker containing plasmid. DNA fragments can be 5' labeled with T4 polynucleotide kinase or 3' labeled using Klenow. As a single end-

40. Moss, T., DNA–protein interactions: principles and protocols, Methods in molecular biology (Humana Press, Totowa, N.J., 2001).

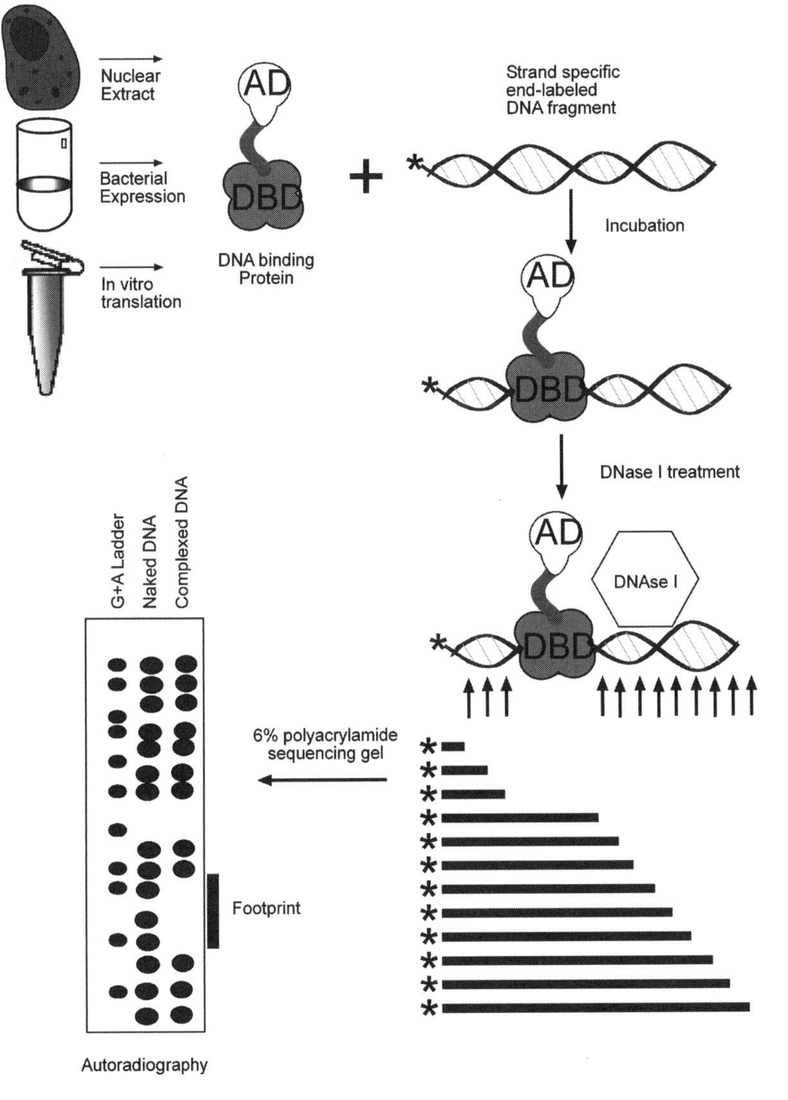

Fig. 4-6. In vitro DNase I footprinting.

labeled fragment allows one to visualize interactions on one strand only, it is usual to repeat the experiment with the same fragment labeled on the other strand. DNase footprinting requires an excess of DNA-binding activity relative to the probe as the higher percentage of occupancy will result in a clearer pattern on the gel. When utilizing crude extracts, an excess of a nonspecific competitor (poly[dIdC], calf thymus DNA) should be added to improve the specific protein–DNA interaction. Also, the amount of DNase I used in digestion must be empirically determined to optimize the detection of a footprint. The digestion is of course not complete with a goal of achieving an average of one cut per strand of DNA. The protein and DNA are allowed to interact, usually in the presence of glycerol to stabilize the complex, on ice for a short period of time (15 minutes to an hour). Subsequently, the DNA is digested with DNase I (0.001–0.1 U), the reaction terminated, and the products extracted. The digested DNA is loaded onto a polyacrylamide sequencing gel and compared to a G+A sequencing ladder generated from the probe.

2.1.3. Gel-Shift (Electophoretic Mobility Shift Assay)

The gel-shift, or electrophoretic mobility shift assay (EMSA), provides a simple and rapid method for detecting DNA-binding proteins. In many ways it is similar to the DNase I footprinting in that it is used widely in the study of sequence-specific DNA-binding proteins. The difference between the two assays is that EMSA examines a putative DNA binding site (using a 20–50 bp fragment) and hence can be used to test a hypothesis. It may also be used to quantify the affinity of a protein for a response element. The assay is based on the observation that complexes of protein and DNA migrate through a nondenaturing polyacrylamide gel more slowly than free DNA fragments or double-stranded oligonucleotides. (As discussed in Chapter 5, EMSAs can also be used to examine RNA–protein interactions.)

The gel-shift assay is performed by incubating a purified protein, or a complex mixture of proteins (such as nuclear or cell-extract preparations or in vitro translation), with a ^{32}P end-labeled DNA fragment containing the putative protein binding site (Fig. 4-7). The binding conditions for an EMSA are very similar to that of DNase footprinting. Low ionic strength buffers with glycerol, cofactors such as Mg^{2+} and an excess of nonspecific DNA (poly[dIdC], calf thymus DNA) are often used. The DNA probe may be a cloned fragment (50–300 bp) or a synthetic oligonucleotide (15–30 bp) and is typically end-labeled using T4 polynucleotide kinase. The reaction products are then analyzed on a nondenaturing polyacrylamide gel. Since it is essential that electrophoresis be done under gentle conditions to minimize dissociation of DNA and protein, temperature must be held constant while resolving. Generally, this may be accomplished by prerunning your gel before loading or by performing the electrophoresis in the cold room. Subsequently, the gel is removed, dried, and exposed to X-ray film.

There are many ways to utilize gel shifts to examine DNA–protein interactions. The specificity of the DNA-binding protein for the putative binding site is established by competition experiments using DNA fragments or oligonucleotides containing a binding site for the protein of interest (specific competitor). In addition, a gel shift in the presence of an excess amount of an unrelated DNA sequence (nonspecific competitor) may be performed to confirm speci-

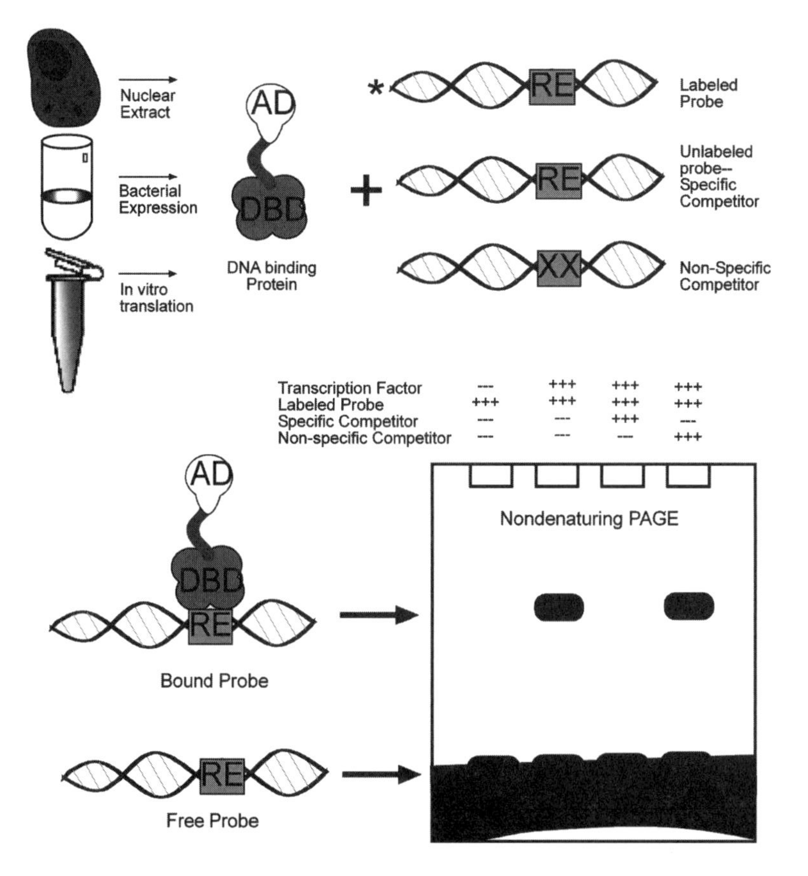

Fig. 4-7. Basics of the gel-shift (electromobility shift assay; EMSA)

ficity. Relative-binding affinity can be determined by titration of unlabeled, specific competitor. The more avid the binding between the *trans*-acting protein and the DNA, the more specific is the competitor required to displace and inhibit the gel shift. Gel shifts may be used to examine mutants of the RE. Point mutants may be end-labeled and used as a probe, or they may be used as a specific competitor. If a particular transcription factor is suspected, a gel shift in the presence of an antibody (supershift assays) may be used to confirm its presence in the shifted band, as will be discussed in a subsequent section.

Gel-shift and DNase footprinting experiments basically examine the same phenomena and utilize similar approaches. The advantage of gel-shift assays is that they are much easier to perform and tend to be more quantitative and reproducible. Among the disadvantages is the relatively small fragment of DNA being examined, and hence further preliminary experiments may be required. In addition, certain DNA–protein complexes either do not cause a change in mobility or they fall apart under the electrophoresis conditions. Thus, DNase footprinting is more amenable to a wider range of complexes.

2.1.4. Reporter Assay

Of the means to examine *cis*-acting elements, perhaps the easiest and the most powerful method is the reporter assay. A reporter gene can be any coding region that results in a product that can be easily measured. Similar to the gel-shift assay, the reporter assay is very versatile and can be used to assess the *cis*-element that is regulating the transcription of a gene. However, the reporter assay also gives information on the *effect* of the protein–DNA interaction. A *cis*-acting element can exert either a positive or a negative effect on gene expression; this will be reflected in the reporter assay but not the EMSA or DNase footprinting.

In order to perform a reporter assay, the only skills required are basic molecular biology (subcloning, to produce the reporter vector) and cell-culture techniques (including transfection). A variety of reporter assays will be discussed throughout this book but the subsequent discussion will examine the simplest approach, to determine if a particular DNA sequence contains a regulatory region that affects the transcriptional activity of a target gene. The goal is to show that an RE explains the effects on mRNA and transcription that were observed in prior studies.

A summary of a reporter assay to examine a particular fragment of DNA for the presence of a *cis*-element is shown in Fig. 4.8. First, a genomic clone of your gene of interest must be obtained. Generally, the 5' untranslated region (5' UTR) is the area of focus, but be aware that enhancer sequences can be found throughout the gene. In the past, this clone would have been obtained by screening a genomic library with a portion of the target gene. This was a time-consuming and arduous process. Fortunately, in the days of genome projects and public databases, it is easy to obtain the sequence from a computer search and utilize PCR to subclone the 5' UTR. Also, several private organizations have

The control plasmid is also called the transfection efficiency control plasmid. Since its activity is not affected by the treatment conditions, it is used to estimate the amount of DNA taken up by the cells. Data is generally expressed as target reporter/control reporter ("relative reporter activity"). It is also appropriate to express data relative to protein content to address extraction efficiency.

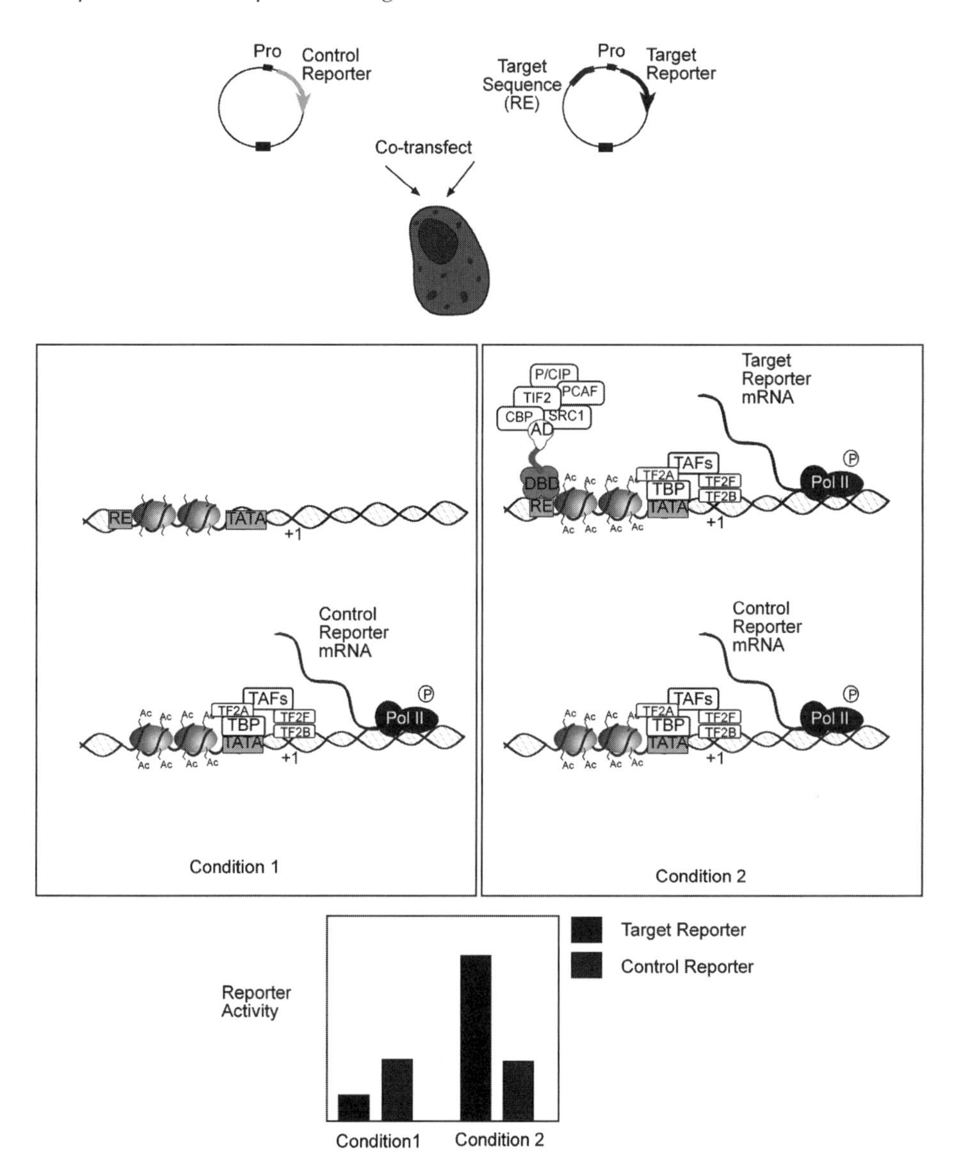

Fig. 4-8. Reporter assay to assess activity of *cis*-sequence.

clones that can be obtained for a reasonable fee (*see* Appendix for websites) and clone-by-phone (or email) from colleagues is almost always possible.

The next step is to subclone the genomic fragment into the reporter plasmid. The choice of reporter genes to use is dependent on the cells being examined, the sensitivity and quantifiability required, and the cost and ease of use

Table 4-4
Commonly Used Reporter Genes

Reporter gene	Substrates	Advantages	Limitations
Alkaline phosphatase (AP)	PNPP, FADP	High sensitivity, large linear range	Endogenous activity in most cells
Secreted AP (SEAP)	PNPP, ADP	Can detect in media (no cell lysis), low endogenous activity	
Chloramphenicol acetyl transferase (CAT)	^{14}C- or ^{3}H chloramphenicol (Cm)	Widely accepted, visualization of activity	Low sensitivity, high cost, radioactive
β-Galactosidase (βGAL)	ONPG, X-Gal	Easy assay, can be performed *in situ*	Low sensitivity, endogenous activity
β-Glucuronidase (GUS)	X-gluc, 4-MUG	Wide variety of assays	Endogenous activity
Green fluorescent protein (GFP)	None needed	No substrate required, *in situ* and in vivo experiments	Low sensitivity
Human growth hormone (hGH)	Iodinated (RIA)	Secreted, low background	Low sensitivity, expensive
β-lactamase	PADAC, CCF2	Easy colorimetric or fluorescent assays	
Luciferase	Luciferin	Fast and easy assays, high sensitivity, large linear range	Specialized equipment, expensive reagents

PNPP, *p*-Nitrophenyl phosphate; ADP, adenosine triphosphate; ONPG, *o*-nitrophenyl β-D-galactopyranoside; 4-MUG, 4-methyl umbelliferyl-bD-glucuronide; RIA, radioiodine assay; PADAC, pyridinium-2-azo-*p*-dimethylaniline chromophore; CCF2, coumarin cephalosporin fluorescein.

(*see* Table 4-4). Also, for transient transfection experiments, a control reporter plasmid should be included; thus, the choice of the reporter gene and assay system to examine *cis*-acting elements must be compatible with that of the control reporter. The control reporter plasmid contains a basal promoter that theoretically is not affected by the test conditions. Of course, this needs to be verified in preliminary experiments. The most popular choice is *firefly luciferase* or chloramphenicol acetyl transferase (CAT) for the target reporter and β-galactosidase or *renilla luciferase* for the control reporter.

The ability to transfect eukaryotic cells with plasmid DNA is essential in many molecular techniques, not the least of which is reporter assays. The methods used vary depending on cell type, growth conditions, and level of expertise. Many currently available methods of transfecting cells with plasmid DNA include diethylaminoethyl (DEAE)–dextran, calcium phosphate, or lipid methodologies, and commercial products are readily available as well. A general rule of thumb is the more differentiated a cell, the less efficiently it is transfected, regardless of the method used. For a very sensitive reporter gene such as luciferase, this is less of an issue. The poorly differentiated cell lines may be the easiest places to start, but keep in mind that they may not recapitulate the tissue-specific gene expression seen in previous experiments. Some commonly used mammalian cell lines that transfect with relatively high efficiency (10–20% of cells receiving DNA) include COS, CV-1, 293, Hepa, and Jurkat. For more differentiated cell lines that have poor transfection efficiency (<2%), transient assays may be unattainable and a stable transfection procedure may be attempted. In stable transfection, the target reporter is engineered with an antibiotic selection marker (i.e., hygromycin transferase) or is co-transfected with a plasmid encoding this enzyme. Under antibiotic selection for extended periods of time, the cells that did not receive plasmid (>98%) will be killed and the survivors will have the target reporter stably integrated into its genome.

Transfection connection: methods of nucleic acid delivery into eukaryotic cells: http://www.the-scientist.com/yr1997/sept/profile1_970929.html.

Stable transfection may be used for hard-to-transfect cell lines. However, these cells have a finite life-span due to loss of plasmid or genetic drift.

Once the initial target reporter plasmid is shown to be regulated under the test conditions, as just described, various deletion constructs may be utilized to further refine the search for *cis*-acting elements. Overlapping fragments of DNA will assure that a site is not lost as a result of restriction enzyme digestion or steric considerations by being too close to the promoter or plasmid sequences. Generally, the search for *cis*-acting elements will include computer-based analysis and gel-shift assays to define a motif that is being affected. Once identified, the RE may be mutated and reporter assays repeated. If the initial reporter does not show a change in activity, alternative hypotheses must be entertained. It is possible that the *cis*-element in not within the fragment as they may be found further upstream, in introns, coding regions, or 3' UTR. It is also possible that the cells being transfected do not contain the proper machinery to regulate this particular target gene, and alternative cell types should be sought.

2.1.5. Equilibrium DNA Binding Assays

Although currently not used frequently, membrane filtration has a long history for the examination of protein–nucleic acid binding [40]. The principle of the filter binding assay is quite simple. Under a wide range of buffer conditions, nucleic acids pass freely through nitrocellulose or nylon membranes, whereas proteins and their bound DNA/RNA are retained. Thus, if a protein within a nuclear extract (or produced by other means) binds to a specific DNA sequence, the complex will remain on the membrane. If the DNA is labeled, the amount bound can be easily ascertained. The technique can be used to analyze both equilibrium or kinetic binding events (if the free and bound probe are separated rapidly), and cold competitors similar to those described earlier can be employed to examine specificity.

40. Moss, T., ed. DNA–Protein Interactions: Principles and Protocols, Humana Press, Totowa, NJ. 2001.

Filter binding assays have several advantages over gel-shift or DNase footprinting. It is extremely rapid and reproducible and may be used to determine equilibrium dissociation constants (K_d). A disadvantage is the fact that not all protein–DNA complexes withstand the filtering and washing processes, and complex mixtures of proteins often lead to a high amount of background (nonspecific) binding of radiolabeled DNA.

Fluorescent polarization (FP) techniques have become the preferred method to determine equilibrium dissociation constants for a protein and DNA fragment. This is a solution-based assay that does not require the separation of free and bound probe and hence is more amenable to a variety of protein–DNA complexes. Also, the affinity between the macromolecules need not be high, since separation is not required. The DNA probe in this case is fluorescently labeled (rhodamine or fluorescein, for example). FP assays take advantage of the fact that fluorescent molecules rotating in solution depolarize an incoming beam of polarized light. The faster a molecule tumbles in solution, the more depolarization occurs and when bound to a larger complex such as a protein the depolarization lessens. A polarimeter is used to measure the amount of polarized light that is emitted. Once again, unlabeled competitors can be included to assess specificity. Although rapid and quantitative, FP assays suffer from low sensitivity and may be affected by multiple proteins in a crude extract. Thus, this technique is generally reserved for quantitation of protein–DNA affinity after the *trans*-acting protein can be identified and purified (*see* next section).

2.3. Examining *Trans*-Acting Proteins

The identification of a *cis*-acting element in a target gene is generally the beginning of a more detailed analysis of protein–DNA and protein–protein interactions. As shown in Fig. 4-1, the *cis*-element (RE) is the base on which a large scaffolding of proteins is assembled. The most proximate event that occurs is an interaction between a protein (i.e., transcription factor, soluble receptor) and the RE through a DNA-binding domain (DBD). In the following discussion, the identification of the DBD-containing protein that is found associated with an RE is described (Sections 2.3.1.–2.3.3.). Similar methods, by which REs for known transcription factors are analyzed, will also be discussed (Sections 2.3.4, 2.3.5.). The subsequent events, such as recruitment of co-regulatory proteins to activation domains (AD), will be described in a later chapter dealing with protein–protein interactions.

2.3.1. Supershift Assay

The "supershift" assay is perhaps the simplest method to identify a protein associated with a particular response element and is a slight variation of the basic gel-shift or EMSA (*see* Section 2.2.3.). The premise of the supershift is that antibodies binding to a protein in the DNA complex (the shifted band in an EMSA) will further retard its migration in a native polyacrylamide gel. A hypothesis is made regarding which protein is binding to a RE based on computer analysis or biological conjecture. The only additional reagent required is the antibody, although not every antibody will suffice. The antibody must recognize an epitope that is accessible in the native, multiprotein complex, and hence may not be the same preparation that is used for Western blotting. An excess of this antibody is added to the protein extract along with the DNA response element; the appropriate amount of antibody may need to be empirically determined. Subsequently, the mixture is resolved on a native gel as stated earlier, and the supershifted band visualized by autoradiography. Owing to the large size of the antibody, the supershifted band may have difficulty entering the gel and may be found in the well. Alternatively, the antibody may ablate the interaction between the protein of interest and DNA, or to a required dimerization partner, and result in a loss of band. Both cases (supershift or ablation) confirm the presence of the hypothesized transcription factor. In addition to the DNA specificity controls, an antibody control must also be added, utilizing preimmune serum, purified IgG, or an antibody to an unrelated protein to assess specificity.

2.3.2. Affinity Purification

In the search for a protein that interacts with a DNA fragment, affinity purification is one of the most powerful approaches and utilizes classical biochemical techniques. Although it suffers from a lack of sensitivity, affinity purification can be relatively unbiased and may identify novel DNA–protein interactions. Following identification of a DNA response element, the fragment is affixed to a resin. Biotinylation is the method of choice for modifying the DNA and streptavidin–agarose is the preferred matrix. This may also be performed with RNA, as discussed in the next chapter. Subsequently the affinity resin is packed into a column and cellular extracts are poured through. Following extensive washing, elution of the bound proteins can be achieved through application of excess unmodified DNA fragment or detergent to the column. The eluted proteins are resolved by SDS-PAGE and identification may be performed by Western blotting or partial peptide sequencing (these methods are described in more detail in later chapters).

There are several variations on this classic approach. The avidin–agarose can be used to precipitate the DNA–protein complex instead of using a column. A particular useful variation is to crosslink the protein–DNA complexes. To produce covalent crosslinks between proteins and DNA, various methods can be employed including ultraviolet or γ-irradiation as well as chemical or physical (vacuum drying) means. Photoaffinity crosslinking is also frequently applied because of its higher specificity. In this case, the DNA is synthesized with photoactivatable deoxynucleotides such as 5-iodo- or 5-azido-2'deoxyuridine-5'monophosphate [40]. The 5-iodo or azido-containing DNA (also biotinylated in this case) is allowed to bind to the protein, and a UV light source is applied (300 nm or longer). Column chromatography is used to separate the covalent protein–DNA complex. Formaldehyde crosslinking can be reversed so that the protein may be recovered and examined by Western blot or sequencing.

2.3.3. Southwestern Blotting

Southwestern blotting entails resolving mixtures of proteins (nuclear extracts or partially purified preparations) by SDS-PAGE and probing the blot with radiolabeled DNA containing the RE [40]. The proteins are transferred by electroblotting to an immobilized membrane, as they would in a standard Western blot. However, the DNA binding activity of the proteins may be affected by the SDS and refolding is probably required. Prior to transfer the gel may be equilibrated in SDS-free buffer; however, the lack of detergent may affect the efficiency of transfer. Alternatively, the proteins may be renatured *in situ* by incubating the blot in 6 *M* urea and gradually decreasing urea concentration with successive changes of media. The membrane is incubated with blocking buffer containing nonfat dry milk or bovine serum albumin to decrease nonspecific binding. Additionally, poly(dIdC), salmon sperm DNA, or calf thymus DNA may be included. The radioactive DNA probe is incubated with the membrane (overnight, 4°C) followed by a series of washes with dilutions of the blocking buffer. Since several DNA-binding proteins require zinc, it may be prudent to add $ZnCl_2$ to all solutions (loading, electrophoresis, transfer, binding, and wash buffers) [40]. Following autoradiography, the apparent molecular weight of the DNA binding protein can be assessed.

SDS-PAGE stands for sodium dodecyl sulfate-polyacrylamide gel electrophoresis.

40. Moss, T., DNA–protein interactions : principles and protocols, Methods in molecular biology (Humana Press, Totowa, N.J., 2001).

Southwestern blotting requires very little specialized expertise to perform and can be used to rapidly examine multiple protein samples for changes in DNA binding activity or to assess the relative affinity of mutant DNA fragments. However, this approach requires that the *trans*-acting protein is able to interact with DNA as a monomer and no additional cofactors are required. Also, not all proteins are equally responsive to the refolding, and, hence, DNA binding activity within the membrane may be insufficient.

2.3.4. One-Hybrid Cloning of DNA Binding Proteins

The Clontech website is an excellent resource for information and resources related to one-, two- and three-hybrid assays (www.clontech.com). An informational bulletin on the one-hybrid may be found at http://www.clontech.com/archive/JAN03UPD/pdf/MM_OneHybrid.pdf.

The yeast one-, two- and three-hybrid systems have been versatile tools to examine protein–DNA and protein–protein interactions. Of these systems, the two-hybrid system is the best known and is discussed in detail in a subsequent chapter; thus, a detailed description of the various plasmids and approaches will not be supplied here.

The yeast one-hybrid assay is a library screening approach, whereby *trans*-acting (DNA-binding) proteins can be cloned. It is perhaps the most powerful and comprehensive method to identify novel protein–DNA associations. To understand the one-hybrid assay, it is necessary to describe the frequently employed two-hybrid assay. In the yeast two-hybrid vernacular there are "bait" and "prey" constructs with the former containing a DNA binding domain (DBD, i.e., Gal4 DBD) and the latter containing an activation domain (AD, i.e., Gal4 AD). Proteins are expressed from these plasmids as chimeras that now contain exogenous DBD and AD. A third plasmid is also transfected into yeast cells, the reporter. This reporter is generally β-galactosidase (but it can also be an amino acid synthesis enzyme) and its also contains the *cis*-element recognized by the DBD contained in the bait construct. If the proteins cloned into DBD- and AD-containing plasmids interact with each other, the DNA binding activity is linked to the *trans*-activation domain and the reporter gene is expressed. Novel protein–protein interactions can be identified by placing a known protein into the bait plasmid and preparing a cDNA library in the prey (or vice versa). Colonies that express the reporter are isolated and the interacting protein identified by sequencing.

The yeast one-hybrid screening approach takes advantage of the cDNA-AD chimeric library. Several such libraries are commercially available from a variety of species and tissues. However, synthesis of a cDNA library has become much more automated and is not as daunting as it once was. First, this *cis*-element (the "bait" in the one-hybrid) is inserted into the reporter plasmid (encodes the nutritional marker such as HIS3 or the enzyme β-galactosidase). Next, the library from the tissue of choice is constructed in the prey (AD) plasmid. Third, yeast cells are cotransfected with the library and the *cis*-element reporter. To screen for positive one-hybrid interactions, the transformation mixture may be spread on selective medium (i.e., His- media if the HIS3 reporter was used) or non-selective media (β-gal) containing a chromagen (X-gal) and incubated at room temperature. With amino acid selection reporters, the colonies that survive are propagated and the DNA binding protein identified by sequencing a rescued plasmid or PCR amplified insert. For β-gal reporters, a simple blue-white screen is performed and the DNA binding proteins identified as described above.

As with the yeast two-hybrid, the one-hybrid suffers from a high rate of false positives. For example, a protein that associates with the basal transcription machinery may increase reporter activity but it is not directly associating with the target DNA sequence. Thus, any interaction found in this approach must be verified by using any of the methods described above to examine DNA–protein interactions.

2.3.5. Selection and Binding Assay

Methods used for characterization of cis-elements for a given trans-acting *protein:*
1. Selection and binding assay
2. Chromatin immuno-precipitation (ChIP) and ChIP cloning

Identification of a *trans*-acting protein that affects a particular gene may result in a further characterization of this protein. For example, one question that may be addressed is "what other response elements does this protein associate with?" Using methods already discussed, it is possible to generate

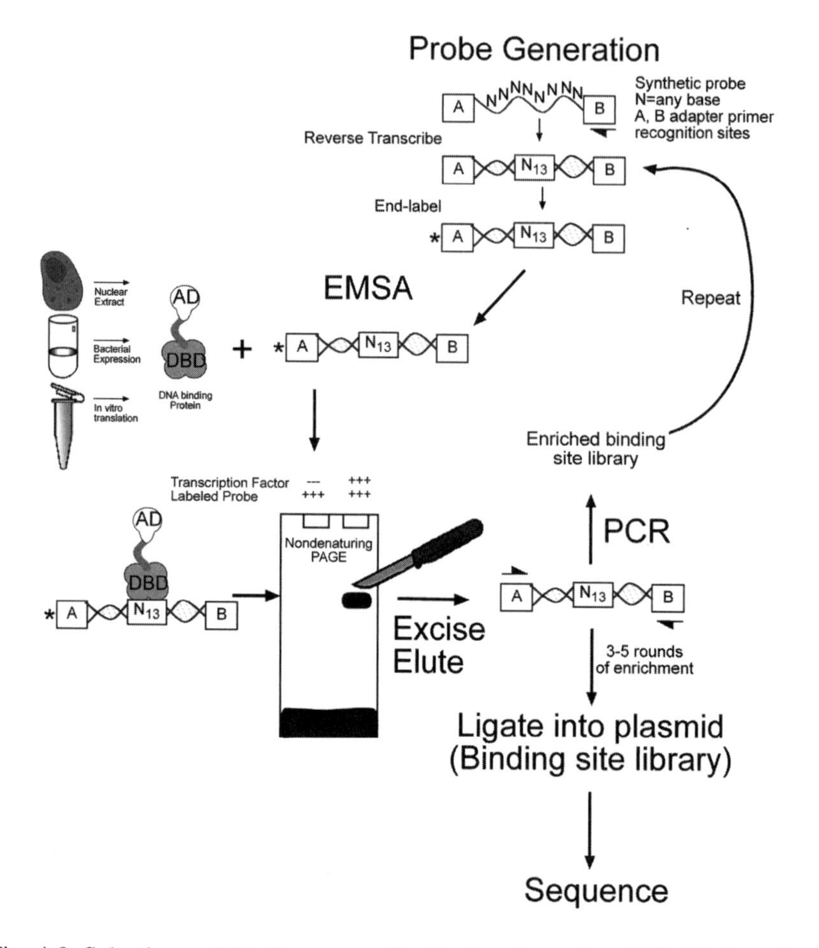

Fig. 4-9. Selection and binding assay (SeaBay) to determine DNA binding sites for a given *trans*-acting protein.

41. Blackwell, T. K. et al. Science 250 (1990) 1149–1151.

a "binding site library" for a given transcription factor or soluble receptor. The interaction of the protein of interest with the DNA will be the criteria used to screen the library, and PCR will be used to enrich the selected oligonucleotides (*see* Fig. 4-9). The exciting feature of the selection and binding sites (SeaBay or SAAB) technique [41] is its ability to examine RNA or DNA binding sites for any protein that binds selectively to a particular sequence. There need not be any indication of which genes this sequence is found or if the protein–DNA interaction has a biological role. The SeaBay technique is an ideal screening tool for detecting a new protein–DNA interaction and may also identify potential new target genes for a transcription factor.

It is beyond the scope of this chapter to discuss how the protein of interest is isolated and expressed. For simplicity, we will also assume that the protein binds as a monomer or homodimer. Of course, if the accessory proteins for a DNA-binding complex are known, they can be added to the incubations, or the protein can be expressed following transfection and nuclear extracts prepared. The first step in the SeaBay procedure is to produce the DNA library. The simplest method to produce these sequences is to synthesize an oligonucleotide that contains primer recognition sites (Fig. 4-9 A,B) flanking a stretch of random bases (n). Following reverse transcription using the B primer as the primer, potentially 4^n different probes are produced (where n = number of random bases). Following end-labeling, this library is screened using a standard EMSA assay. Following autoradiography, the retarded band is excised from the gel, eluted from the polyacrylamide, and PCR amplified using the appropriate primers (A+B). The PCR products are once again end labeled and the enrichment process repeats three to five times. Finally, the amplified products are cloned and sequenced. The subsequent REs can be aligned to determine a consensus sequence. Additionally, the RE identified can be used to search the databases to determine if any genes known to be regulated by the *trans*-acting protein contain this sequence.

2.3.6. Chromatin Immunoprecipitation (ChIP) and ChIP Cloning

Chromatin immunoprecipitation (ChIP) assays are relatively new tools to examine DNA–protein interactions in vivo (Fig. 4-10). They can be used to examine chromatin structure, identify transcription factors involved in the regulation of a specific gene, or be employed in the search for target genes of an individual transcription factor. Using ChIP to identify a transcription factor involved in gene regulation is very similar to affinity chromatography described above (Section 2.3.2.), although it has the advantage of examining DNA–protein interactions in vivo. ChIP cloning can be used to identify target genes that are under the direct control of a particular *trans*-acting protein. In most instances other techniques such as SAGE and microarrays will identify both direct targets as well as indirect targets of the particular transcription factor.

In brief, ChIP involves using formaldehyde to chemically crosslink bound proteins to underlying DNA [42]. Formaldehyde may be added directly to cell culture media or to homogenized tissue samples (1% v/v). Fixation is relatively rapid (10–15 min) and is stopped by the addition of excess glycine. Adherent cells are trypsinized, all types of samples are collected by centrifugation then swelled in the presence of detergent (i.e., 0.5% NP40) and, following homog-

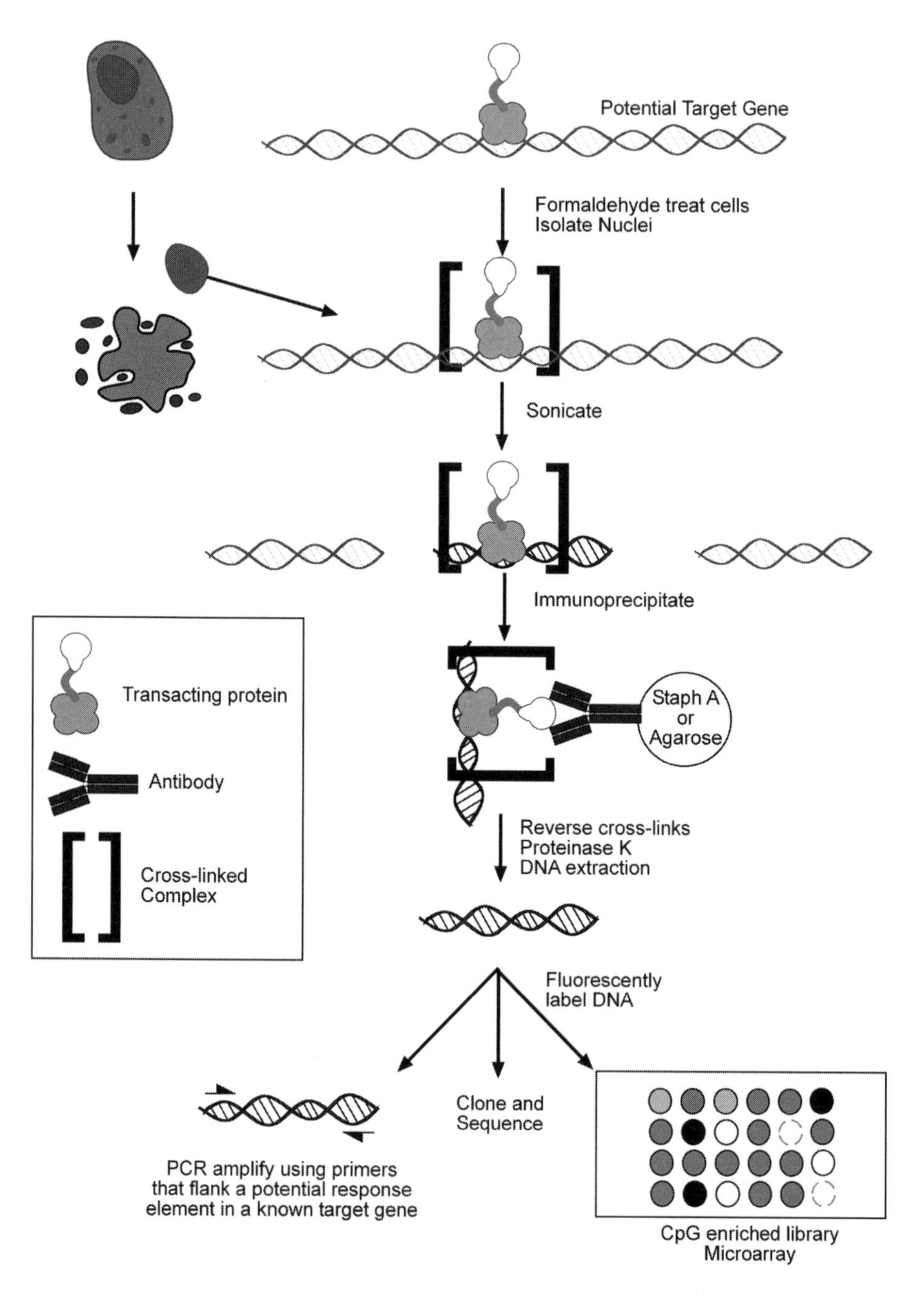

Fig. 4-10. Chromatin immunoprecipitation (ChIP) assay.

enization, the nuclei are isolated and resuspended. The crosslinked chromatin is sonicated to achieve the desired size of DNA fragments. Sonication time and power must be determined empirically to obtain DNA fragments with an average length of 500–1000 bp. Protein complexes of interest are immunoprecipitated with an antibody against one of the crosslinked proteins. Immunoprecipitations involve many considerations of antibody design and use; these will be discussed in more detail later in the book (Part II). However, blocking nonspecific antibody–protein and DNA interactions is critical. Sheared DNA (calf thymus, salmon sperm) and bovine serum albumin should be used to decrease the background binding to staph A cells, protein (A/G) agarose, or any other reagent used to precipitate the antibody. Following this, antibodies are added that are directed against a *trans*-acting protein as well as an antibody control (no antibody, preimmune serum, or antibody against a non-DNA binding protein). This antibody control will be examined in parallel throughout the experiment. Immunoprecipitation, washing, and elution of the complexes is performed as described in Part II. Crosslinks are reversed by the addition of NaCl to a final concentration of 200 m*M* and RNA removed by RNase A digestion. The DNA fragments are precipitated in ethanol, proteinase K digested and extracted into phenol:chloroform:isoamyl alcohol.

At this point, at least three approaches may be used to identify the DNA associated with the immunoprecipitated protein. Following reversal of the crosslink, PCR may be performed using primers that span the suspected RE in a target gene. This is an excellent way to confirm that a RE identified above is being affected by a transcription factor in the intact cell. In fact, this is perhaps the best way to show that a particular protein is involved in regulating gene expression by directly associating with the promoter region of a gene. The second approach to identify the DNA targets is to clone the immunoprecipitated fragments into a suitable vector and

42. Wells, J., Farnham, P. J. Methods 26 (2002) 48–56.

sequence the resultant clones. This is a brute force method to identify target genes and is in many ways similar to electronic subtraction or serial analysis of gene expression (SAGE). That is, a side-by-side comparison is necessary between DNA sequences from the control antibody and the clones obtained by immunoprecipitation using the *trans*-acting protein antibody. Finally, a short cut to sequencing is to label the DNA with a fluorescent dye and hybridize to a promoter-enriched DNA array. The test and control fragments would be labeled with different dyes (Cy5/Cy3), and analysis would be similar to the gene expression microarrays. The most common source of probes for such an array are derived from CpG-enriched libraries [43], which may be amplified and spotted onto glass slides. As stated in the chapter introduction, CpG islands often correspond to promoter regions of genes. Thus, this type of a library would be ideal for screening ChIP-derived DNA fragments for REs.

43. Cross, S. H. et al. Nat. Genet. 6 (1994) 236–244.

2.4. Regulation of Transcriptional Elongation

Intrinsic factors that affect transcriptional elongation include cis-elements and chromatin structure; extrinsic factors generally affect polII phosphorylation.

The regulation of transcriptional elongation is not commonly considered when mRNA accumulation is examined. However, genes that are under very strict cellular control, such as proto-oncogenes, cytokines and cell cycle regulators, utilize this mechanism as a rheostat to fine-tune the amount of a transcript produced. The proto-oncogene *c-myc* is an excellent example; its transcriptional elongation is affected by tumor promoters such as phorbol esters, as well as by vitamin D and estrogen. Transcriptional elongation is affected by two types of regulators, intrinsic and extrinsic factors. Intrinsic factors are sequences within the gene itself (*cis*-elements, chromatin structure) that cause a pause or arrest in pol II elongation. Often these *cis*-elements are found within introns and are recognized by *trans*-acting proteins. Another intrinsic regulator is chromatin structure; a highly compacted configuration would hinder the rate of transcriptional elongation. The examination of chromatin structure is discussed below (Section 2.4.5.). Extrinsic factors are those that affect the activity of pol II, possibly by altering the phosphorylation of the carboxy terminal domain (CTD) of this protein or by affecting the amount and activity of pol II accessory proteins. The approaches to examinination of transcriptional elongation are for the most part the same as those for the rate of transcription. In the present section, the focus will be on determining if transcriptional elongation is affected, followed by examination of intrinsic factors that affect this process.

It is prudent, perhaps, to discuss why you may want to examine transcriptional elongation as a mechanism of regulation of a target mRNA. The previous discussion of enhancer elements and promoters dealt primarily with the initiation of transcription. Under some circumstances, you may find that mRNA accumulation is different upon treatment due to transcriptional effects, but transcription initiation is not affected. Two lines of evidence may suggest that transcription initiation is not affected: (1) *cis* elements in the promoter cannot be found by any of the methods described previously; and (2) chromatin has not been differentially remodeled around the promoter under the applied treatment conditions (*see* Section 2.4.5.). Alternatively, you may find that transcriptional initiation is affected, but not to the same degree as the mRNA accumulation (assuming, of course, that mRNA stability is also unaffected by the treatment). Thus, this lends credence to examining transcriptional elongation as a regulated event under your treatment conditions.

2.4.1. Nuclear Run-On Assays and S1 Nuclease Mapping

The nuclear run-on assay is perhaps the best way to directly examine if a particular treatment is affecting transcriptional elongation, as well as if it affects the overall rate of transcription. In fact, there is very little difference in the approaches used. The predominant alteration that allows for the examination of transcriptional elongation is probe design. In the classical run-on assay, a probe that is found in a 3' exon is commonly employed. To examine transcriptional elongation, probes that are found toward the 5' end of the gene are used. In fact, several probes can be used to pinpoint where a stall in elongation is occurring. Thus, regulation of pause or arrest sites is performed in a manner identical to the determination of a transcript start site (*see* Fig. 4-4A).

Although not as commonly employed, S1 nuclease mapping can also be used to examine transcription elongation [44]. Once again, this is very similar to transcription start-site determination (Fig. 4-4C). A probe is designed to span a potential pause site in the gene. To avoid an overwhelming signal from fully elongated mRNA found in the cytoplasm, it may be necessary to isolate nuclei and extract RNA from this compartment. Following hybridization of the probe with the nuclear RNA, S1 nuclease digestion and polyacrylamide gel elec-

44. Epshtein, V., Nudler, E. Science 300 (2003) 801–805.

trophoresis is performed. A difference in the size of the protected fragment between treatments may indicate that elongation has paused.

2.4.2. Reporter Assays

To examine transcription initiation, reporter constructs are designed with prospective enhancer elements (5' UTR sequences, synthetic RE), coupled to a basal promoter (SV40, thymidine kinase [TK]) and reporter gene. If transcription elongation is the goal of study, this construct needs to be modified slightly. The first difference would be to have a promoter that is constitutively active and is not affected by the treatment conditions. Generally strong basal promoters like SV40 and cytomegalovirus (CMV) work well, although certain treatments and conditions affect their activity and alternatives may be necessary. It is possible to use the target gene's natural promoter if you have shown in previous experiments that pol II recruitment is similar under your treatment conditions. Following the exogenous promoter, the region of the target gene immediately downstream of its natural promoter is cloned. Often pause and termination sites are found in the first exon or intron, although they could be found throughout the sequence. Finally, the reporter gene itself is inserted and this plasmid is examined identically to that described earlier for the examination of *cis*-acting elements. If a treatment affects transcriptional elongation, the inclusion of the DNA with the pause site will give a different amount of reporter gene to be expressed. The site of the pause may be dissected identically to that for an enhancer using reporter assays (EMSA, among others).

2.4.3. In Situ *Hybridization*

A relatively novel approach for examining transcription elongation is to perform *in situ* hybridization [45]. This is somewhat analogous to the nuclear run-on assay in that the probes are designed to hybridize to the 5' portion of the transcript.

45. Wilson, A. J. et al. Cancer Res. 62 (2002) 6006–6010.

However, this is performed in cells that have been fixed and does not require radioactivity or isolation of nuclei. In this technique, probes are designed to be complementary to the 5' transcript (i.e., first exon) and to a full-length transcript (3' portion of the mRNA). Each probe may be labeled with a different chromo- or fluorophore. Cy5 or Cy3 dyes used in microarray experiments are appropriate, as are FITC and rhodamine. The choice of dyes may be dependent on the microscope capabilities and the laboratory's expertise. The cells under the treatment conditions are fixed and *in situ* hybridization with the probes performed (a detailed explanation of *in situ* hybridization will be presented in Chapter 5). Cells are scored for hybridization to the 5' probe relative to the 3' probe. If transcription elongation is being halted, one would expect to see cells with higher intensity of 5' probe fluorescence than its 3' probe counterpart. The quantitative scoring of fluorescence intensity is not trivial and requires much cell biology expertise, as well as microscopy software. Nonetheless, it has the advantage of being an approach to examine the in vivo processes of transcription.

2.4.4. ChIP Analysis of RNA Pol II and Accessory Proteins

Chromatin immunoprecipitation assays can be used to study the protein–DNA interactions that are associated with transcriptional elongation much in the same way they are used for *trans*-acting proteins and enhancer elements. The key determinants for the study of transcriptional elongation are the antibodies used to immunoprecipitate the complex and the primers used to detect the associated DNA. A short list of potential types of antibodies that may be employed to study elongation are shown in Table 4-5 and include RNA pol II and its associated proteins. Thus, the analysis would be to immunoprecipitate the protein–DNA complexes using RNA pol II and PCR amplify using a series of primer pairs. If

Table 4-5 Antibodies Used to Examine Transcriptional Elongation in ChIP Assays

Antibody directed against
RNA pol II
Pol II phosphorylation site in CTD
Histone H4 acetylation site
Transcription elongation associated proteins (pTEF, TFIIH, cyclin H)

stalling/halting is occurring in a particular region, the RNA pol II should be found more frequently at that location.

2.5. In Vivo Footprinting to Study Chromatin Structure

The in vivo analysis of DNA–protein analysis and chromatin structure can provide crucial information on how a gene is being regulated. Clearly, gene promoters and DNA structure are best studied in vivo and footprinting assays such as those described herein are accurate predictors of the transcriptional activity of a target gene. A "closed" chromatin structure, or DNA tightly associated with histones, generally signifies a silent gene. In order for transcriptional regulation to occur, a more "open" state must be achieved so that transcription factors and RNA polymerase can gain access.

To assess the chromatin configuration of the intact gene, there are many approaches that can be used [40]. Two of the most popular in vivo footprinting assays are described, DNase (or MNase) cleavage in permeabilized cells and DMS/piperidine directed cleavage (Fig. 4-11). In vivo fooptrinting with DNase I is very similar to its in vitro counterpart (*see* Fig. 4-11A); however, DNase I cannot penetrate cells without first permeabilizing the membranes. Most cells can be efficiently permeabilized with Nonidet P40 (NP40) or lysolecithin. It is preferable to treat whole cells instead of isolated nuclei since polyamine buffers will result in the loss of transcription factors upon nuclei isolation. Nonetheless, to examine the location of histones within a particular gene, DNase I footprinting with isolated nuclei is acceptable. DNase I binds to the minor groove of DNA and cuts the phosphodiester backbone on both strands, but because of its size (31 kDa) it is unable to cut DNA under and around a bound protein. In fact, because of steric hindrance the DNase molecule cannot cleave in the immediate vicinity of a histone and often a hypersensitive site

40. Moss, T., ed. DNA–Protein Interactions: Principles and Protocols. Humana Press, Totowa, NJ. 2001.

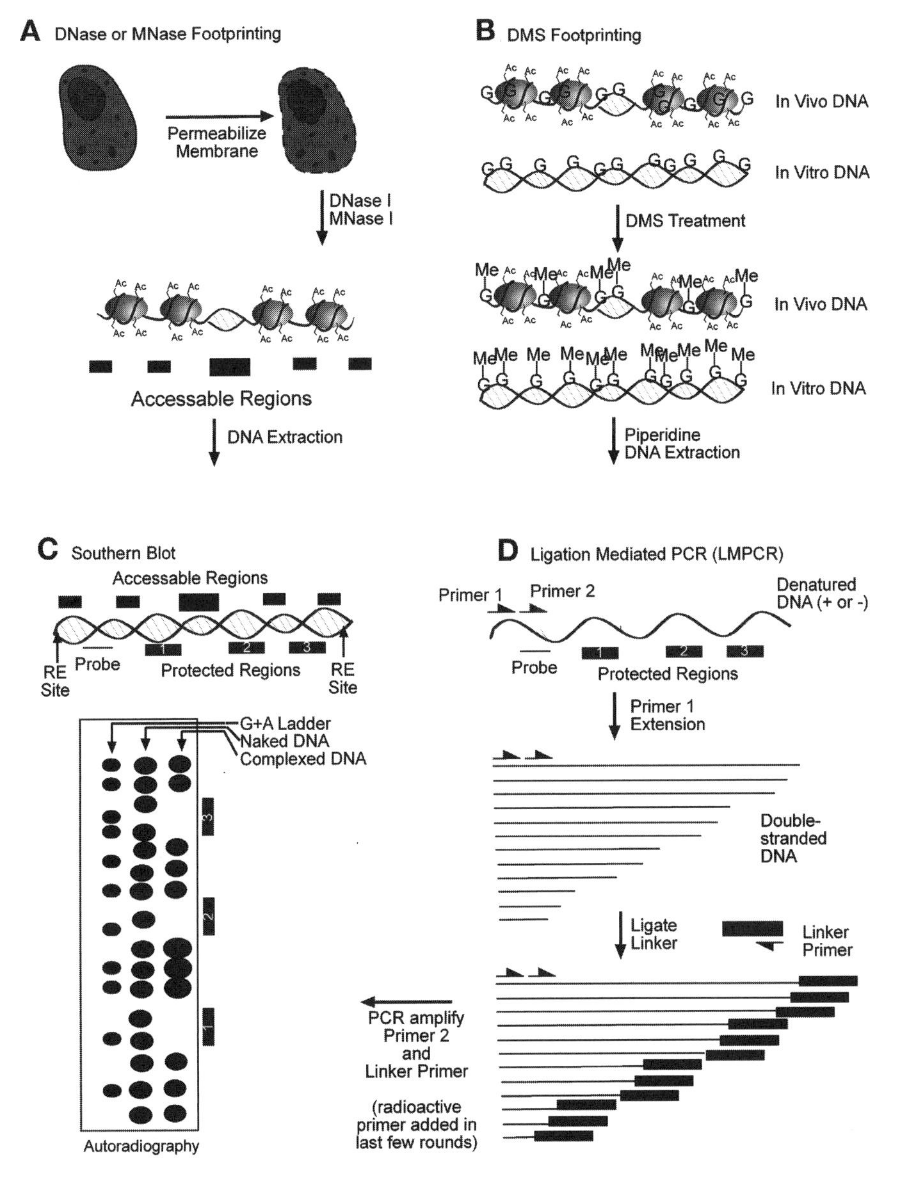

Fig. 4-11. Approaches to in vivo footprinting.

10 bp from a nucleosome is observed. Linker DNA (the DNA between nucleosomes) has a high propensity of hypersensitive sites, which is fortunate since DNA wrapped around histones within the nucleosome may not have a distinct footprint. Hence, the nucleosomal boundaries are discernible even though they are not as clear as in vitro footprinting with *trans*-acting proteins. Also, a bound

protein may bend the DNA in such a way as to render a site hypersensitive to this nuclease. To reveal the most precise footprint, the amount of DNase I added to permeabilized cells must be determined empirically. Concurrently with DNase I treatment of permeabilized cells, naked DNA should be used as a comparison. This is important since DNase I does have some preference for particular DNA sequences and structures, regardless of whether the template is from in vitro or in vivo sources. Following DNase I digestion, the footprint may be observed using Southern blotting or ligation-mediated PCR (LMPCR) as outlined in Fig. 4-11, C and D, respectively. The advantage of LMPCR as a detection method is its enhanced sensitivity, although both procedures should result in similar footprinting patterns.

Other nucleases may also be used for in vivo footprinting in permeabilized cells. MNase I and exonuclease III are both commonly used. The molecular size and cleavage specificity of these enzymes are different from DNase I and may result in a slightly different footprinting pattern.

An alternative to using permeabilized cells is to assess the local reactivity to modifying agents of DNA in living cells as compared to that of purified DNA. Dimethylsulfate (DMS) is an alkylating agent that methylates guanine residues at the N7 position. In purified DNA virtually every guanine has the same probability of reacting with DMS. However, in the intact cell DMS methylation of guanine residues that are bound by proteins would not proceed with the same efficiency. Being a small molecule, DMS can be added to tissue culture media where it easily enters the cell. Hence, this method is perhaps more reflective of the true transcriptional state and does not suffer from permeabilization or isolation artifacts. DMS is added to the media (0.1–0.5%, v/v) and incubated for a short period of time (5–10 min). The cells are washed with fresh media, trypsinized, pelleted, and DNA purified. Subsequently, hot piperidine (2 *M*) is used to cleave the glycosylic bond of methylated guanines. Processing of the DNA for Southern blotting or LMPCR is identical to that described for DNase I footprinting. Similar to DNase I footprinting, DNA structure results in both DMS footprints and hypersensitive sites.

DMS footprinting has the advantage of detecting DNA–protein contacts under more physiological conditions. However, this approach may not detect all DNA–protein interactions. First, the DNA contact site must contain guanine residues to be detected. Second, the accessibility of some DNA sites to DMS is unaffected by protein association. Last, the reactivity of DMS may affect weak DNA–protein contacts and these would be lost in the subsequent analysis.

2.6. Histone Modification

An important consideration regarding the structure of chromatin in a localized area is whether histones are posttranslationally modified. A variety of well-conserved posttranslational modifications occur in the histone tail domain, including acetylation, phosphorylation, methylation, ribosylation, ubiquitinylation, and glycosylation; the most-studied of these is the acetylation of histone tail domains on specific lysine residues. Conventional wisdom holds that these modifications alter the strength of the interaction between DNA and the histone core, thereby changing the accessibility of this DNA to transcription factors, leading to changes in gene expression. In fact, lysine residues of histone tails are commonly acetylated in promoter regions of genes that are transcriptionally active. Thus the study of histone acetylation should be considered as part-and-parcel of the examination of chromatin structure (i.e., footprinting) as well as DNA methylation (Section 2.7.).

As stated in previous discussion, chromatin immunoprecipitation (ChIP) is a powerful and versatile tool for studying such DNA–protein interactions within a native chromatin environment. Chromatin analysis using ChIP requires high-titer, residue- and modification-specific histone antibodies. (See the marginal note for a URL to a list of antibodies and their suppliers.) The basic ChIP protocol for examining histone modification is the same as that shown in Fig. 4-10. Following treatment under conditions in which the gene of interest is transcriptionally affected, nuclear protein–DNA complexes are crosslinked with formaldehyde. Nuclei are then isolated and the DNA is sheared by sonication to produce chromatin fragments approx 500–1,000 bp long. After DNA fragmentation, chromatin immunoprecipitation is achieved using the appropriate histone modification-specific histone antibody. The crosslinking is then reversed and the DNA is purified and quantified. Finally, the enriched DNA sequences are examined for regions of interest, generally by PCR. The choice of primer location may be dependent on previous footprinting assays to assure that the primers span only one nucleosome. Alternatively, in the absence of footprinting information, a series of overlapping primers may be employed.

A list of suppliers of antibodies for examination histone modifications by ChIP assays can be found in http://www.the-scientist.com/yr2002/jan/profile1_020107.html or The Scientist 16(1):47, Jan. 7, 2002. Upstate Biotechnology has a wide array of histone modification-specific antibodies as well as ChIP kits (www.upstatebiotech.com)

2.7. DNA Methylation Within CpG Islands

Methyl group tags in the DNA of humans and other mammals play an important role in determining whether some genes are or are not expressed. In addition to genomic imprinting, methylation of the 5 position of cytosine is involved in embryogenesis, X chromosome inactivation, cellular differentiation, and establishment of tissue-specific patterns of gene expression [46]. Importantly, DNA methylation of tumor suppressor genes has been linked to cancer. This epigenetic effect has been observed in GC-rich regions and CpG islands (5'-CG-3') that are prevalent in the promoter region and first exon of genes. As a result of CpG hypermethylation, chromatin structure in the promoter can be altered, causing a compacted, less-active state or, alternatively, affecting the interaction with the transcriptional machinery.

46. Sutherland, J., Costa, M. In: Cellular and Molecular Toxicology, Vanden Heuvel, J. P. et al., eds. Elsevier Science, Amsterdam, The Netherlands. 2002, vol. 14, pp. 299–332.

The term "epigenetic" refers to reversible, heritable changes in gene function that occur without a change in the sequence of DNA.

The study of CpG methylation is generally performed in one of two ways: by using methylation-specific restriction enzymes or by modifying unmethylated DNA with bisulfite followed by analysis of the sensitivity (sequencing, single nucleotide primer extension [SNuPE], or microarray). These methods are depicted in Fig. 4-12 and described in the following subsections.

2.7.1. Methylation Analysis Using Restriction Enzyme Digestion

This is a classical method of methylation analysis based on the inability of some restriction enzymes to cut methylated DNA. Since in eukaryotic DNA only cytosine in the CG context can be methylated, the restriction enzymes with CG sequences within their restriction sites come into question. The most common enzyme pair used to examine methylation of DNA is *Hpa*II–*Msp*I (CCGG). Both enzymes recognize the CCGG sequence; however, *Hpa*II is unable to cut DNA when the internal cytosine is methylated (CCGG vs

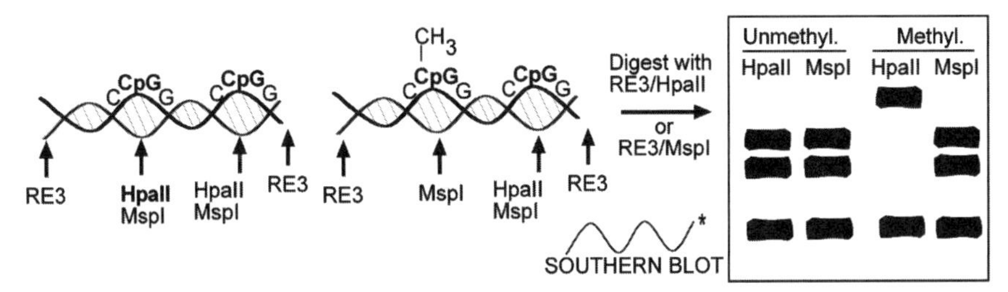

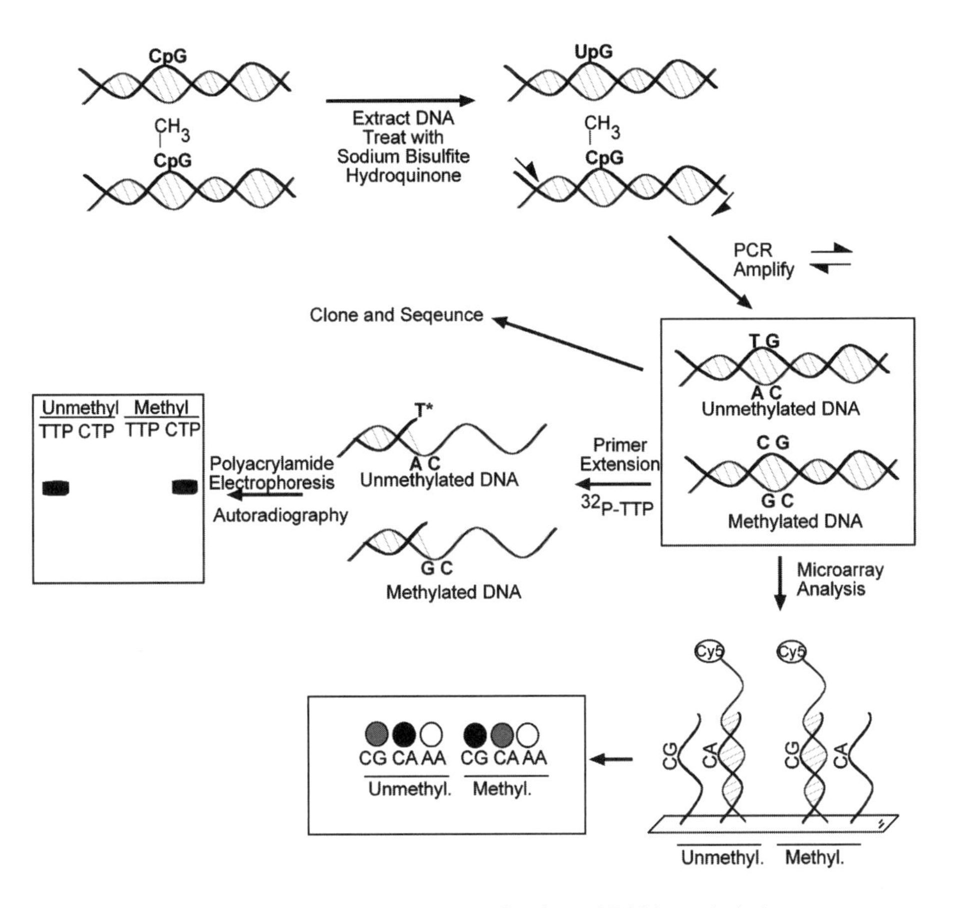

Fig. 4-12. Approaches to examination of DNA methylation.

CmCGG). A third restriction enzyme is required that will digest the DNA flanking the potential CpG site and results in a 500–1000-bp band. It is advantageous to have less than 10 *Hpa*II–*Msp*I sites within this fragment to simplify the banding pattern observed. Genomic DNA is isolated and parallel samples are digested with this third enzyme and either *Hpa*II or *Msp*I. Southern blotting is generally performed to examine the banding pattern when the fragment generated from the third enzyme alone is used as the probe. Alternatively, if a specific site is suspected as being differentially methylated, the digestion products may be amplified by PCR using primers that flank the site. Although technically simple, one shortcoming is that not all methyl-cytosine are located within CCGG sequences, and hence potential methylation sites will be overlooked.

2.7.2. Analysis of Methylation by Bisulfite Sequencing

Alternatives to restriction enzyme digestion rely on cytosine's differential chemical reactivity to reagents such as sodium bisulfite, hydrazine, or permanganate relative to methyl-cytosine. Under appropriate conditions, unmethylated cytosines in DNA react with sodium bisulfite to yield deoxyuracil, which behaves as thymidine in subsequent template-directed polymerization. In contrast, methylated cytosines are unreactive to bisulfite and behave as cytosine upon polymerization. Genomic DNA (1–2 μg) is denatured in NaOH, split into two reactions, and one reaction receives sodium bisulfite (3 *M*) and hydroquinone (10 m*M*). Both reactions are subjected to PCR using primers that flank the region of interest. The PCR products can be purified, precipitated, and cloned into a plasmid for sequencing or may be used directly for primer extension sequencing (direct sequencing). Comparison of the sequence obtained with and without bisulfite treatment is used to determine the presence of methyl–cytosine in the genomic sample.

2.7.3. Single-Nucleotide Primer Extension

SNuPE is a powerful method that can be used for the precise analysis of methylation in a certain position. (This technique is similar to one method used for single-nucleotide polymorphism (SNP) analysis.) Genomic DNA is treated with bisulfite as stated earlier, PCR amplified using primers that flank the site of interest, purified, and divided into two aliquots. Primers are designed to anneal immediately before (5') the suspected methyl–cytosine site. A one-cycle PCR is performed using this primer; however, the only nucleotides supplied are ^{32}P-CTP in one reaction and ^{32}P-TTP in the parallel reaction. If the cytosine immediately 5' of the primer was methylated, the extension reaction will incorporate the radioactive CTP, but not TTP. If unmethylated, the bisulfite will introduce a mutation (to adenosine) and hence primer extension will occur with

TTP but not CTP. The reactions are denatured and analyzed following polyacrylamide gel electrophoresis and autoradioagraphy.

2.7.4. Methylation-Specific Oligonucleotide Microarray

A novel use of oligonucleotide arrays to examine methylation sites has been described (methylation-specific oligonucleotide microarray (MSO) [47, 48]). With MSO, bisulfite treated DNA is PCR amplified and labeled with a fluorescent dye (Cy5). An oligonucleotide array is prepared that contains probes specific for a particular CpG site. Two probes are needed for each site, one that would hybridize with the methylated DNA (*xx*CG*xx*, where *x* is any base), the second recognizes the unmethylated (*xx*CA*xx*) form. Following hybridization of Cy5-labeled DNA, washing, and scanning, the analysis is simply based on the amount of fluorescence associated with each probe.

47. Gitan, R. S. et al. Genome Res. 12 (2002) 158–164.

48. Balog R. P. et al. Anal. Biochem. 309 (2002) 301–310.

Theoretically, a DNA chip can be generated that contains thousands of oligonucleotides designed to discriminate between methylated and unmethylated sequences in these gene promoters. To date, the method has only been tested and optimized on a small set of probes. Cross-hybridization has been a problem, since the discrimination between probes is only one nucleotide. Thus, the sequence composition for each oligonucleotide probe is critical in the assay. The specificity of a probe drops greatly when it contains more than four consecutive T or G residues. In addition, some probes may have inherently diminished hybridization signals, probably a result of decreased duplex stability of targets and probes [47].

Another alternative [47] is to use oligonucleotide probes that are designed so that their 3' termini end just before interrogating CpG sites, as described above. These probes are also arrayed on glass slides by attaching to the surface via their 5' ends.

47. Gitan, R. S. et al. Genome Res. 12 (2002) 158–164.

Unlabeled bisulfite-modified targets are prepared, essentially following the steps described. After target-probe annealing, *in situ* polymerase reaction is performed to allow extension to only one base (either Cy5-ddTTP or Cy3-ddCTP) away from an oligonucleotide primer. The extended molecule is identified on the basis of the incorporation of different fluorescent dyes for either the converted or unconverted nucleotides, that is, unmethylated or methylated cytosine residues, at the interrogating CpG sites. Quantitative analysis of methylation is determined by two-color fluorescence analysis.

3. Summary

The analysis of regulation of gene expression at the transcriptional level is the preoccupation of thousands of researchers. The number of techniques required to fully understand whether, and how, a particular gene is regulated in this fashion may seem daunting at first. However, most of the approaches require a few basic tools, including cell culture and isolation of nuclei, polymerase chain reaction, handling and use of modifying enzymes (restriction enzymes, RNases, DNases, T4 polynucleotide kinase), and electrophoretic separation of nucleic acids and proteins. Thus, mastery of one approach will increase the likelihood that success can be achieved in others as well. The key is to know what approach to use to answer the central hypothesis in the most definitive way, utilizing the resources and expertise at your disposal. Table 4-6 gives a brief summary of the various approaches with their strengths and weaknesses.

Table 4-6
Summary of Methods Used to Examine Regulation of Gene Expression at the Transcription Level

Method	Comments
Transcription analysis	
RNA polymerase inhibitors	Cells are treated with a RNA polymerase inhibitor such as actinomycin D and the effects on regulation of the target gene are examined. Straightforward technique, but chemical inhibitors have nonspecific effects.
Nuclear run-on assays	This is the best method to determine if a treatment or condition regulates a gene at the transcription level. Nuclei are prepared from the different treatment conditions and initiated transcription complexes are allowed to continue elongation in the presence of a radioactive nucleotide. Detection is by dot or slot blot. Several controls are required to assure specificity of the probes.
Mapping transcription start sites	Not used to determine if transcription is occurring but will explain which promoter is being used. Several methods are available including S1 nuclease mapping, rapid amplification of cDNA ends, primer extension, and nuclear run-on analysis. All use common techniques, although S1 nuclease mapping is the most often utilized.
Analyzing *cis*-sequences	
Computer search	Various public domain databases are available that contain consensus sequences for promoters, enhancers, and other features. The regulatory region of the target gene is compared to these databases. This is a rudimentary, first step that may lead to new hypotheses, although it rarely proves that a *cis*-element is involved in transcriptional regulation.
In vitro footprinting	DNA containing the target gene's regulatory region is incubated with purified or crude protein extracts and digested with a nuclease such as DNase I. Regions that are associated with proteins will be protected from cleavage, resulting in a footprint when resolved on a polyacrylamide gel. Detects many types of protein–DNA contacts and can identify multiple response elements in a single construct.

(continued)

Table 4-6 (Continued)

Method	Comments
Gel-shift assays	Short pieces of the target gene regulatory region are end-labeled and incubated with purified or crude extracts. When resolved on a nondenaturing polyacrylamide gel, complexed DNA will move more slowly through the gel, resulting in a retarded band. Simple experiments that can rapidly detect protein–DNA interactions, as well as measure affinity and specificity of this interaction. Not all complexes result in a gel shift, and assay is only semiquantitative.
Reporter assays	The regulatory region of a target gene is cloned upstream of an easy-to-measure gene product (reporter gene). When transfected into an appropriate cell, the reporter gene will respond to stimulus similarly to the endogenous target gene. Various mutation or deletion constructs can be used to find the *cis*-elements that are required for regulation. Simple, yet powerful assays to find *cis*-elements that can be used to find positive and negative regulatory elements. Does not necessarily mean a gene is transcriptionally regulated as the reporter construct may also be regulated posttranscriptionally.
Equilibrium DNA binding	Purified or crude protein extracts are incubated with end-labeled DNA. The free and bound DNA are separated and the amount of radioactivity associated with protein is quantified. Can measure equilibrium dissociation constants (K_d) for the protein–DNA interaction, but little other information is obtained.
Examination of *trans*-acting proteins	
Identification of a trans-*acting protein for a given response element*	
Supershift assays	A standard gel-shift assay is performed in the presence of an antibody to one of the constituents. Owing to the mass of the antibody, the protein–DNA complex is further retarded on the native polyacrylamide gel, resulting in a supershift. Simple method to determine if a particular protein is part of the complex associated with a *cis*-element. Unfortunately, not all antibodies are appropriate for supershift assays as they may not recognize the *trans*-acting protein in this context.

Affinity purification	Biochemical approach to identify a *trans*-acting protein once a *cis*-element is described. The DNA is affixed to a resin and protein extracts are passed through a column packed with this affinity matrix. After the protein is eluted, it may be probed by Western blot or partially sequenced. Utilizes simple techniques, but is not sensitive or rapid and often numerous proteins are eluted. Nonetheless, it is possible to identify novel DNA-binding proteins using this approach.
Southwestern blots	Proteins from cellular extracts are resolved by SDS-PAGE, blotted and re-folded. Subsequently, the end-labeled *cis*-element is incubated and the mass of the interacting protein assessed. Rapid method to gain some information on the *trans*-acting protein, although difficult to determine the identity of this protein based solely on this data. Many proteins do not regain DNA-binding activity when refolded, and *trans*-acting factors that require heterodimerization are not detected.
One-hybrid cloning	An adaptation of the yeast two-hybrid cloning technique. In the one-hybrid, the "bait" construct is the *cis*-element cloned upstream of a reporter gene, while the "prey" plasmid is a library of cDNAs expressed as a hybrid with an activation domain. If a protein interacts with the *cis*-element the reporter will be expressed and the colony can be isolated for identification of the *trans*-acting protein. A very powerful method to find novel *trans*-acting proteins that may occur in low abundance, and it utilizes a system that is commonly used by molecular biology laboratories. Suffers from a high rate of false positives, and yeast may not recapitulate what occurs in a mammalian cell.
Characterization of cis-*elements for a given* trans-*acting protein*	
Selection and binding assay	A library of response elements flanked by PCR primer sites is prepared, usually by synthesizing a degenerate probe. This library is screened with the protein of interest in a gel-shift assay. The retarded band is excised, eluted, and amplified. After three rounds of this enrichment, the resultant products may be cloned and sequenced. The consensus binding site may be derived from this examination, although whether the interaction has biological activity is not known.

(continued)

Table 4-6 (Continued)

Method	Comments
ChIP assays and ChIP cloning	Chromatin immunoprecipitation (ChIP) assays have fast become the method of choice for the examination of protein–DNA interactions within the context of the endogenous gene. Protein–DNA complexes are crosslinked in vivo using formaldehyde, the DNA sheared by sonication, and immunoprecipitated using an antibody against one of the constituents. The DNA that is associated with the protein is assessed by PCR, sequencing, or microarray. Requires technical expertise in many assays, but once mastered is an excellent way to examine *trans*-acting proteins associating with endogenous gene promoters.
Regulation of transcriptional elongation	
Nuclear run-on and S1 nuclease mapping	The examination of events associated with transcriptional elongation (stalling, pausing, arresting) are virtually identical to determination of transcription start sites. The difference is in the choice of probes with elongation using probes designed 3' to the start site.
Reporter assays	Similar to reporter assay to look at transcription initiation, except the region 3' of the transcription start site is examined.
In situ hybridization	Novel approach to look at transcriptional elongation in the intact cell. *In situ* hybridization is performed using probes at the 5' end of the transcript (close to the initiation site) and compared the signal derived from probes to the 3' end. Difficult to quantify, but does show what is occurring in vivo.
ChIP analysis of RNA pol II and accessory proteins	Antibodies against members of the transcriptional elongation complex (RNA pol, TFIIH, and so on) are used in ChIP protocols. Can show whether a pause has occurred in the intact gene.
In vivo footprinting and chromatin structure	
DNase footprinting	The cells are permeabilized and treated with DNase I. Footprinting is then examined as stated for in vitro footprinting assays. Ligation mediated PCR (LMPCR) is often for its increased sensitivity. Generally gives a good indication of chromatin structure, although it is often difficult to optimize.

DMS footprinting	Dimethyl sulfate is used to methylate guanine residues and because of its small size does not require permeabilization. DNA that is associated with protein would be less accessible to the adduction. Subsequently, extracted DNA is treated with piperidine, which cleaves at the methylated residues. DMS footprinting has the advantage of detecting DNA–protein contacts under more physiological conditions. However, this approach may not detect all DNA–protein interactions.
Histone modification	
ChIP assays	Antibodies against specifically modified residues of histone molecules are used in ChIP protocols. Acetylated lysine residues are often examined to determine if histone acetyl transferase (HAT) or deacetylase (HDAC) activity is present. In conjunction with other assays, can give a good impression of the chromatin structure of the target gene.
DNA methylation and CpG islands	Restriction enzyme digested DNA is extracted and digested with pairs of restriction enzymes, one that is sensitive to methylation of cytosine and the other that is unaffected by this modification. Simple method to detect methylation but is only applicable for specific methylation sites.
Bisulfite modification	Unmethylated cytosines remain reactive to bisulfite whereas methyl-cytosine is inactive. Following bisulfite treatment and PCR amplification, the modified cytosines are mutated to adenosine. This mutation is then detected using a variety of methods. More universal in its ability to detect methylation than restriction enzyme digestions but technically more challenging.

103

5

Posttranscriptional Processing of Messenger RNA

1. Concepts

1.1. Introduction to RNA Processing

Posttranscriptional processing is a very important process in eukaryotic mRNA accumulation, although it is often overlooked as a primary means of regulating gene expression [1,30]. Another key point is that in mammalian cells, transcription, processing, and stability are intertwined, and it is often difficult to dissociate an effect on one process from the others. When transcription of bacterial rRNAs and tRNAs is completed, they are immediately ready for use in translation with no additional processing required. Translation of bacterial mRNAs can begin even before transcription is completed owing to the lack of the nuclear-cytoplasmic separation that exists in eukaryotes. An additional feature of bacterial mRNAs is that most are polycistronic, which means that multiple polypeptides can be synthesized from a single primary transcript. This does not occur in eukaryotic mRNAs. In contrast to bacterial transcripts, eukaryotic RNAs (all three classes) undergo significant posttranscriptional processing and are transcribed from genes that contain introns. The sequences encoded by the intronic DNA must be removed from the primary transcript prior to the RNAs being biologically active. The process of intron removal is called RNA splicing. Additional processing occurs to eukaryotic mRNAs. The 5' ends of all eukaryotic mRNAs are capped with a unique 5'-to-5' linkage to a 7-methyl-

1. Alberts, B. et al. In: Molecular Biology of the Cell. Garland Publishing, New York, NY. 1994, pp. 401–476.

30. Vanden Heuvel, J. P. In: Cellular and Molecular Toxicology, J. P. Vanden Heuvel et al., eds. Elsevier, Amsterdam, The Netherlands. 2002, vol. 14, pp. 57–79.

Gene Expression Control at the mRNA Level by J. P. Vanden Heuvel
From: *Regulation of Gene Expression*
By: G. H. Perdew et al. © Humana Press Inc., Totowa, NJ

guanosine residue (m^7G). Messenger RNAs also are polyadenylated (poly[A]) at the 3' end.

Although not discussed here, eukaryotic tRNAs and rRNAs are extensively processed. In addition to intron removal in tRNAs, extra nucleotides at both the 5' and 3' ends are cleaved, the sequence 5'-CCA-3' is added to the 3' end of all tRNAs, and several nucleotides undergo modification [1]. There have been more than 60 different modified bases identified in tRNAs. Both prokaryotic and eukaryotic rRNAs are synthesized as long precursors termed preribosomal RNAs. In eukaryotes a 45S preribosomal RNA serves as the precursor for the 18S, 28S and 5.8S rRNAs. In the following section, we will limit our discussion to eukaryotic mRNA and will emphasize the important connections between transcription initiation, pre-mRNA elongation and mRNA processing. These events are summarized in Fig. 5-1.

1. Alberts, B. et al. In: Molecular Biology of the Cell. Garland, New York, NY, 1994, pp. 401–476.

1.2. Capping of mRNA

The addition of m^7G to RNA is the first posttranscriptional event to occur, and it proceeds as a result of three enzymes: RNA triphosphatase (RTP), guanyltransferase (GT), and 7-methyltransferase (MT) [49,50]. GT and MT each bind exclusively to the phosphorylated carboxy terminal domain (CTD) of RNA pol II. The end result is the addition of a "cap" with a unique 5' to 5' linkage (Fig. 5-2). Although the capping enzymes are distinct from the transcriptional machinery, they are coupled to the transcription process [51] (Fig. 5-1). The GT and methylase bind to the phosphorylated form of the pol II CTD and carry out their enzymatic functions before the transcript has reached 30 nucleotides in length. CTD phosphorylation by the general transcription factor TFIIH occurs shortly after initiation and is important for the transition to productive elongation. The addition of the 5' cap may serve three distinct functions. First, it may signal the switch from transcription initiation to that of tran-

49. Bentley, D. Curr. Opin. Cell Biol. 14 (2002) 336–342.

50. Cramer, P. et al. FEBS Lett. 498, (2001)179–182

51. Proudfoot, N. Trends Biochem. Sci. 25 (2000) 290–293.

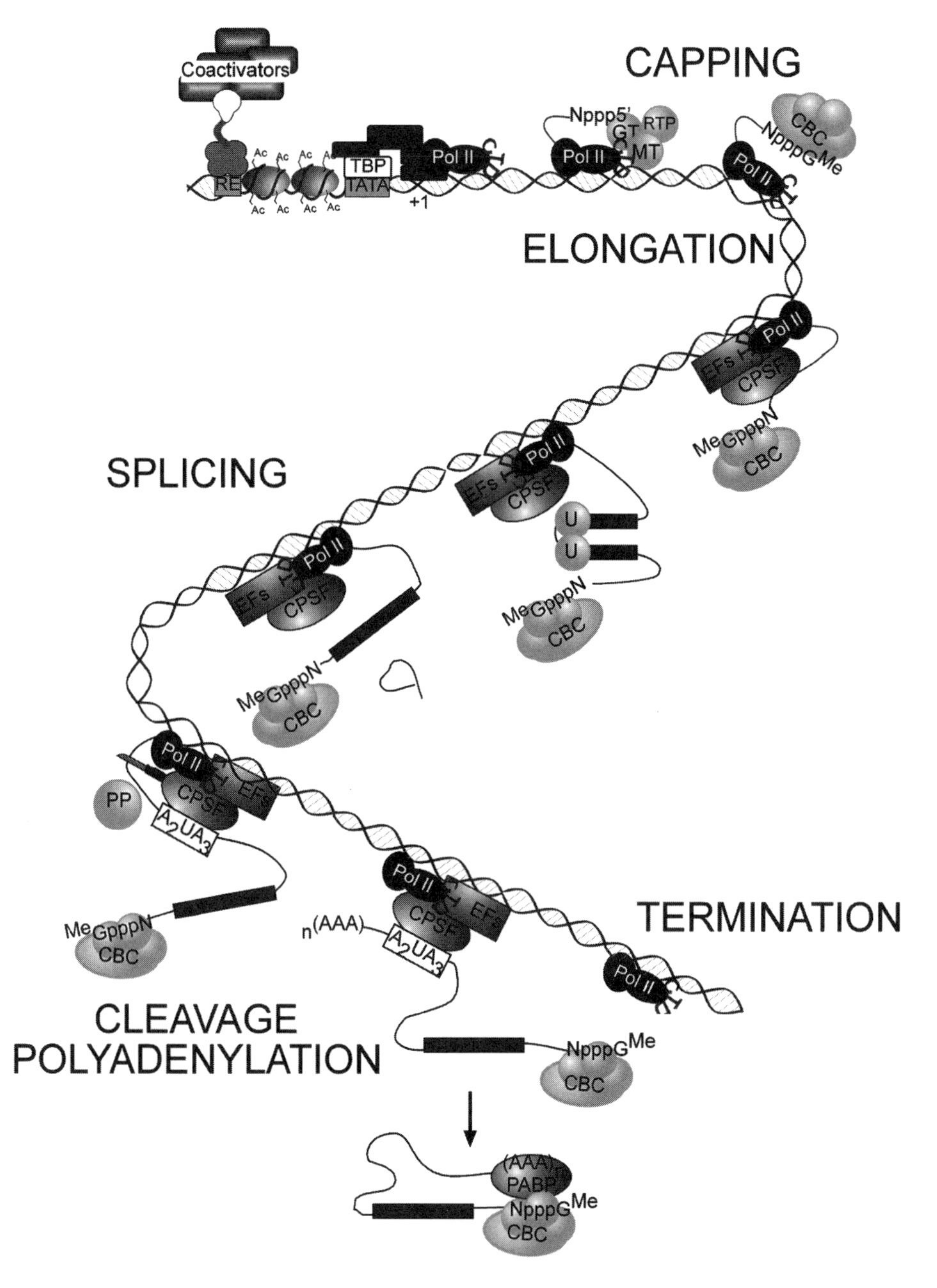

Fig. 5-1. RNA processing and relationship to transcription.

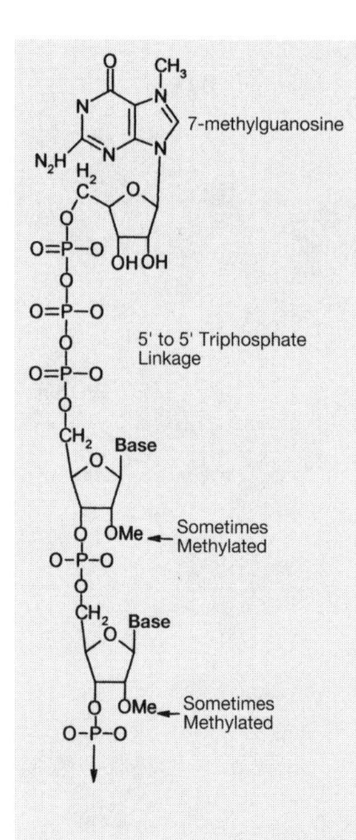

Fig. 5-2. Capping of mRNA.

script elongation. Stalling of transcription elongation may result from defective capping of mRNA. Thus, the link between capping and transcription elongation may serve as a checkpoint to ensure that pol II commits to productive elongation only after the transcript has been capped [49]. The cellular transcription factor Spt5 binds to GT and stimulates capping and thereby affects elongation. The HIV Tat transcription factor PTEFb stimulates transcription elongation by recruiting another CTD kinase and also binds GT and enhances capping in vitro [49]. Second, capping may be required for efficient RNA processing. The cap probably needs to be in place by the time the first intron is spliced, because the cap binding complex (CBC) stimulates removal of this intron. Splicing of this intron is enhanced by interaction between U1 small nuclear ribonucleoprotein (snRNP; *see* next subsection) and CBC, which, like other mRNA-binding proteins, associates with its target co-transcriptionally. Third, owing to the 5' to 5' linkage of the m^7G at the capped end of the mRNA, the transcript is protected from exonucleases and, more importantly, is recognized by specific proteins of the translational machinery. Thus, capping affects downstream events including mRNA stability and translational efficiency.

1.3. Splicing of RNAs

The next RNA processing step to occur is that of intronic splicing, although as mentioned previously this step occurs concurrently with transcriptional elongation. Most eukaryotic genes contain introns, with approx 90% of the gene composed of noncoding sequences. The typical human gene contains an average of eight exons and most genes express more than one mRNA by alternative splicing. Not surprising is the fact that disruption of splicing may result in a variety of diseases and pathologies [52]. Splicing of nuclear mRNAs requires the formation

52. Faustino, N. A., Cooper, T. A. Genes Dev. 17 (2003) 419–437.

of an internal "lariat structure" and is catalyzed by specialized RNA–protein complexes called small nuclear ribonucleoprotein particles (snRNPs, pronounced "snurps") [53]. The RNAs found in snRNPs are identified as U1, U2, U4, U5, and U6. The genes encoding these snRNAs are highly conserved in vertebrate and insects and are also found in yeast and slime molds indicating their importance. Analysis of a large number of mRNA genes has led to the identification of highly conserved consensus sequences at the 5' and 3' ends of essentially all mRNA introns (Fig. 5-3). The U1 RNA has sequences that are complimentary to sequences near the 5' end of the intron. The binding of U1 RNA distinguishes the GU at the 5' end of the intron from other randomly placed GU sequences in mRNAs. The U2 RNA also recognizes sequences in the intron, in this case near the 3' end. The addition of U4, U5, and U6 RNAs forms a complex identified as the splicesome that then removes the intron and joins the two exons together.

53. Newman, A. Curr. Biol. 8 (1998) R903–R905.

As with mRNA capping, splicing is also closely connected with transcription and may in fact occur simultaneous to mRNA elongation [51]. The splicesome is a large complex that includes the snRNPs just described, as well as splicing regulatory proteins (SR proteins). In general, SR proteins possess RNA-binding domains (RRMs) that target the protein to exon enhancer sequences and arginine/serine-rich region (RS) that may provide for protein–protein interaction. In vitro splicing reactions can be affected by the phosphorylation status of the pol II CTD and SR proteins interact directly with the heptad repeats of this protein. Overexpressing various SR proteins affects the splicing patterns initiated at specific promoters. This is consistent with a model in which SR protein interaction with CTD are set up early in the transcription initiation process and affect splicing patterns as the mRNA chain is elongated [51].

51. Proudfoot, N. Trends Biochem. Sci. 25 (2000) 290–293.

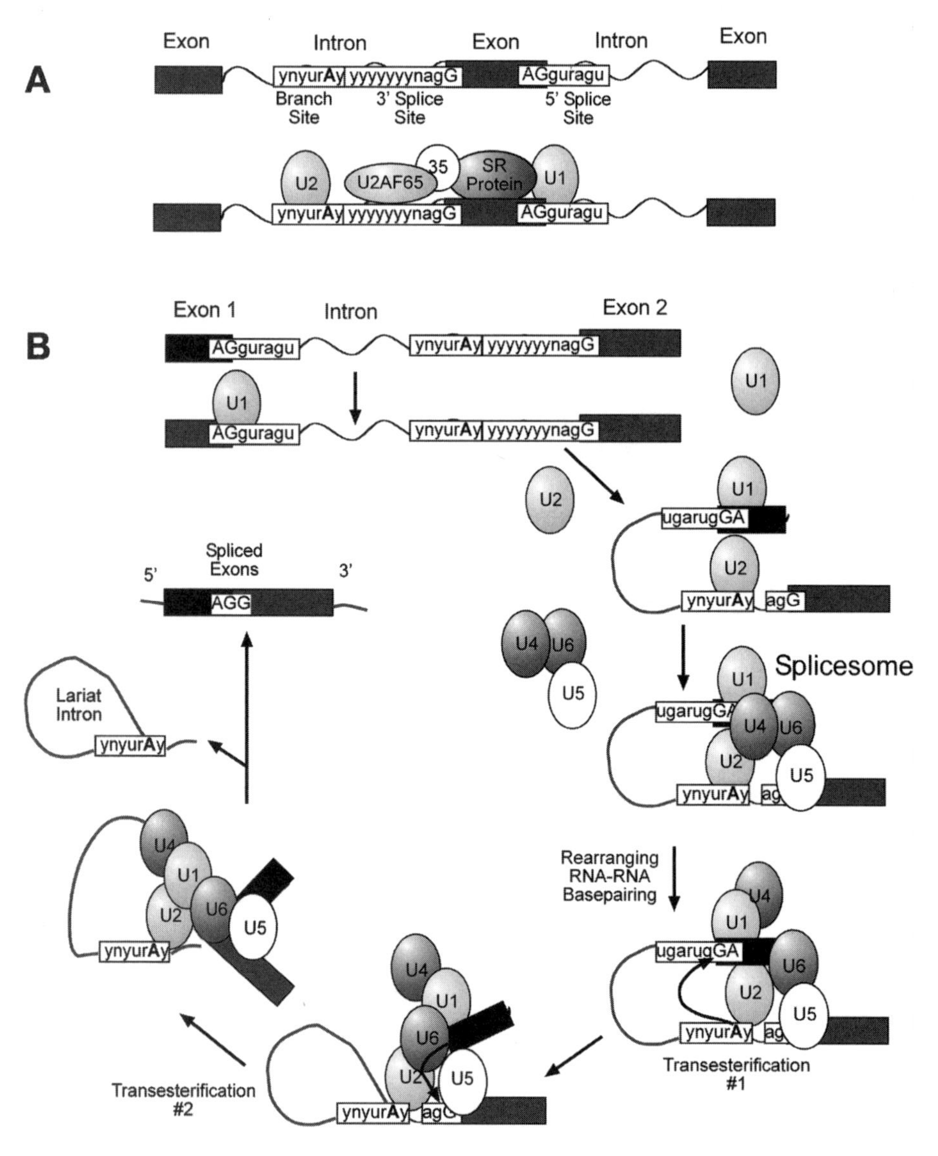

Fig. 5-3. Messenger RNA splicing.

1.4. Polyadenylation

A specific sequence, AAUAAA, is recognized by the endonuclease activity of cleavage polyadenylation stimulatory factor (CPSF) that cleaves the primary transcript approx 11– 30 bases 3' of the sequence element. A stretch of 20–250 A residues is then added to the 3' end by the polyadenylate polymerase (PP) activity. The poly(A) tail has been conclusively shown to play a role in mRNA stabilization and decay, as will be discussed below. However the addition of the 5' cap and poly(A) tail may also be involved in nuclear export of the transcript.

The CTD of pol II affects the polyadenylation of mRNA and the processing of mRNA regulates transcriptional elongation. Shortly after transcription initiation, CPSF binds to pol II and is carried along during the elongation process. Deletion of the CTD inhibits the polyadenylation process, once again linking transcription to mRNA processing. The recognition of the poly(A) signal site triggers the termination of the transcription process. Without a poly(A) signal site, long RNA transcripts are produced as the result of pol II remaining attached to the elongating transcript [51].

51. Proudfoot, N. Trends Biochem Sci. 25 (2000) 290–293.

1.5. RNA Transport

As briefly mentioned earlier, a major difference between eukaryotic and prokaryotic mRNA production is the fact the former segregates RNA and protein synthesis. This segregation allows for a level of regulation of gene expression not available to prokaryotes [54]. Although all RNAs must be transported, our focus will remain with mRNA. Prior to mRNA export from the nucleus, the introns must be removed. The addition of the m^7G cap at the 5' end and poly(A) at the 3' end may enhance the rate of transport. The general scheme for mRNA nuclear export is shown in Fig. 5-4.

In the nucleus, the pre-mRNA is associated with heterogeneous nuclear ribonucleoproteins (hnRNPs), a family of approx 20 proteins [54]. Some hnRNPs such as hnRNPA1 contain both nuclear localization sequences (NLS) and export sequences (NES), while others including hnRNPC are retained in the nucleus. Several other proteins are needed for efficient nuclear export of mature mRNA. Tap and its coregulator p15 bind to mRNA as a heterodimer and shuttle between the nucleus and the cytoplasm. The

54. Cullen, B. R. Mol. Cell. Biol. 20 (2000) 4181–4187.

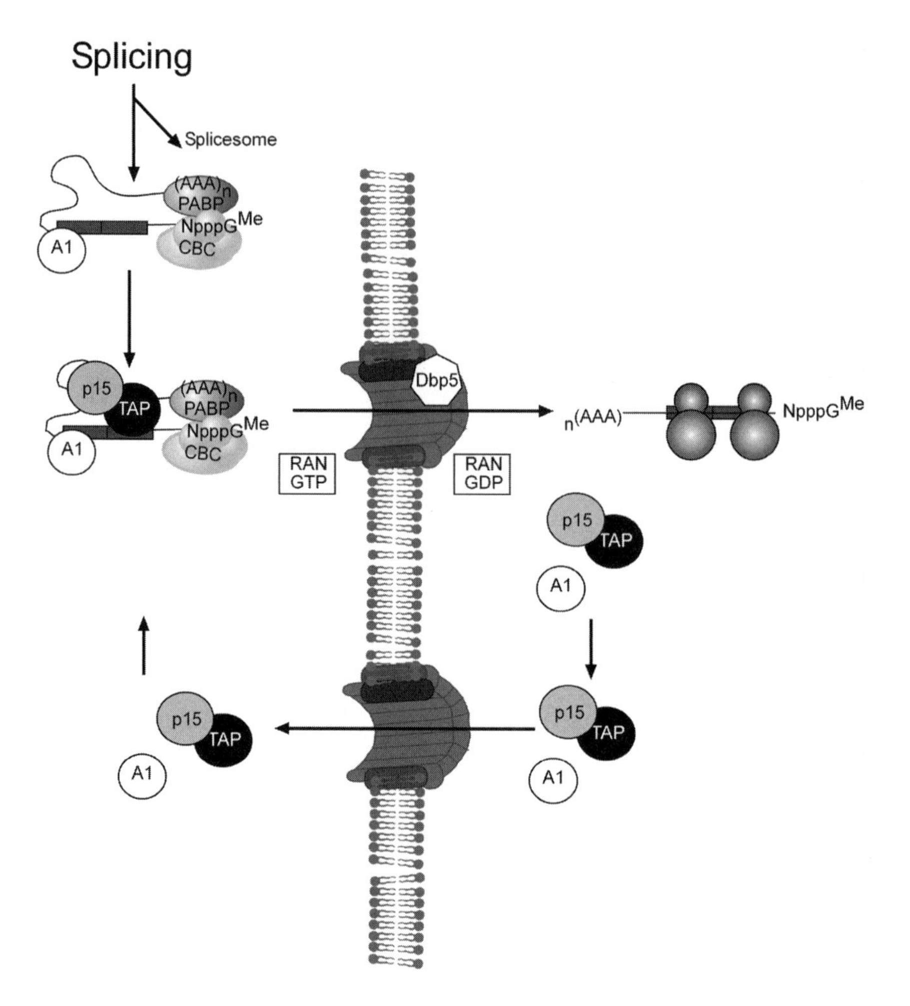

Fig. 5-4. RNA transport.

carboxy terminus of Tap is recognized by several nucleoporins and transportin, which are part of the nuclear pore complex (NPC). The function of p15 in this complex is less clear, but its presence may increase Tap binding to mature mRNA. The helicase Dbp5 is also required for poly(A) mRNA transport and may function by unwinding mRNA as it travels through the NPC. This may also assist in the removal of nonshuttling hnRNPs for retention in the nucleus.

The end result of this complicated mechanism of nuclear export is that only properly processed RNA will be allowed to exit the nucleus for later translation. Retroviruses have evolved ways to express both sliced and unspliced transcripts in the cytoplasm by circumventing some of these controls [54]. The regulation of mRNA transport by chemicals, or by disease states such as cancer, has not been studied extensively. However, it is potentially a plausible path to regulation of gene expression.

54. Cullen, B. R. Mol. Cell. Biol. 20 (2000) 4181–4187.

1.6. Introduction to mRNA Stability

Modulation of mRNA stability is an important posttranscriptional mechanism of modulating gene control. For example, inherently unstable mRNAs can reach steady-state levels more rapidly than their stable counterparts. This property would allow for cytoplasmic concentrations and translation to be altered quickly in response to a change in transcription rate [55]. The half-life of mRNAs is often controlled in a cell cycle-dependent manner, and important genes such as *c-myc* are controlled to a large extent by transcript stability. The importance of mRNA decay in gene expression is underscored by the fact that there are large differences in the half-lives of transcripts, ranging from 15 minutes to several hundred hours [56]. This variation may result in many-fold differences in the levels of specific proteins being expressed. Despite the importance of mRNA stability in gene expression, these processes have been largely ignored. The predominant reason for this oversight is the fact that the tools available for examining transcription, such as reporter assays, have not become as universally accepted, and thus examining decay rates can be quite difficult.

55. Staton, J. M. et al. J. Mol. Endocrinol. 25 (2000) 17–34.
56. Jacobson, A., Peltz, S. W. Methods 17 (1999) 1–2.

1.7. *Cis*-Acting Elements in mRNA Stability

Similar to transcription, mRNA decay is a precise process dependent on a variety of specific *cis*-acting sequences and *trans*-acting factors [57]. Entry into the pathways of mRNA decay is triggered by at least three types of initiating events: poly(A) shortening, arrest of translation at a premature nonsense codon, and endonucleolytic cleavage. However,

57. Jacobson, A., Peltz, S. W. Annu. Rev. Biochem. 65 (1996) 693–739.

other potential mRNA sequences may play a role. After poly(A) shortening or premature arrest of translation and endonuclease cleavage, the cap is removed and exonuclease activity proceeds. In general the mRNA 5' cap remains unchanged after mRNA transport and the 5'-to-5' phosphodiester bond makes the mRNA resistant to most ribonucleases.

Most eukaryotic mRNAs contain a 3' poly(A) tract, which is gradually shortened in the cytoplasm in a transcript- and cell-specific manner. For some mRNA, the shortening of the poly(A) tail is the rate-determining event in transcript decay. Shortening by 3'-to-5' exonuclease is followed by decapping at the 5' end and entry of exonucleases that proceed in the 5'-to-3' direction. Exonuclease digestion in the 3'-to-5' direction is, in general, inhibited by a poly(A) tail of at least 30 nucleotides. Deadenylation-independent decapping occurs in some mRNAs that contain nonsense codons. This mechanism, which occurs by the protein decapping protein 1 (Dcp1), exists to ensure that aberrant mRNAs are not translated.

The presence of specific-sequence elements can affect the rate of decay of certain mRNAs. These rate-modifying elements include adenosine+uridine and hairpin structures in the 3' and 5' UTR or in the coding region. The best characterized destabilization sequence, the AU-rich element (ARE), is characterized as a pentamer of AUUUA, repeated once or several times in the 3' UTR. AREs are found in many labile cytokines, transcription factors, and oncogenes (*see* Table 5-1). The 3' terminal stem loop, represents a stable structure that affects the stability of cell cycle regulated histone genes. These transcripts are not polyadenylated.

Evidence that poly(A) stimulates translation initiation, that some destabilization sequences must be translated in order to function, and that premature translation termination promotes rapid mRNA decay supports a close linkage between the elements regulating mRNA decay and components of the protein synthesis apparatus.

1.8. *Trans*-Acting Factors in mRNA Stability

Subsequent to transport via the nuclear export machinery, mRNAs are bound by the cytoplasmic cap-binding complex, eIF4F, which is composed of three subunits: eIF4E, eIF4A, and eIF4G. The subunit eIF4E interacts directly with the m^7G cap while eIF4A contains helicase activity and eIF4G interacts with the polysomal complex. Other proteins bind with high affinity to the 5' region of mRNA, including iron regulatory proteins (IRPs) that block entry of the small ribosomal subunit and repress translation. The decay of mRNA commences while it is still attached to the translation apparatus, indicating how eIF4F and IRPs affect mRNA turnover.

Table 5-1
AU-Rich Elements

Group	Motif	Examples
I	WAUUUAW	c-fos, c-myc
IIA	AUUUAUUUAUUUAUUUAUUUA	GM-CSF, TNF-α
IIB	AUUUAUUUAUUUAUUUA	Interferon-α
IIC	WAUUUAUUUAUUUAW	Cox-2, IL-2, VEGF
IID	WWAUUUAUUUAWW	FGF2
IIE	WWWWAUUUAWWWW	u-PA receptor
III	U-rich, non-AUUUA	c-jun

An important, functional mutation in a gene is one that adds a premature stop codon. These mutant transcripts must be recognized and removed before they lead to disease. Nonsense mutations in *BRCA1* and *NF1* lead to human diseases, including breast cancer and neurofibromatosis. Several proteins that may affect nonsense-mediated mRNA decay are Upf1p, a helicase, and Upf2p and Upf3p, RNA binding proteins. Upf3p or Upf2p recognize the nonsense mutation and bind to the transcript that must be removed. Subsequently Upf1p interacts with the mRNA and unwinds the RNA making it more accessible to nuclease digestion.

The AU-binding proteins (AUBPs or AUFs), are 30–45 kDa proteins that appear to be involved in both stabilization and destabilization of mRNA [58]. The human AUBP HuR has been postulated to chaperone mRNA from the nucleus and decreases susceptibility to nucleases. AU-binding factor 1 (AUF1) has been implicated in many cytokine-regulated genes [55] and binds to AU containing transcripts in the nucleus or cytoplasm. This protein interacts with two other RNA binding proteins eIF4G and poly(A) binding protein (PABP).

PABP is a highly conserved 71-kDa protein that binds with high affinity to poly(A) sequences. Stabilization of mRNA by PABP requires current translation and affects the turnover of transcripts

58. Kren, B. T., Steer, C. J. FASEB J. 10 (1996) 559–573.

55. Staton, J. M. et al. J. Mol. Endocrinol. 25 (2000) 17–34.

that rely on polyadenylation dependent pathways. One model suggests that PABP holds the 5' and 3' end of the transcript in close proximity, presumably by protein–protein interactions with eIF4G.

2. Methods and Approaches

Often, the analysis of posttranscriptional processing or turnover of mRNA is not performed until other possibilities are exhausted. For example, one may find that mRNA accumulation far exceeds what can be explained by transcriptional regulation. A logical progression of studies is shown in Fig. 5-5. Albeit rare, there may be instances that arise where the primary hypothesis is that mRNA stability is being affected. If mRNA accumulation is slow after treatment, there may be a suggestion of nontranscriptional events (although some transcriptional events are delayed). Perhaps a better indication of mRNA-stability effects can come from examining the primary sequence of the target gene or by scouring the literature for effects of your treatment on stability of other genes. For example, the target gene may contain *cis*-elements that affect mRNA stability such as AU-rich regions (Table 5-1).

In contrast to this bias toward transcription in forming hypotheses, the methods used to examine stability and/or processing of transcripts are not as well established. Although not inherently more difficult than methods described previously, the molecular biology protocol books and other general laboratory references are silent on the issue. Thus, often one has to adapt methods from the literature or develop new approaches. In the following discussion, we will focus on some of the more straightforward and often-used methods.

2.1. Stability Analysis

2.1.1. Chemical Transcription Inhibitors

The use of transcription inhibitors to examine mRNA stability is virtually identical to that described in Chapter 4. Cells are treated with inhibitors of RNA polymerase and the rate of disappearance of the mRNA examined over time. The goal of these studies is to determine the elimination rate (k_e) and ultimately the half-life ($t_{1/2}$) of a given target gene. As shown in Table 4-2, there are two commonly employed eukaryotic RNA polymerase inhibitors, Actinomycin D and α-amanitin. Cells are pretreated with actinomycin D (1–5 μg/mL) or α-amanitin (1–5 μg/mL) or vehicle control prior to the treatment (or establishment of the condition or state). At various times after treatment, RNA is extracted and the target gene examined by any means as discussed in Chapter 2. If mRNA stability is affected by treatment, the rate of disappearance of the target gene will differ from that of an appropriate control cell. The half-life can

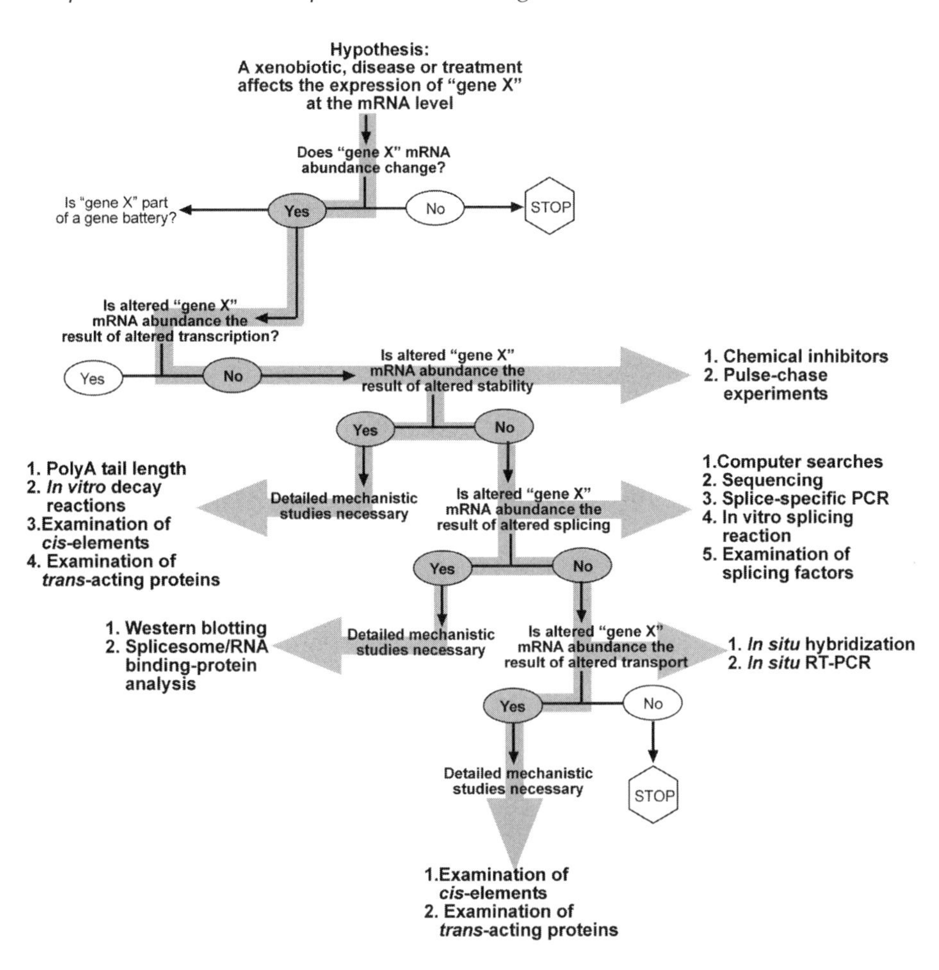

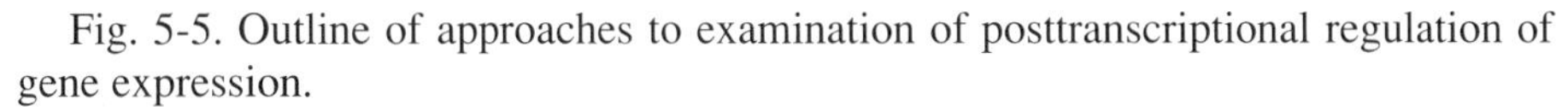

Fig. 5-5. Outline of approaches to examination of posttranscriptional regulation of gene expression.

be determined from a semilogarithmic transformation of the expression data plotted against time or by nonlinear regression. (For this analysis, it is assumed that first-order elimination takes place. $[mRNA] = [mRNA]_0 e^{-kt}$; $\ln 2 / k_e = t_{1/2}$).

2.1.2. Pulse-Chase Labeling

Because of the ease of use of the transcription inhibitors to measure mRNA half-life, the pulse-chase mRNA labeling approach has virtually vanished. However, as mentioned in Chapter 4, chemical inhibitors are wrought with non-specificity and thus pulse-chase labeling is the most accurate and definitive means to measure mRNA half-life. Cells are labeled for 3–6 hr with a radioactive nucleotide (e.g., [5,6-^{3}H]uridine 100–200 μCi/mL [59]). The labeling media is removed and the chase is initiated by adding media containing an excess of the cold nucleotide (5 m*M* uridine or 5 m*M* each uridine and cytosine). Treatment is initiated at the time of addition of the chase media. The chase continues for various times and cells harvested and cytoplasmic RNA isolated. Dot or slot blot analysis is performed essentially as described for nuclear run-on analysis. The mRNA half-life is determined as described in Section 2.1.1.

59. Paek, I., Axel, R. Mol. Cell. Biol. 7 (1987) 1496–1507.

In addition to requiring a large amount of radioactivity, the kinetics of mRNA synthesis and turnover must be appreciated. That is, if a target gene is synthesized slowly, it may require a longer pulse time to assure a high specific activity of the mRNA. Similarly, if the turnover is slow, a long period of chase will be required.

2.2. Mechanistic Studies on mRNA Stability

Messenger RNA decay is a precise process dependent on a variety of specific *cis*-acting sequences and *trans*-acting factors [57]. Entry into the pathways of mRNA decay can be triggered by a variety of events, although none has been as well studied as poly(A) shortening. For many mRNAs, the shortening of the poly(A) tail is the rate-determining event in transcript decay. Exonuclease digestion in the 3'-to-5' direction is inhibited by a poly(A) tail of at least 30 nucleotides. In addition, shortening of the poly(A) tail by 3'-to-5' exonuclease is followed by decapping at the 5' end and entry of exonucleases that proceed in the 5'-to-3' direction. Since decapping is generally the result of poly(A) shortening, measurement of this event will not be discussed.

57. Jacobson, A., Peltz, S. W. Annu. Rev. Biochem. 65 (1996) 693–739.

Internal, *cis*-acting elements that may be present in the mRNA dictate endonuclease cleavage or contain binding sites for *trans*-acting proteins. RNA-binding proteins that affect stability of the transcript can be thought of as analogous to DNA-enhancer element binding proteins. When a transcription factor binds to a stretch of DNA, it recruits enzymatic activity and can serve as a scaffold for other proteins to associate. The same may be true for RNA-binding proteins. Specific binding to a *cis*-element in RNA may affect endonuclease activity or it may bring other factors to the site (such as the splicesome). In addition, the structure of the RNA may be profoundly affected by interaction with proteins that may in turn affect its ability to be transported, degraded, and translated. The identification of *cis*- and *trans*-acting factors in mRNA stability will be examined.

2.2.1. Poly(A) Tail Length

The poly(A) tail has been conclusively shown to play a role in mRNA stabilization and decay as well as in nuclear export of the transcript. In general, the length of the poly(A) tail cannot be examined using procedures such as a traditional Northern blot, since the length of the tail relative to the rest of the transcript is too small. Thus, RNase mapping experiments are needed (Fig. 5-6). A mapping probe (antisense to the target gene) is designed that is 200–600 bp from the poly(A) tail. The RNA sample is denatured (65°C), and hybridized to the complementary mapping probe. The samples are then incubated with RNase H (1–10 U) and the fragments resolved on agarose/formaldehyde (or urea) gels. RNAs are transferred to nylon, UV crosslinked, and a standard Northern blot performed using a probe between the RNase H mapping probe and the poly(A) tail. Generally, autoradiographs of the blots will show a smeary band, the length of which is indicative of the length of the poly(A) tract. Transcripts that lack the poly(A) tail can be generated by incubating the RNA with oligo$(dT)_{15}$ and treating with RNase H.

Some variations on this method have been used. For example, if the length of the transcript is relatively small, it may be possible to examine the tail length directly through the use of a Northern blot. The poly(A) tail may be removed from the RNA prior to analysis using oligo(dT) as mentioned above, and may serve as a comparison. The rate of nuclear adenylation versus cytoplasmic deadenylation may be compared by performing the RNase H mapping experiments with RNA isolated from the two compartments.

2.2.2. In Vitro Decay Reaction

The examination of posttranscription processing of RNA in simplified in vitro systems has been championed by the Brewer and Ross labs (*see* for example [60–64]). In these assays, cell extracts or purified enzymes and proteins are incubated with mRNA substrates. This system allows for the study of *trans*-acting proteins as well as exonucleases involved in degradation. In this section, the

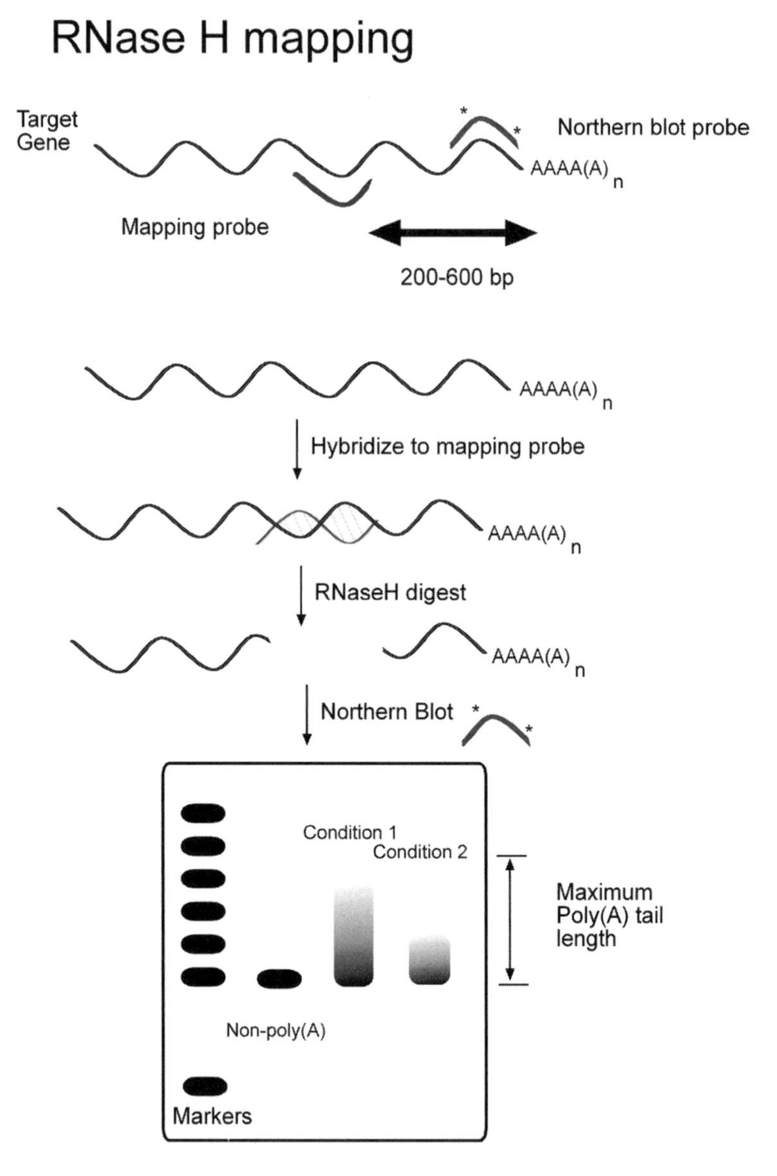

Fig. 5-6. RNase H mapping to determine poly(A) tail length.

isolation of extracts and examination of nuclease activity will be described. In subsequent sections, the use of in vitro reactions for analysis of *cis*- or *trans*-elements will be explored.

The first step in the development of an in vitro assay is isolation of a crude extract, usually the polysome fraction. Polysomal proteins include the core ribosomal proteins as well as translation initiation factors, RNases, and cytosolic proteins that bind the ribosome [65]. To isolate polysomes, cells are homogenized on ice by shearing (e.g., Potter–Elvehjem homogenizer), and the lysate is centrifuged at low speed (8300*g*) to collect nuclei and membranes. The supernatant (S8) is layered over a sucrose cushion and centrifuged at high speed (130,000*g*, S130). The polysomes are present in the pellet and after washing are resuspended.

The next component of the reaction is the mRNA, the production of which is dependent on the questions being addressed. The standard design of the plasmid used to generate the transcript and the flexibility of in vitro decay reactions is shown in Fig. 5-7. In the simplest scenario, the target gene is cloned into a suitable plasmid. This plasmid should contain a recognition sequence for a phage RNA polymerase such as SP6, T7, or T3. Other factors to consider are the direction of the RNA relative to the promoter (unlike creating probes for Northern blots, the plus strand is to be transcribed) and whether a heterologous polyadenylation site is needed. Many plasmids contain a poly(A) signal or a tract of adenosines to improve translation in in vitro systems. The plasmid is added to the appropriate reaction buffer containing the phage RNA polymerase; in most instances a radioactive nucleotide is added, although the RNA can be end labeled subsequently (*see* Fig. 5-7B). To this transcript the sample containing RNase activity is added. At various times after addition of the RNase activity, the samples are removed and resolved on polyacrylamide gels followed by autoradiography. Using this simple approach one can test polysomes from different sources to determine if a treatment is affecting production of RNases or other RNA binding proteins. Also, purified RNases or column fractions can be tested for activity. Various cofactors can be added to the reaction (such as magnesium) or inhibitors added to test for specificity of degradation.

60. DeMaria, C. T. and Brewer, G. Prog. Mol. Subcell. Biol. 18 (1997) 65–91.

61. Ross, J. Microbiol. Rev. 59 (1995) 423–50.

62. Ross, J. Sci. Am. 260 (1989) 48–55.

63. Peltz, S. W. et al. Crit. Rev. Eukaryot. Gene Expr. 1 (1991) 99–126.

64. Wilson, G. M. and Brewer, G. Prog. Nucleic Acid Res. Mol. Biol. 62 (1999) 257–91.

65. Ross, J. Methods 17 (1999) 52–59.

More information on in vitro transcription reactions can be found at http://www.epicentre.com/pdftechlit/018pl092.pdf

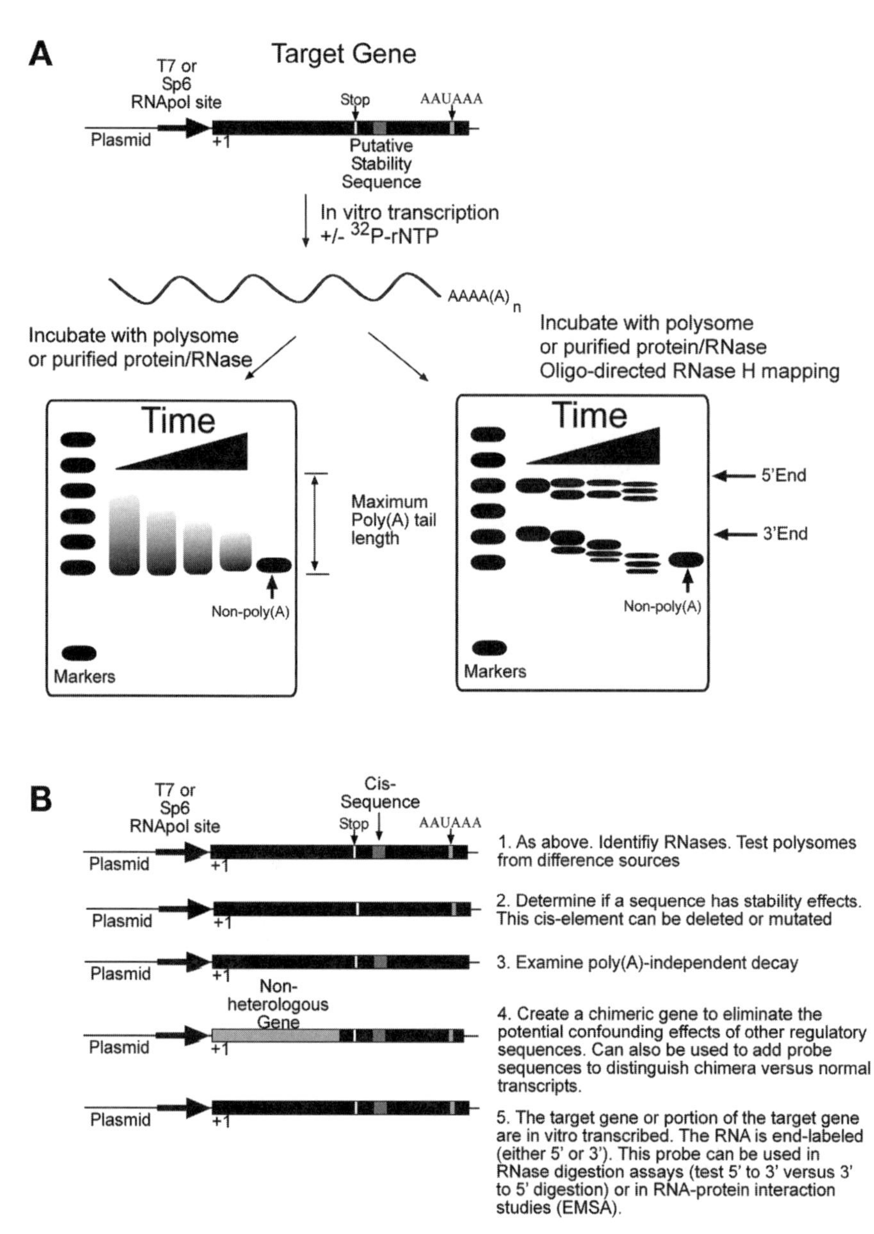

Fig. 5-7. In vitro decay reactions. **(A)** Basic design of in vitro decay reaction. The RNase-dependent digestions (purified enzyme or polysome extract) can be determined for a gene that is in vitro transcribed, usually in the presence of a radioactive nucleotide. Similar methods are used to examine the rate of deadenylation of the transcript using RNase H mapping. **(B)** The in vitro decay reactions are very flexible: a variety of tests may be performed based on the design of the plasmid to be transcribed.

There are many possible variations of the in vitro decay reaction. For example, RNase H mapping may be performed essentially as described in the previous section, using an in vitro transcribed target gene. Having the target gene in a plasmid allows for site-directed mutagenesis to examine the effects of *cis*-elements on decay rates. Also, chimeric transcripts can be made to either take advantage of a known activity found in another transcript or to negate confounding factors within the target gene.

More information on in vitro mRNA capping can be found at http://www.epicentre.com/pdftechlit/047pl092.pdf

The RNA prepared as described above is amenable to examination of specific posttranscriptional modifications [65]. For example, the 5'-to-3' degradation of mRNA by certain types of nucleases is inhibited by stem-loop structures. A good example is a stretch of 18 consecutive G residues placed in the target gene. This transcript is labeled on either the 5' end or the 3' end (5' label with polynucleotide kinase, 3' label with RNA ligase) and then digested with the nuclease. The release of acid-soluble radioactivity is used to assess the direction of the degradation. The transcript may be capped in vitro and compared to the unmodified transcript. This can be done by reducing the concentration of GTP in in vitro transcription reactions and adding a cap analog such as diguanosine triphosphate in either its unmethylated, monomethylated, or trimethylated form.

65. Ross, J. Methods 17 (1999) 52–59.

2.2.3. *Examination of* cis*-Elements in mRNA*

2.2.3.1. Computer Searches

Analogous DNA binding protein approaches can be found in Chapter 4.

The search for *cis*-elements and RNA binding proteins is not as straightforward as analogous searches for DNA equivalents. There does not seem to be a simple search algorithm to determine which *cis*-elements may be used to regulate mRNA stability, as the sequences often do not have a consensus primary sequence. Rather, RNA splicing, regulation of transcription, translation, and RNA:DNA, RNA:RNA and RNA:protein interactions are determined by the RNA structure. Thus, the search for *cis*-elements and other regulatory elements should begin with an idea of the secondary and tertiary RNA structure. However, it is extremely difficult to crystallize large RNA molecules for precise structural determination. Computational methods are valuable, because determination of secondary structure, particularly for long-chain RNA molecules, is difficult by experimental means.

S-Fold: http://www.bioinfo.rpi.edu/applications/sfold/index.pl
M-Fold: http://www.bioinfo.rpi.edu/applications/mfold/old/rna/form1.cgi
X-RNA: http://rna.ucsc.edu/rnacenter/

There are several tools on the web that may help in examining the secondary structure of an mRNA (*see* the marginal notes for URLs). As mentioned in the discussion of in vitro decay reaction, hairpin structures may help prevent RNase activity. Thus, a computer search (such as that shown in Fig. 5-8 for *c-myc*) may indicate regions that increase stability, such as hairpins. Note that an mRNA molecule can manifest countless conformations. For this reason, the computation algorithms generally report enthalpy estimates so that the probability of a particular shape can be estimated. As a tool to discover new *cis*-elements, "squiggle plots" have a limited use. However, once a region is discovered to be important in mRNA decay, through any means outlined below, these plots may confirm that a particular secondary structure may contribute. In addition, these algorithms are used in the design of antisense oligonucleotides or small inhibitor RNA molecules, both of which require accessible regions of mRNA to hybridize.

Webgene: http://www.itb.cnr.it/sun/webgene
Softberry: http://www.softberry.com/berry.phtml?topic=polyah&group=programs&subgroup=promoter

Regarding the lack of tools for examination of *cis*-elements in mRNA, two exceptions may exist: finding splice sites and identifying polyadenylation signals. Splice sites will be discussed subsequently. In the marginal note are URLs for web-based tools for the identification of 3' polyadenylation.

2.2.3.2. Reporter and Chimeric Transcripts

The use of reporter genes for examination of *cis*-elements in DNA is very common. The 5' UTR of the target gene, or fragments thereof, are subcloned into a reporter vector. The DNA fragments are

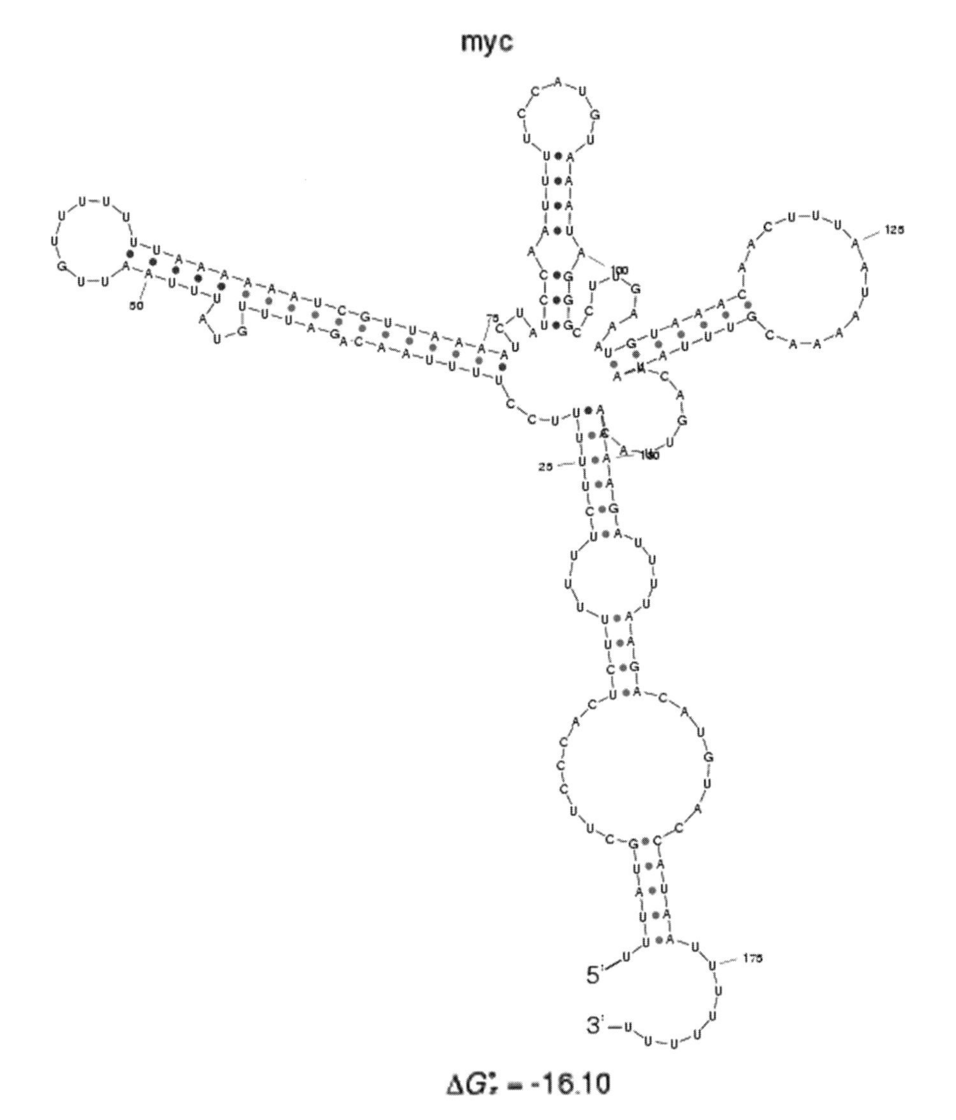

Fig. 5-8. Secondary structure of c-myc 3' UTR. The last 180 bp of mouse c-myc mRNA was examined using S-fold http://www.bioinfo.rpi.edu/applications/sfold/index.pl. Shown is one of the possible configurations.

placed upstream of the reporter gene's promoter and transcription start sites. Enhancer elements will affect the rate of transcription, which will in turn affect reporter activity. A similar approach may be taken with 3' UTR sequences that affect mRNA decay. A reporter construct with a constant amount of basal transcription is used, such as a luciferase under the control of the SV40 or cytomegalovirus (CMV) promoter. (*See* Table 4-4 for some of the common reporter genes.) The mRNA sequence under question is placed to the 3' end of the reporter gene, after the stop codon. Since the amount of luciferase enzyme will be proportional to the amount of transcript, any stabilizing or destabilizing sequence will affect the reporter activity measured. Unfortunately, to our knowledge, no commercially available reporter plasmids have a multicloning site on the 3' end of the reporter gene. Often the restriction enzyme sites are limited, unless one constructs a more suitable plasmid.

It is also possible to examine 5' sequences thought to control mRNA stability and translation using a reporter gene as well. For example, iron responsive elements (IREs) are stem loop structures found in the 5' UTR of ferritin (Ft) and mitochondrial aconitase (m-Aco) mRNAs. Iron regulatory proteins (IRP1 and IRP2) are RNA-binding proteins that affect the translation and stabilization of specific mRNAs by binding to IREs. The IRE (or other potential *cis* element) would be placed downstream of the transcription start site and before the reporter gene start codon. Other 5' regulatory sequences such as internal ribosomal entry sites (IRES), which would affect translational efficiency, can also be examined in this manner.

An analogous approach may be used where a chimeric construct is made, much like that shown in Fig. 5-7B. A plasmid is produced that contains a chimeric gene (i.e., β-globin) without its mRNA decay regulatory elements. The region(s) of the target gene thought to control mRNA stability are inserted. Following transfection (for in vivo) or transcription with viral polymerase (for in vitro decay reactions), mRNA is isolated. RT-PCR is performed using primers specific for the chimeric gene. The method has an advantage in that translation of the transcript is not required, and hence mRNA stability is examined more precisely. Also, this approach may be used to determine mRNA half-life and the effects of particular *cis*-elements. The plasmids are transfected, RNA polymerase inhibitors are employed, and at various times mRNA is extracted and examined by RT-PCR.

2.2.3.3. RNA Electrophoretic Mobility Shift Assay (REMSA)

The theoretical underpinnings of an RNA gel shift are identical to that described previously for DNA, although the former requires a few extra precautions. The most direct approach to preparing a probe for use in an REMSA has been described above (*see* Fig. 5-7). The region of the gene suspected of containing the *cis*-element is placed in a plasmid containing an RNA polymerase binding site (T7, SP6, T3). Following linearization of the plasmid (in particular if using SP6), in vitro transcription is performed in the presence of [α-^{32}P]UTP or CTP. Alternatively, the probes may be synthetic and end-labeled with T4 polynucleotide kinase. This probe may be incubated with mixtures from the in vitro decay reactions or with cytosolic proteins, usually in the presence of glycerol (to stabilize the complex) and RNasin. After incubation, the reactions are incubated with RNase T1, which cuts unprotected RNA downstream to guanosine residues. (RNase A1 may be substituted for RNase T1.) Radiolabeled RNA and RNA–protein complexes are resolved on nondenaturing, RNase-free Tris-borate EDTA (TBE), polyacylamide gels. Protein–RNA complexes are visualized by autoradiography. The same controls and caveats to DNA EMSA assays must be observed with REMSA; specific and nonspecific unlabeled competitors would be included, and the binding buffers and conditions must be optimized to reduce nonspecific binding.

2.2.3.4. RNA Footprinting/RNase T1 Protection Assays

Although not commonly employed, RNase T1 may be used in a footprinting assay to identify protein–RNA contact sites (*see* for example [66]; Fig. 5-9). Since RNase T1 is a large molecule, protection against RNase T1 may be somewhat exaggerated. To further define the minimal protein–RNA contact site, footprint analysis using lead hydrolysis may be performed. The RNA is chemically synthesized (generally examining 25–75-bp RNA molecules), and labeled at the 5' end using T4 polynucleotide kinase and [α-^{32}P]ATP. RNAs can be labeled at the 3' end using T4 RNA ligase and 5'-[^{32}P]pCp. Reaction mixtures contain labeled RNA and the protein source, either a complex mixture or a purified protein. Mixtures are preincubated with RNase T1 or lead acetate to a final concentration of 0.05–0.5 U/mL or 10 m*M*, respectively. Reactions are further incubated for 10 min and then stopped and analyzed by 10% polyacrylamide/urea gel electrophoresis. An RNase T1 digest of the element is in a parallel lane to provide size markers; since RNase T1 cleaves at G residues, the location of the protected bands may be determined.

66. Zhao, Z. et al. Nucleic Acids Res. 28 (2000) 2695–2701.

RNA footprinting, in particular when it pertains to ribosomal position on the transcript, is sometimes referred to as "toeprint" analysis.

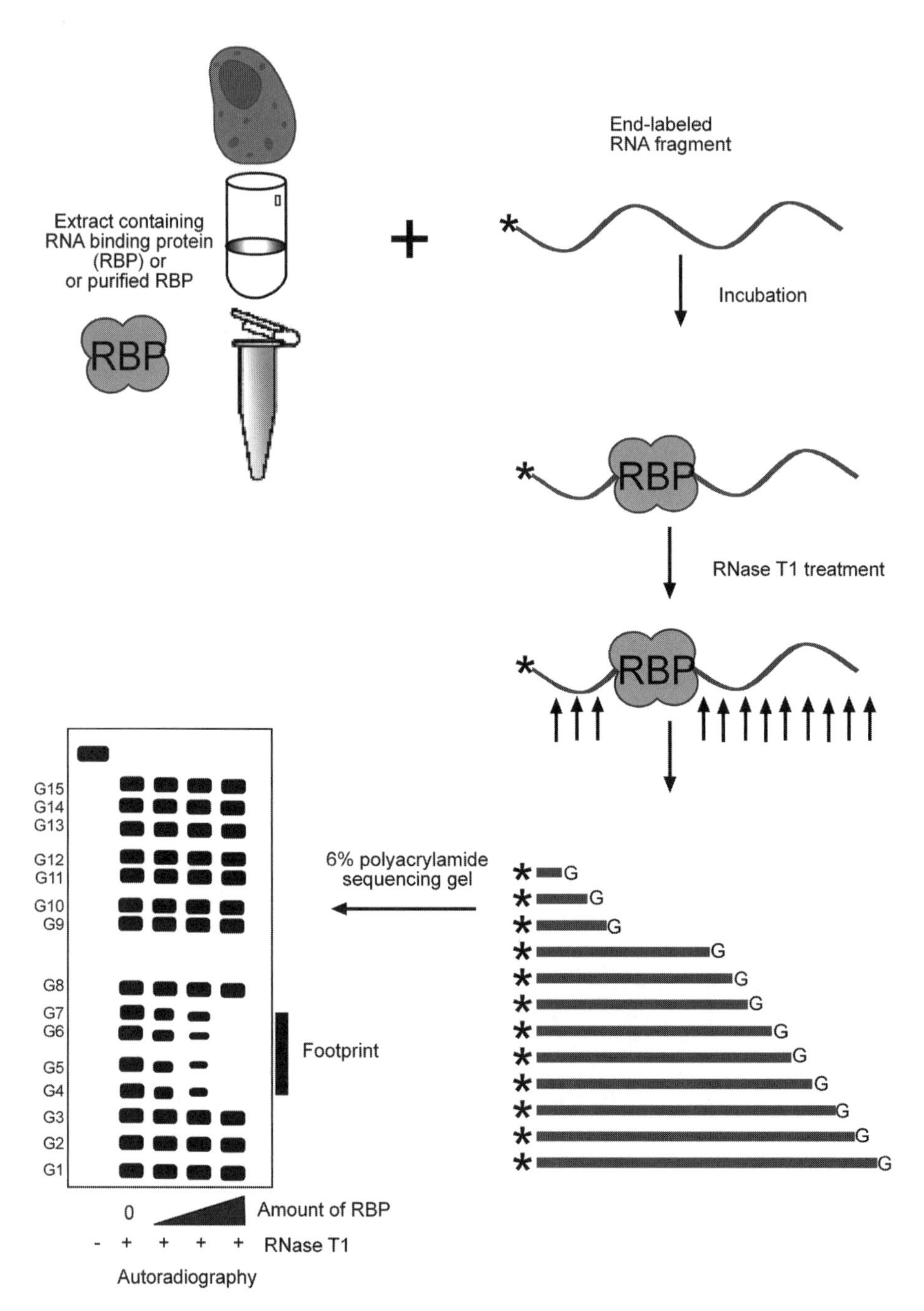

Fig. 5-9. RNA footprinting assay.

2.2.4. Examination of Trans*-Acting Proteins in mRNA*

2.2.4.1. REMSA and Supershift Assays

The following approaches used to identify a *trans*-acting RNA-binding protein depend on the identification of a *cis* element in the target gene. One may then take educated guesses as to what the protein may be. For example, AUF1 is an important protein in mRNA turnover. This protein may be bacterially expressed, purified, and examined using REMSA to see if it does indeed bind to the *cis* element. Alternatively, supershift assays with an antibody to AUF1 may be attempted. The term "supershift" refers to decreasing, by use of an antibody, the mobility of the protein–probe complex as it runs through the gel. REMSA would be performed as stated in the previous section, with the exception that an antibody to a suspected *trans*-acting protein would be included.

2.2.4.2. UV Crosslinking

In UV crosslinking, a covalent bond is formed by irradiation between the RNA and closely interacting proteins. Note however, that the reaction is inefficient and not all proteins and RNA bases react the same. The radiolabeled probe is incubated with the source of RNA binding proteins, as if an REMSA assay is being performed. After UV irradiation (i.e., 2500 μJ for 3 min), the RNA–protein complex is digested with RNase T1 (or RNase A). Since these complexes are stable, they may be resolved on SDS-PAGE gels and detected by autoradiography. The approximate size of the RNA-binding protein may be determined.

2.2.4.2. Northwestern Blots

In yet another "directional blot" the interaction between an mRNA sequence known to affect stability with an unknown protein can be examined. Cytoplasmic proteins can be separated by SDS-PAGE and transferred to nitrocellulose. Following blocking (as would be performed in a Northern blot, i.e., Denhardt's solution), the ^{32}P-labeled RNA *cis*-element is added. Following incubation and washing, the filter is exposed to X-ray film. The size of the interacting protein may be determined. It is possible to isolate the protein, digest with trypsin, and sequence the peptides to determine its identity. To improve the resolution and decrease the chance of sequencing a contaminating protein, crude proteins may be first fractionated by centrifugation or chromatography. Northwestern blots are then performed on the fractions.

2.2.4.3. Affinity Purification

The identification of a *cis*-element of RNA provides a very useful reagent for the purification of a *trans*-acting protein. That is, this RNA segment can be used to create an affinity resin to purify enough protein for sequences and identification. For example, the *cis*-element may be synthesized with a biotinylated U on

67. Snee, M. et al. J. Cell Sci. 115 (2002) 4661–4669.

the 3' end [67]. These oligoribonucleotides are incubated with streptavidin-coated beads and washed. The affinity matrix is incubated with cellular extracts, washed, and eluted with SDS. The RNA binding proteins are analyzed by SDS-PAGE and Coomasie staining. Gel slices or proteins embedded in nitrocellulose may be trypsinized and analyzed by Edman microsequencing.

2.3. Splicing Analysis

The analysis of splice sites and splice variants is much simpler than might have been implied in the introduction to this chapter. In fact, most of the techniques to be used are commonplace. Splice sites (mRNA *cis*-elements) may be discovered by computer searches and can be detected via sequencing or PCR. To determine the mechanism for the altered splice, the search for *trans*-acting proteins can be made with methods described above. The only subtlety to be described is an in vitro splicing assay, the rest will be outlined briefly.

2.3.1. Computer Searches

A listing of programs that can be used to find exons and splice sites can be found at: http://restools.sdsc.edu/biotools/biotools16.html

There are several programs that can be used to find potential splice sites in your target gene. Generally, the genomic sequence (or area around a potential splice) is examined for splice donor/acceptor sites (*see* for example, SPL at http://www.softberry.com/). Alternatively, the sequence is compared to the cDNA database and introns are identified as a gap in the alignment. It is also possible to analyze a cDNA sequence and find exon–exon borders (as in RNASPL at http://www.softberry.com/). Commercially available programs are generally equipped to do many of these searches as well, such as GeneQuest (DNAStar, Madison, WI).

Splice variants in your target gene can also be identified and analyzed by comparing the sequence to the various genome projects and to mRNA and ESTs already deposited in the public databases. It is now possible to trace the target mRNAs back to the region of the genome from which they were tran-

Gene on Genome | Annotated mRNA | Links

Maps
This gene MYC covers 6002 bp, from 128418563 to 128424564 (33), on the direct strand of chromosome 8.

MOLECULES
Transcripts
According to our analysis, this gene produces, by alternative splicing, 5 types of transcripts, predicted to encode 5 distinct proteins. It contains 6 confirmed introns, 6 of which are alternative. Comparison to the genome sequence shows that 5 introns follow the consensual [gt-ag] rule, 1 is fuzzy or ill defined.

Transcript size	5' UTR	Completeness	3' UTR	# exons	Transcr.unit
variant a 2352bp	510bp		477bp, polyA	3	5352bp
variant b 1251bp	380bp	1 exon inferred	no	2	2338bp
variant c 2042bp	1160bp		63bp	3	5044bp
variant d 1236bp	471bp		no	3	3467bp
variant f 813bp	344bp	1 exon inferred	no	1	813bp

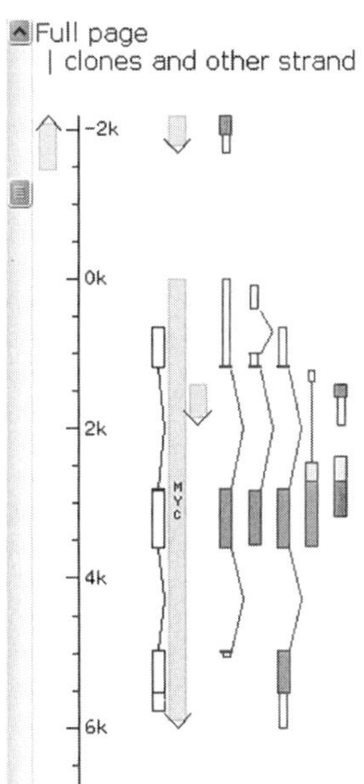

Fig. 5-10. The Aceview program to predict splice sites based on genome information and submitted cDNA and EST data. Human *c-myc* was used as an example. *See* http://www.ncbi.nlm.nih.gov/IEB/Research/Acembly/index.html.

scribed and to try to reconstruct, by assembling the other sequences that have been submitted, the most likely image of the original transcripts. The AceView program is extremely useful in this regard (http://www.ncbi.nlm.nih.gov/IEB/Research/Acembly/index.html). As shown in Fig. 5-10 using human *c-myc* as an example, information about potential splice sites can be obtained, even if they do not follow the typical GT–AG rule. Also, the AceView program allows for a visualization of the alternative splices and the ultimate coding region.

2.3.2. Sequencing of Transcripts

The basic premise of this analysis is similar to that of the AceView program. The cDNA for the target gene is cloned and sequenced from a series of individuals [68]. These sequences are compiled and aligned and splice variants can be determined.

68. Auerbach et al., S.S. Nucleic Acids Res. 31 (2003) 3194–3207.

Briefly, cDNA is prepared and PCR amplification is performed; the primers may contain restriction endonuclease sites to assist in later cloning steps. The design of the primers is dependent on the target gene itself. For example, it may be possible to amplify the entire cDNA for the target gene (generally less than 2000 bp). Alternatively, primers may be selected that span a particular intron/splice junction. PCR should be performed using a high fidelity polymerase and with the fewest cycles possible to reduce the errors in the resulting sequence. Following amplification, the fragments are purified and cloned into an appropriate sequencing vector. If a variant is observed, the effects on the coding sequence of the target gene should be assessed.

2.3.3. Splice-Specific RT-PCR

Although an exhaustive way to detect splice variants, the sequencing approach is not amenable to high-throughput analysis and is not cost effective for examination of a known splice variation. Also, this method assumes that splicing within a cell is "all-or-none" and that only one type of transcript is produced per individual. If high-throughput or quantitative analysis is required, splice site-specific RT-PCR would be performed. In these assays, primers would be selected that have one of the following characteristics:

1. Span the splice site. In this assay, the spliced and unspliced transcripts would give differently sized products. This is only appropriate if the excised piece is within a particular range that can be easily detected. If the intron is <50 bp or >2000 bp, it may be difficult to detect.
2. One primer is wholly within the spliced sequence. Amplification would only be seen in the pre-mRNA and not in the fully processed version.
3. The 3' end of one of the primers contains the splice junction. This is very similar to the second approach; however, in this case the primer has only 2–3 bases that are contained in the intron. This is the method of choice for very small introns and the reactions are optimized much like PCR detection of single-nucleotide polymorphisms. Quantitation of the amount of splicing that occurs can be performed by real-time PCR or a competitive PCR approach, as outlined in Chapter 2.

2.3.4. In Vitro Splicing Assays

The generation of pre-mRNA substrates for splicing assays can be performed essentially as described in the in vitro stability section. Plasmids containing the gene of interest are in vitro transcribed in the presence of radioactive nucleotides (i.e., [α-^{32}P]CTP). However, in this case, the mRNA will contain the intronic sequences or the region suspected of containing a splice site. Following purification, the radioactive pre-mRNA is incubated with nuclear extracts (i.e., HeLa cell nuclear extracts), plus ATP and creatine phosphate [69]. Reactions are allowed to proceed (30°C for 1–3 h) followed by standard chloroform–phenol extractions. RNAs are separated on denaturing polyacrylamide gels and examined by autoradiography or phosphorimaging.

The splicing reactions can be highly adaptable. For example, heterogeneous nuclear ribonucleoproteins (hnRNPs) can be added to the nuclear extracts and the extent of processing assessed. Alternatively, the nuclear extracts may be depleted of a particular protein via immunoprecipitation, followed by the in vitro splicing reactions.

69. Caputi, M. et al. EMBO J. 18 (1999) 4060–4067.

2.3.5. Splicing Factors

The analysis of splicing factors proceeds as described for other *trans*-acting proteins in mRNA regulation. In this case the *cis*-elements may contain the splicing acceptor or donor sites or any other sequence believed to be regulatory. For example, the *cis*-element may be affixed to an agarose bead, nuclear extracts applied, and protein analyzed by SDS-PAGE [69]. Other assays such as REMSA and supershift assays may be performed with the pre-mRNA sequences.

69. Caputi, M. et al. EMBO J. 18 (1999) 4060–4067.

2.4. mRNA Transport

2.4.1. *In Situ* Hybridization

Perhaps the only tools for the examination of mRNA localization are *in situ* hybridization (ISH) or *in situ* RT-PCR. Both techniques are capable of analysis and localization of specific mRNAs within cells or tissues. There are several advantages to these methods. Traditional methods of mRNA analysis (i.e., Northern blot, RPA, RT-PCR) do not provide detailed spatial information. For example, it is unknown where in the tissue or within the cell the transcript is observed. Also, messages that are expressed in a limited number of cells will be greatly diluted in the total RNA pool during extraction, thereby reducing the sensitivity.

An excellent source of technical information on in situ hybridization can be found at the Ambion website, http://www.ambion.com/techlib/tn/104/14.html.

The proper preparation of the cells or tissue for *in situ* hybridization is critical. Preparation and storage must be performed to maintain the cellular architecture and to inhibit the RNases. For tissues, fixed, paraffin embedded, or cryostat-cut sections may be used. To prepare 5–20-μm sections made from fixed, paraffin-embedded tissue for ISH, the paraffin is first removed by submerging the slides in xylene and rehydrated in decreasing concentrations of ethanol, ending in nuclease-free water. Finally, the section is equilibrated in the reaction buffer for the subsequent proteinase K digestion step. Proteinase K disrupts membranes and the tissue matrix allowing the reagents to infiltrate the tissue section, and giving the probe access to the target RNA. Cryostat-cut sections are thaw mounted onto positively charged slides prior to fixation. Cells in culture may be grown on slides or mounted subsequently. The slides are immersed in 4% paraformaldehyde, washed, and then subjected to proteinase K treatment and ethanol dehydration. At this point, the slides are available for either ISH or *in situ* RT-PCR.

The probes for ISH may be radioactive or fluorescent, depending on the microscopic capabilities and expertise of the researcher. In either case, experiments should be designed with positive and negative controls in mind. Positive controls include probes for target genes known to be expressed in the cells being examined. This will assure that the slides have been prepared properly and that the techniques are appropriate. A negative control is generally the sense-strand probe for your target gene. The negative control will help examine the nonspecific binding.

For radioactive probes, the RNA molecules are generally 250–1500 bases long. The insert may be cloned into a vector that contains phage RNA polymerase sites (T7, SP6, or T3) on both ends of the multicloning site to produce either the sense or antisense probes. In vitro transcription in the presense of [α-^{35}S], [α-^{32}P] or [α-^{33}P] UTP or CTP is performed and the probes are purified to remove unincorporated nucleotides. For fluorescent ISH, the

A list of sources for fluorescent nucleotides can be found at http://www.the-scientist.com/yr2001/may/nucleotides.html.

probes may be enzymatically synthesized with the appropriate nucleotide analog (Cy5, Cy3, fluorescein, rhodamine). Alternatively, fluorescent probes may be directly synthesized with the modified ribonucleotides. (Although not as frequently employed currently, the probes may be biotin or digoxygenin labeled. In these instances the detection is indirect as other reagents are required.)

The processing of the slides is very similar to that of Northern blots. The probe is applied to the cells in hybridization buffer and the cover slip is sealed (rubber cement, fingernail polish, or mounting adhesive). After an overnight hybridization at 60°C, the slides are washed (4X–0.1X SSC), then treated with RNase A to remove the unhybridized probe. Following washing and ethanol dehydration (50–100% ethanol), the slides are ready for detection. Fluorescent probes may be directly visualized, whereas radioactive ISH must be examined by autoradiography or phosphor imaging. If desired, a representative slide can be stained with H&E, or with cresyl violet to visualize structural elements.

2.4.2. *In Situ PCR*

The advantages of *in situ* PCR relative to ISH is the increased sensitivity and selectivity. However, there are more steps to the process and it is often difficult to optimize. (Detailed protocols may be found elsewhere [18].) The cells are fixed and dehydrated (50–100% ethanol) and proteinase K treated as stated above. Next, the slides are treated with RNase-free DNase overnight in a humid chamber at 37°C. Subsequently, they are washed and air dried. A reverse transcriptase solution is applied that contains the enzyme, nucleotides, target gene reverse primer, and RNasin. The solution incubates at 42°C for 0.5–2 h. The slides are washed and a PCR mix is applied that contains a fluorescent-labeled forward primer, reverse primer, nucleotides, and Taq DNA polymerase. A cover slip is sealed and the slides are placed in the PCR thermocyler. (Special

18. Vanden Heuvel, J. P. In: PCR Protocols in Molecular Toxicology, Vanden Heuvel, J. P., ed. CRC Press, Boca Raton, FL. 1997, pp. 41–98.

in situ PCR units are available. Alternatively, the slides may be placed directly on top of the heating element of the thermocycler.) Following washing, the slides are ready to be visualized by fluorescent microscopy.

3. Summary

The analysis of posttranscriptional processing or turnover of mRNA is generally not performed until transcription has been ruled out as a possible mechanism of altered mRNA levels following a particular treatment or condition. In addition to this bias toward transcription in forming an hypothesis, the methods used to examine stability and/or processing of transcripts are not as well established. However, the methods are not inherently more difficult than those used to examine gene transcription.

Table 5-2
Posttranscriptional Processing of mRNA

Method	Comments
mRNA stability/half-life determination	
RNA polymerase inhibitors	Cells are treated with a RNA polymerase inhibitor such as actinomycin D and the effects on regulation of the target gene are examined at various times after treatment. Straightforward technique, but chemical inhibitors have nonspecific effects.
Pulse-chase analysis	mRNA is labeled by treating with radioactive nucleotides. At the time of treatment, a chase media is added that has excess nonradioactive nucleotide, and RNA is extracted at various times. This may be the most definitive measure of mRNA half-life. However, it may be difficult to label the target mRNA to high enough specific activity, especially if it is slowly synthesized. Also, if the mRNA degrades slowly, the chase may require excessive time in culture.
Poly(A) tail length	
RNase H mapping	This method takes advantage of the fact that RNase H digests RNA:DNA hybrids. A probe is designed to the 3' end of the mRNA, close to the start of the poly(A) tail. Following hybridization with this probe, the RNA is incubated with RNase H. The resulting RNA fragments are analyzed by Northern blotting.
In vitro decay reactions	
In vitro decay reactions	In vitro transcribed radioactive mRNA is prepared and incubated with a protein mixture, most often a polysome fraction. The mRNA produced by in vitro transcription can be designed to test certain aspects of mRNA processing. For example, *cis*-elements may be inserted or deleted. Following incubation with the polysome, the RNA products are resolved on a polyacrylamide gel and degradation assessed by autoradiography.

(continued)

Table 5-2 (Continued)

Method	Comments
Examination of *cis*-elements in mRNA	
Computer searches	Not as straightforward as finding *cis*-elements in DNA. The most common approach is to estimate the structure of the mRNA in solution. Several computer programs are available to perform this analysis. Hairpin sequences are often associated with affecting mRNA stability.
Reporter assays	The regulatory region of a target gene is cloned downstream of an easy-to-measure gene product (reporter gene). A basal promoter is used that will have a constant amount of transcription. When transfected into an appropriate cell, the reporter gene will respond to stimulus similarly to the endogenous target gene. Various mutation or deletion constructs can be used to find the *cis*-elements that are required for regulation. Simple, yet powerful assay to find *cis*-elements and can be used to find positive and negative regulatory elements.
mRNA gel-shift assays (REMSA)	Short pieces of the target gene regulatory region are synthesized in vitro and incubated with purified or crude extracts. When resolved on a nondenaturing polyacrylamide gel, complexed mRNA will move more slowly through the gel, resulting in a retarded band. Extra care must be given to handling RNA compared to DNA.
RNA footprinting	RNase T1 cleaves RNA at guanosines that are unprotected by an RNA-binding protein (RBP) In vitro transcribed RNA is incubated with the RBP and digested with RNase. The resulting products are resolved on a polyacrylamide gel and the footprint is observed following autoradiography.
Examination of *trans*-acting proteins	
Identification of a trans-acting protein for a given response element	
Supershift assays	A standard REMSA is performed in the presence of an antibody to one of the constituents. Owing to the mass of the antibody, the protein–RNA complex is further retarded on the native polyacrylamide gel resulting in a supershift. Simple method to determine if a particular protein in part of the complex is associated with a *cis*-element. Unfortunately, not all antibodies are appropriate for supershift assays as they may not recognize the *trans*-acting protein in this context.

UV crosslinking	The radioactive *cis*-element is incubated with the protein source and UV irradiated. The RNA binding protein is now covalently labeled, due to the radioactive RNA, and its relative molecular mass can be determined by SDS-PAGE. This is a relatively low-resolution technique but may be used to purify enough protein for sequencing.
Northwestern blots	Proteins from cellular extracts are resolved by SDS-PAGE, blotted, and refolded. Subsequently, the end-labeled *cis*-element is incubated and the mass of the interacting protein assessed. Rapid method to gain some information on the *trans*-acting protein, although difficult to determine the identity of this protein based solely on this data. Many proteins do not regain RNA binding activity when refolded and *trans*-acting factors that require heterodimerization are not detected.
Affinity purification	Biochemical approach to identify a *trans*-acting protein once a *cis*-element is described. The RNA is affixed to a resin and protein extracts are passed through a column packed with this affinity matrix. After the protein is eluted, it may be probed by Western blot or partially sequenced. Utilizes simple techniques, but not sensitive or rapid and often numerous proteins are eluted. Nonetheless, it is possible to identify novel RNA-binding proteins using this approach.
Examination of mRNA Splicing	
Computer searches	Various programs are available to predict intron/exon boundaries and mRNA splice sites. Also, the human genome project has afforded a more global look at human splice variants.
Sequencing of splice variants	The coding region of the target gene cDNA is PCR amplified, cloned, and sequenced. If enough individuals are examined, the existence of a splice site variation may be observed. Alternatively, a particular region may be amplified and sequenced. Can be used to identify new splice variants but is not high throughput.
Splice-specific PCR	Primers are designed that can discriminate between the spliced and unspliced transcript. Can be rapid, sensitive, and quantitative.

(continued)

Table 5-2 (Continued)

Method	Comments
In vitro splicing	Pre-mRNA is transcribed in vitro in the presence of a radioactive nucleotide. This transcript is incubated with nuclear extracts and the extent of splicing examined by denaturing gel electrophoresis and autoradiography. Can be adapted to examine various RNA-binding proteins through immunodepletion and/or reconstitution of the nuclear extracts.
Examination of splicing factors	Since splice sites are essentially *cis*-elements, splicing factors may be examined using any of the methods described for examination of *trans*-acting proteins.
Examination of mRNA transport	
In situ hybridization	The two methods to examine mRNA localization are very similar. Cells are fixed and permeabilized. With *in situ* hybridization, the cells are incubated with labeled antisense RNA, much like a Northern blot. Localization of the labeled probe is performed by microscopy.
In situ RT-PCR	Fixed and permeabilized cells are incubated with DNase and reverse transcription is performed. Subsequently, a labeled primer and PCR reagents are added. Localization of the labeled probe is performed by microscopy.

Study Questions

The following questions are designed to help integrate the information found in Part I. They do not necessarily have definitive answers. Each individual and laboratory has a different set of skills and infrastructure that may dictate the method or approach to be utilized. You should be able to justify your rationale for choosing certain approaches, however. The answers should be in an outline form without specifics. Observations from another laboratory have been made regarding the regulation of a particular gene ("Gene X") in mouse liver following treatment with a xenobiotic. The following questions relate to the understanding of this event.

1. Your laboratory has been using mouse hepatocyte cell lines to examine the effects of this xenobiotic on gene expression, but you have not previously determined if Gene X is affected. Outline your studies to determine if Gene X mRNA accumulates in your model system. (If this proves to be the case, the model system will be used in the subsequent studies.)
2. What two experiments would you perform to prove that the effect of the xenobiotics on Gene X mRNA was transcriptional or posttranscriptional?
3. Questions 3–6 assume that the regulation is transcriptional. Outline the studies you would perform to identify the cis-element in Gene X that is responsible for the transcriptional regulation. Assume that there are plasmids available containing the genomic sequence of Gene X. The ultimate goal is to find one small stretch of DNA that is responsible for the xenobiotics effects.
4. Once a cis-element is obtained, outline the studies to be performed that would identify the trans-acting protein.
5. In most cases, trans-acting proteins such as transcription factors will recruit chromatin remodeling factors to the target gene. Describe two studies that could be used to examine the chromatin structure around the cis-element discovered in Question 3.
6. Although not examined frequently, methylation of the DNA around the promoter may also be affected by the transcription apparatus. How could you examine the alternation in the methylation of CpG islands within Gene X's promoter following xenobiotic treatment?
7. Questions 7–10 assume that posttranscriptional regulation of Gene X is suspected. First, how can it be shown definitively that mRNA turnover is being affected by the xenobiotic? (The absence of a transcriptional effect is not sufficient to prove posttranscriptional mechanisms.)

8. If turnover is affected, how would you identify the cis-element in the mRNA that is responsible for this event?
9. Once a cis-element is obtained, outline the studies to be performed that would identify the trans-acting protein that is responsible for the altered mRNA stability.
10. The degradation of mRNA can follow many paths. Using in vitro systems, how would you determine the nucleases responsible for Gene X's mRNA decay and whether the effects of the xenobiotic are attributable to poly(A)tail length.
11. Questions 11–12 assume that subsequent studies show that the effects of the xenobiotic on Gene X protein production are more dramatic than production from either the transcriptional or posttranscriptional mechanisms. For example, a twofold difference in Gene X mRNA is seen, but a fivefold difference in protein is observed. How would you prove the hypothesis that the xenobiotic is affecting the splicing of Gene X mRNA which in turn affects translation efficiency?
12. Similarly, outline one study that proves whether Gene X mRNA transport is affected by the xenobiotic.
13. The biology of Gene X has not been well characterized. Very little is known about its biological role or whether regulation by the xenobiotic is associated with toxicity. Gene transcript profiling provides one way in which the regulation of Gene X can be compared to thousands of genes with known function. This may assist in determining its biological function. Outline studies you could perform to identify genes that are regulated similarly to Gene X.
14. Assume that Gene X can regulate gene expression, for example, if this protein is a signaling molecule or transcription factor. How could you determine the target genes of this protein?

References

1. Alberts B, Bray D, Lewis J, Raff M, Roberts K, and Watson JD. Control of gene expression. In: Molecular Biology of the Cell, Garland Publishing, New York, NY. 1994, pp. 401–476.
2. Pugh BF. Control of gene expression through regulation of the TATA-binding protein. Gene 255(1): 1–14, 2000.
3. Dillon N and Sabbattini P. Functional gene expression domains: defining the functional unit of eukaryotic gene regulation. Bioessays 22(7): 657–65, 2000.
4. Huang L, Guan RJ, and Pardee AB. Evolution of transcriptional control from prokaryotic beginnings to eukaryotic complexities. Crit Rev Eukaryot Gene Expr 9(3–4): 175–82, 1999.
5. Kornberg RD. Eukaryotic transcriptional control. Trends Cell Biol 9(12): M46–9, 1999.
6. Chin JW, Kohler JJ, Schneider TL, and Schepartz A. Gene regulation: protein escorts to the transcription ball. Curr Biol 9(24): R929–32, 1999.
7. Roeder RG. Role of general and gene-specific cofactors in the regulation of eukaryotic transcription. Cold Spring Harb Symp Quant Biol 63: 201–18, 1998.
8. Franklin GC. Mechanisms of transcriptional regulation. Results Probl Cell Differ 25: 171–87, 1999.
9. Ausubel FM, Brent R, Kingston RE, et al. Current Protocols in Molecular Biology. John Wiley, New York, NY, 1994.
10. Sambrook J, Fritsch EF, and Maniatis T, eds. Molecular Cloning: A Laboratory Manual. Cold Spring Harbor Press, Cold Spring Harbor, NY, 1989.
11. Davis LG, Kuehl WM, and Battey JF. Basic Methods in Molecular Biology. Appleton & Lange, Norwalk, CT, 1994.
12. Kaufman PB. Handbook of Molecular and Cellular Methods in Biology and Medicine. CRC Press, Boca Raton, FL, 1995.
13. Vanden Heuvel JP, Mattes WB, Corton JC, Bell DA, and Pittman G. PCR Protocols in Molecular Toxicology. CRC Press, Boca Raton, FL, 1998.
14. Vanden Heuvel JP. Approaches for the identification of xenobiotic-inducible genes. In: Toxicant-receptor interactions (Denison MS and Helferich WG, eds.), pp. 217–235. Taylor and Francis, Philadelphia, 1998.
15. Lewin B. Genes IV. Oxford University Press, Oxford, UK, 1990.
16. Bishop JO, Morton JG, Rosbash M, and Richardson M. Three abundance classes in HeLa cell messenger RNA. Nature 270: 199–204, 1974.
17. Mattes WB. The basics of the polymerase chain reaction. In: PCR Protocols in Molecular Toxicology, Vanden Heuvel JP, ed. CRC Press, Boca Raton, FL, 1997, pp. 1–40.
18. Vanden Heuvel JP. Analysis of gene expression. In: PCR Protocols in Molecular Toxicology, Vanden Heuvel JP, ed. CRC Press, Boca Raton, FL, 1997, pp. 41–98.

19. Gilliland GS, Perrin K, and Bunn HF. Competitive PCR for quantitation of mRNA. In: PCR Protocols: A Guide to Methods and Applications, Innis MA, Gelfand DH, Sninsky JJ, and White TJ, eds. Academic Press, San Diego, CA, 1990, pp. 60–69.
20. Wan JS, Sharp SJ, Poirier GM, et al. Cloning differentially expressed mRNAs. Nat Biotechnol 14(13): 1685–91, 1996.
21. Liang P and Pardee AB. Differential display of eukaryotic messenger RNA by means of the polymerase chain reaction. Science 257: 967–971, 1992.
22. Sokolov BP and Prockop DJ. A rapid and simple PCR-based method for isolation of cDNAs from differentially expressed genes. Nucleic Acids Res 22: 4009–4015, 1994.
23. Velculescu VE, Zhang L, Vogelstein B, and Kinzler KW. Serial analysis of gene expression. Science 270(5235): 484–7, 1995.
24. Velculescu VE, Vogelstein B, and Kinzler KW. Analysing uncharted transcriptomes with SAGE. Trends Genet 16(10): 423–5, 2000.
25. Mattes WB. Emerging technologies and predictive assays. Introduction and overview. In: Cellular and Molecular Toxicology, Vol. 14, Vanden Heuvel JP, Perdew GH, Mattes WB, and Greenlee WF, eds. Elsevier, Amsterdam, 2002, pp. 463–492.
26. Bilban M, Buehler LK, Head S, Desoye G, and Quaranta V. Normalizing DNA microarray data. Curr Issues Mol Biol 4(2): 57–64, 2002.
27. Hegde P, Qi R, Abernathy K, et al. A concise guide to cDNA microarray analysis. Biotechniques 29(3): 548–50, 552–4, 556 passim, 2000.
28. Quackenbush J. Microarray data normalization and transformation. Nat Genet 32 Suppl: 496–501, 2002.
29. Pennie WD. Custom cDNA microarrays; technologies and applications. Toxicology 181–182: 551–4, 2002.
30. Vanden Heuvel JP. Control of gene expression. In: Cellular and Molecular Toxicology, Vol. 14, Vanden Heuvel JP, Perdew GH, Mattes WB, and Greenlee WF, eds. Elsevier, Amsterdam, The Netherlands, 2002, pp. 57–79.
31. Conaway JW, Shilatifard A, Dvir A, and Conaway RC. Control of elongation by RNA polymerase II. Trends Biochem Sci 25(8): 375–80, 2000.
32. Bregman DB, Pestell RG, and Kidd VJ. Cell cycle regulation and RNA polymerase II. Front Biosci 5: D244–57, 2000.
33. Uptain SM, Kane CM, and Chamberlin MJ. Basic mechanisms of transcript elongation and its regulation. Annu Rev Biochem 66: 117–72, 1997.
34. Wolffe AP and Hayes JJ. Chromatin disruption and modification. Nucleic Acids Res 27(3): 711–20, 1999.
35. Struhl K. Fundamentally different logic of gene regulation in eukaryotes and prokaryotes. Cell 98(1): 1–4, 1999.
36. Mannervik M, Nibu Y, Zhang H, and Levine M. Transcriptional coregulators in development. Science 284(5414): 606–9, 1999.

37. Bevan C and Parker M. The role of coactivators in steroid hormone action. Exp Cell Res 253(2): 349–56, 1999.
38. Workman JL and Kingston RE. Alteration of nucleosome structure as a mechanism of transcriptional regulation. Annu Rev Biochem 67: 545–79, 1998.
39. Robertson KD and Jones PA. DNA methylation: past, present and future directions. Carcinogenesis 21(3): 461–7, 2000.
40. Moss T, ed. DNA–Protein Interactions: Principles and Protocols. Humana Press, Totowa, NJ, 2001.
41. Blackwell TK, Kretzner L, Blackwood EM, Eisenman RN, and Weintraub H. Sequence-specific DNA binding by the c-Myc protein. Science 250(4984): 1149–51, 1990.
42. Wells J and Farnham PJ. Characterizing transcription factor binding sites using formaldehyde cross-linking and immunoprecipitation. Methods 26(1): 48–56, 2002.
43. Cross SH, Charlton JA, Nan X, and Bird AP. Purification of CpG islands using a methylated DNA binding column. Nat Genet 6(3): 236–44, 1994.
44. Epshtein V and Nudler E. Cooperation between RNA polymerase molecules in transcription elongation. Science 300(5620): 801–5, 2003.
45. Wilson AJ, Velcich A, Arango D, et al. Novel detection and differential utilization of a c-myc transcriptional block in colon cancer chemoprevention. Cancer Res 62(21): 6006–10, 2002.
46. Sutherland J and Costa M. DNA methylation and gene silencing. In: Cellular and Molecular Toxicology, Vol. 14, Vanden Heuvel JP, Perdew GH, Mattes WB, and Greenlee WF, eds. Elsevier Science, Amsterdam, The Netherlands, 2002, pp. 299–332.
47. Gitan RS, Shi H, Chen CM, Yan PS, and Huang TH. Methylation-specific oligonucleotide microarray: a new potential for high-throughput methylation analysis. Genome Res 12(1): 158–64, 2002.
48. Balog RP, de Souza YE, Tang HM, et al. Parallel assessment of CpG methylation by two-color hybridization with oligonucleotide arrays. Anal Biochem 309(2): 301–10, 2002.
49. Bentley D. The mRNA assembly line: transcription and processing machines in the same factory. Curr Opin Cell Biol 14(3): 336–42, 2002.
50. Cramer P, Srebrow A, Kadener S, et al. Coordination between transcription and pre-mRNA processing. FEBS Lett 498(2–3): 179–82, 2001.
51. Proudfoot N. Connecting transcription to messenger RNA processing. Trends Biochem Sci 25(6): 290–3, 2000.
52. Faustino NA and Cooper TA. Pre-mRNA splicing and human disease. Genes Dev 17(4): 419–37, 2003.
53. Newman A. RNA splicing. Curr Biol 8(25): R903–5, 1998.
54. Cullen BR. Nuclear RNA export pathways. Mol Cell Biol 20(12): 4181–7, 2000.

55. Staton JM, Thomson AM, and Leedman P. Hormonal regulation of mRNA stability and RNA–protein interactions in the pituitary. J Mol Endocrinol 25(1): 17–34, 2000.
56. Jacobson A and Peltz SW. Tools for turnover: methods for analysis of mRNA stability in eukaryotic cells [editorial]. Methods 17(1): 1–2, 1999.
57. Jacobson A and Peltz SW. Interrelationships of the pathways of mRNA decay and translation in eukaryotic cells. Annu Rev Biochem 65: 693–739, 1996.
58. Kren BT and Steer CJ. Post-transcriptional regulation of gene expression in liver regeneration: role of mRNA stability. FASEB J 10(5): 559–73, 1996.
59. Paek I and Axel R. Glucocorticoids enhance stability of human growth hormone mRNA. Mol Cell Biol 7(4): 1496–507, 1987.
60. DeMaria CT and Brewer G. Cell-free systems for analysis of cytoplasmic mRNA turnover. Prog Mol Subcell Biol 18: 65–91, 1997.
61. Ross J. mRNA stability in mammalian cells. Microbiol Rev 59(3): 423–50, 1995.
62. Ross J. The turnover of messenger RNA. Sci Am 260(4): 48–55, 1989.
63. Peltz SW, Brewer G, Bernstein P, Hart PA, and Ross J. Regulation of mRNA turnover in eukaryotic cells. Crit Rev Eukaryot Gene Expr 1(2): 99–126, 1991.
64. Wilson GM and Brewer G. The search for trans-acting factors controlling messenger RNA decay. Prog Nucleic Acid Res Mol Biol 62: 257–91, 1999.
65. Ross J. Assays for analyzing exonucleases in vitro. Methods 17: 52–59, 1999.
66. Zhao Z, Chang FC, and Furneaux HM. The identification of an endonuclease that cleaves within an HuR binding site in mRNA. Nucleic Acids Res 28(14): 2695–701, 2000.
67. Snee M, Kidd GJ, Munro TP, and Smith R. RNA trafficking and stabilization elements associate with multiple brain proteins. J Cell Sci 115(Pt 23): 4661–9, 2002.
68. Auerbach SS, Ramsden R, Stoner MA, Verlinde C, Hassett C, and Omiecinski CJ. Alternatively spliced isoforms of the human constitutive androstane receptor. Nucleic Acids Res 31(12): 3194–207, 2003.
69. Caputi M, Mayeda A, Krainer AR, and Zahler AM. hnRNP A/B proteins are required for inhibition of HIV-1 pre-mRNA splicing. EMBO J 18(14): 4060–7, 1999.

PART II

Regulation of Protein Levels and Transcription Factor Function

Gary H. Perdew, PhD

CONTENTS

Chapter 6. Dissection of Signaling Pathways

1

Overview

The basic approaches to understanding how a drug, chemical exposure, or disease results in altered protein level or activity will be described in this part, with particular emphasis on transcription factors. In addition, molecular approaches to understanding the role of a given protein in signaling pathways will also be examined. Exposure to a xenobiotic can alter protein activity or levels through a variety of mechanisms. We will explore the various approaches available to determine how a treatment leads to altered protein function by a mechanism other than by changes in mRNA levels. This part examines the experimental approaches that can define the general mechanism(s) that lead to altered protein activity, as depicted in Figure 1-1.

Often overlooked, the first mechanism, which will be discussed in Chapter 2, is altered protein synthesis or rate of protein turnover, which leads to an altered level of a given protein. Chapter 3 will explore the use of proteomics techniques, such as 2-D gel electrophoresis to globally examine and detect changes in the level of expression of specific proteins. A second mechanism that causes altered protein activity is through protein–protein interactions will be discussed in Chapter 4. A variety of different experimental approaches will be described to examine the ability of one protein to bind to another, including both in vitro and in vivo methodologies such as yeast two-hybrid, immunoprecipitations, and far-Western blotting techniques. In Chapter 5, a third protein-mediated mechanism that can be altered by xenobiotic treatment is posttranslational modification, such as phosphorylation, ubiquitination, acetylation, or glycosylation. A flowchart that outlines the possible mechanisms that can lead to altered activity of a transcription factor is shown in Figure 1-2 and provides a framework for the experimental approaches to be discussed in this part.

Chapter 6 explores the experimental approaches that can be used to examine the role of a protein in cellular signaling pathways. Often in the pursuit of determining the mechanism of xenobiotic- or disease-mediated changes in cellular function, a key protein that may mediate the observed effect may need to be blocked or down regulated to establish its role in the process. This can be

Regulation of Protein Levels and Transcription Factor Function by G. H. Perdew
From: *Regulation of Gene Expression*
By: G. H. Perdew et al. © Humana Press Inc., Totowa, NJ

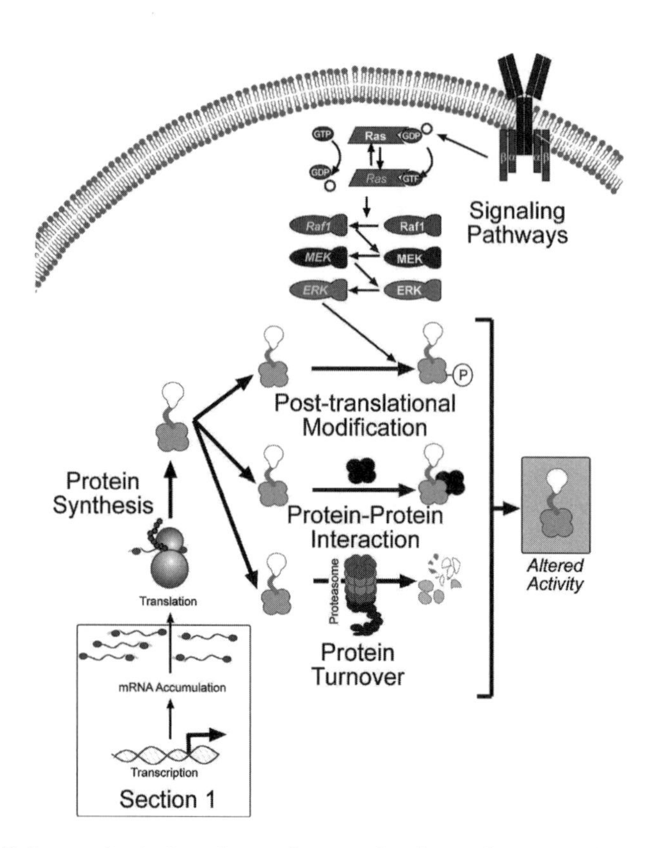

Fig. 1-1. Scheme depicting the major mechanisms that modulate protein activity.

accomplished in cell lines through one of two basic experimental approaches, the first being the disruption of protein–protein interaction. The second approach utilizes techniques that block expression of a specific protein. Part III will expand on the approaches that can be utilized to dissect the role of specific proteins in xenobiotic-mediated changes in gene expression by disrupting expression of specific genes in mouse models.

Examples of How Various Approaches Must Be Integrated to Determine a Mechanism of Action

In order to firmly understand how a protein's activity is altered under a particular set of conditions, it may be necessary to utilize several of the approaches described in this part. In addition, there may be situations where the approaches must be individualized. Another important point to keep in mind is that there could be more than one mechanism of action, thus determining a mechanism

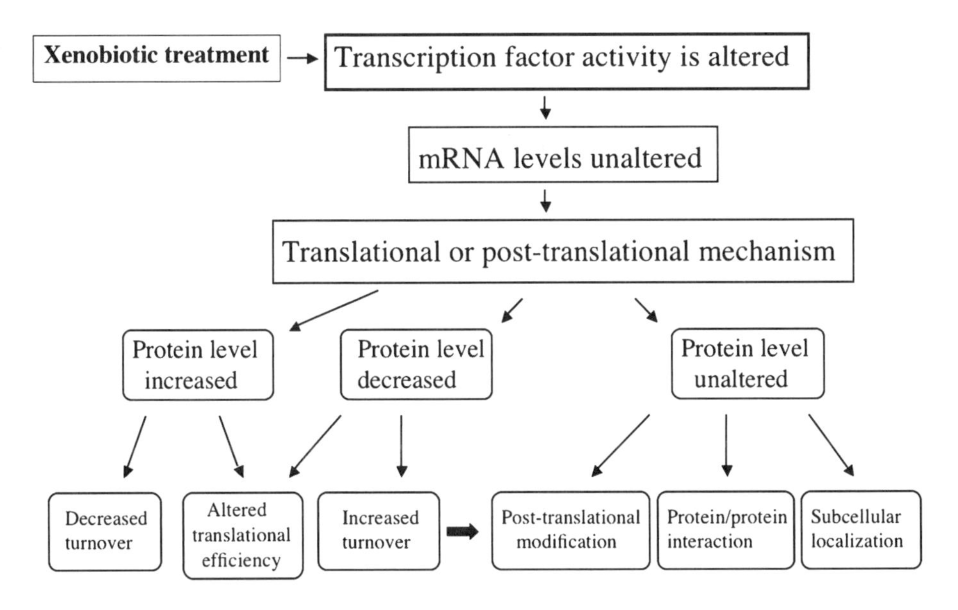

Fig.1-2. Flow chart of the possible mechanisms that can lead to altered transcription factor activity upon treatment of cells with a xenobiotic.

does not preclude consideration of other possibilities. Often it is through alterations in the function of specific protein domains after a given treatment that leads to phenotypic changes. For example, treatment with a protein kinase AKT inhibitor can lead to translocation of the transcription factor Forkhead (FKHR) into the nucleus. Before the actual mechanism can be firmly established, the domains involved in nuclear and cytoplasmic localization need to be characterized. Transcription factors often carry a number of important interaction domains or motifs that mediate specific protein–protein interaction events. Some of these domains can be tentatively identified by amino acid sequence-homology analysis. However, experimental analysis is required to actually determine that a domain or motif is indeed functional. An excellent example of this process is the determination of nuclear import and export signals. Nuclear localization sequences (lysine/arginine-rich sequences, e.g., KKKRKRKK) bind importins that facilitate nuclear uptake, and nuclear export sequences (e.g., LXXLXXLXL) bind to CRM1. Near consensus sequences for these two protein interaction sites have been described, but having these sequences does not necessarily mean that a protein will exhibit these activities, as the overall folding of a protein is important in obtaining the proper motif conformation around these types of sequences. As an example, this can be established through the fusion of the domain containing the putative motif and green fluorescent pro-

Many transcription factors have both an import- and export-signal sequence and undergo nucleocytoplasmic shuttling.

tein. Then the ability of this fusion protein to translocate into the nucleus, or to be exported out of the nucleus, can be assessed on its transient expression. Thus, while the presence of a wide array of possible protein–protein interaction domains can be determined through sequence homology with previously defined domains, it is essential to experimentally determine whether that domain actually exhibits the activity suggested by the sequence homology detected. Now getting back to our example, once the domains or motifs responsible for FKHR nucleocytoplasmic shuttling have been determined, the impact of AKT phosphorylation can be understood and thus the mechanism of action of the AKT inhibitor on FKHR function established. It is this type of approach that is extremely common in molecular toxicology or studies aimed at dissection of disease processes.

A second example is the observation that antibiotic geldanamycin (GA) has been shown to increase PPARα transcriptional activity and decrease Ah receptor transcriptional activity [1]. The mechanism to explain this observation has been determined through a series of experiments. First, it has been established through the use of affinity chromatography, that GA binds hsp90, and, more specifically, to its N-terminal ATP binding site. In general, GA binding to hsp90 leads to disruption of hsp90's ability to chaperone proteins. Second, immunoprecipitation experiments have established that hsp90 binds directly to both PPARα and the Ah receptor. Protein stability studies revealed that GA treatment of cells leads to rapid proteolytic degradation of the Ah receptor and has little effect on PPARα levels [2]. The use of transcriptional reporter-cell assays revealed that the transcriptional activity of the Ah receptor markedly decreased after ligand and GA treatment. In contrast, PPARα transcriptional activity was increased in the presence of GA and ligand. Taken together, these studies would suggest that hsp90 represses PPARα activity and has little effect

1. Sumanasekera, W.K. et al. Biochem. 42 (2003) 10,726–10,735.

2. Chen, H-S. et al. Arch. Biochem. Biophys. 348 (1997) 190–198.

on its stability, while hsp90 appears to be required to protect the Ah receptor from rapid proteolytic degradation within the cell. The mechanistic studies outlined above are just two examples of the type of protein function studies that can be performed through the use of the experimental approaches described in Part II.

2

Protein Synthesis and Turnover

1. Concepts

Chapters 4 and 5 in Part I have detailed the regulation of mRNA synthesis and stability. The next regulatory step in the process of protein production is the rate of protein synthesis or mRNA translation. Protein synthesis can be divided into three discreet steps; initiation, elongation, and termination. An important aspect of overall regulatory control of protein synthesis is regulation of peptide chain elongation by a group of eukaryotic elongation factors (eEFs). The activity of several eEFs is regulated by phosphorylation. For example, the kinase that stimulates eEF2 activity is regulated by calcium ions, calmodulin, and mitogen-activated protein (MAP) kinase. Also, low energy levels in the cell can lead to downregulation of the eEF2 kinase and thus peptide chain elongation activity. Other key contributors to translational control are the eukaryotic initiation factors (eIFs). One rate-limiting member of the eIF-4F translation initiation complex is the mRNA cap-binding protein eIF-4E. This protein binds the cap structure at the 5' terminus of mature mRNAs and recruits mRNAs to the eIF-4F complex, which then scans from the 5' cap through the untranslated region (UTR), unwinding secondary structure to reveal the translation initiation codon, which then leads to ribosome loading. Messenger RNAs with short unstructured 5' UTRs are more easily translated than mRNA with long 5' UTRs, as they require more secondary structure unwinding. Many proteins that are key players in angiogenesis, cell

Overall rate of protein synthesis is tightly regulated by initiation and elongation factors.

Regulation of Protein Levels and Transcription Factor Function by G. H. Perdew
From: *Regulation of Gene Expression*

growth, and apoptosis have long 5' UTRs. Also, an increase in ribosome production could facilitate the translation of these mRNA with long 5' UTRs. Often in tumor cells high eIF-4E activity is observed, which would lead to more efficient translation of key mRNAs that would enhance tumor growth. Yet another example of how the specificity of translation of certain mRNAs could be altered is by upregulation or posttranslational modification of such ribosomal proteins as S6. An increase in phosphorylation of S6 can lead to increased translation of mRNAs that contain a terminal oligopyrimidine tract in their 5' UTR. Constitutive phosphorylation of S6 is commonly observed in tumor cells. Thus, alterations in the levels of key factors that regulate various steps in protein synthesis by exposure to modulatory chemicals could contribute to important disease end points.

While the actual level of a protein within the cell is determined by the combination of the rate of synthesis and degradation, in general, the level of many proteins is primarily determined by the amount of mRNA available for protein synthesis. However, there are many examples where a rapid alteration in protein stability leads to important biochemical events. An excellent example is the transcription factor HIF1α that responds to hypoxia (low O2 levels) within the cell. Upon oxygen deprivation, the half-life of HIF1α is greatly increased through a decrease in the posttranslational hydroxylation of key proline and asparagine residues. Other examples of changes in protein stability can be mediated by protein–protein interactions and phosphorylation. For example, IκB forms a stable p50/RelA heterodimer (NFκB) in the cytoplasm of the cell. Upon activation of IκB kinase complexes through signaling casacades, IκB is phosphorylated. This leads to rapid ubiquitination and subsequent proteolytic turnover. Thus, protein stability is an important parameter that should be tested to thoroughly understand the mechanism(s) behind an increase or decrease in a protein's level in response to a chemical treatment. Even when a change in mRNA is observed in a given experiment, it is still important to consider that several mechanisms may be occurring that lead to an alteration in a protein's level.

In addition to alterations in the synthesis or stability of specific proteins, more global changes in protein synthesis or degradation can occur in cells upon exposure to various treatments (e.g., heat shock). This is often overlooked in studies examining the effect of cell exposure to modulators. It has been well established that growth factors and protease inhibitors can have a global effect on protein levels in cells. More recently, several proto-oncogenes and tumor suppressors have been shown to directly regulate ribosome production or initiation of protein synthesis. It is quite possible that loss of key checkpoints in protein synthesis could contribute to the carcinogenesis process. Interestingly, an overall increase in protein synthesis rates can increase the levels of a normally low abundance regulatory protein(s) to a threshold level that could lead to abnormal cell growth or responses to external stimuli (e.g., c-myc). These observations would suggest that both quantitative and qualitative differences in protein synthesis or protein turnover can be important parameters to consider in the overall assessment of an exogenous modulator's mechanism of action.

Growth factors and heat shock are two examples of global regulators of protein synthesis.

2. Methods and Approaches

2.1. Analysis of Protein Synthesis and Turnover Rates

Protein synthesis can be measured in cultured cells by determining the level of a radioactive amino acid (e.g., leucine, methionine) incorporated into proteins. Total radioactive amino acid incorporation into proteins can be measured by acid precipitation of protein from total cell lysate and expressed as the amount of radioactivity per microgram of protein. A time course in the presence of a possible modulator can be performed to test if an overall effect on protein synthesis has occurred. If the rate of synthesis of a specific protein needs to be assessed, an immunoprecipitation should be performed and the level of radioactive amino acid incorporation determined after sodium dodecyl sulfate polyacrylamide gel electrophoresis (SDS-PAGE) analysis. The relative amount of radioactivity in the protein band of interest can be determined by phosphor imaging.

For studies that have noted a rapid change in protein levels (e.g., 1–2 h), it is logical to assume that the most likely cause for this effect is a change in protein turnover. This is especially true for proteins that have a relatively long half-life, such as 8–24 h. However, many regulatory proteins have a relatively short half-life, for example, the half-life for c-myc is approx 30 min. A rapid change in c-myc levels could be attributed to either a change in synthesis rate or degradation of mRNA and/or protein. There are two basic techniques to monitor protein

Radioactive amino acid pulse-chase cell culture experiments are an effective means to determine a protein's half-life.

turnover rates. Currently, the most commonly used method is termed "pulse-chase." This approach involves incubating cells with radioactive amino acid (e.g., ^{35}S-methionine) for 1–4 h in medium with low levels of the nonradioactive version of that amino acid, which is termed a "pulse." An excess of the non-radioactive amino acid is added to dilute the radiospecific activity and thus serve as a "chase" of the radioactive amino acid from proteins as they degrade and new synthesis occurs. The time frame for the chase time-course is dependent on the turnover rate of the protein being studied. In a typical experiment, cells that have undergone a chase for 2, 4, 8, 12, and 18 h are harvested and the protein of interest is immunoprecipitated. Cell lysates are subjected to SDS-PAGE and the amount of radioactivity in the specific protein band determined. The time-course data is plotted and the half-life is defined as the amount of time to achieve 50% loss of radioactivity in the protein being investigated. A second method for determining protein turnover rates involves incubating cells with a radioactive amino acid for 4 h followed by disruption of protein synthesis with cyclohexamide over a time-course during which samples are taken for analysis. Each cell preparation is lyzed and centrifuged to remove membranes, and the protein of interest is immunoprecipitated and analyzed as described for pulse-chase analysis. However, this method is not used very often because the high level of toxicity of cyclohexamide can lead to nonspecific effects.

2.2. Use of Polyacrylamide Electrophoresis and Protein Blotting to Determine Specific Protein Levels

Perhaps the single most important technique for analysis of proteins SDS-PAGE. There are several variations of this standard technique that can be quite useful. First, an investigator has the option of using either standard format (e.g., 15 3 15 cm) or a mini-gel format (e.g., 8 3 8 cm). This can be an issue if one needs to obtain better separation between two bands. Second, the use of gradient polyacrylamide gels (e.g., 7–15% acrylamide) can also improve the resolving power of the gel system. A third potential consideration is the basic type of gel-buffer running system, two systems being most widely utilized. Tris-glycine SDS-PAGE is the most extensively used system and offers excellent resolution, as the stacking and lower gels use different pH buffers. However, the pH can affect the storage life of precast gels. The second gel system utilizes a tricine-buffer system, whose gels have the same pH buffer in both the stacking and lower gel and thus should have a longer shelf life. An additional advantage of this system is enhanced stacking potential, which is particularly important for low-molecular-weight proteins (e.g., 1–2 kDa). One disadvantage of this system is the relatively high cost for tricine. After electrophoresis, the protein is transferred to either nitrocellulose or PVDF membrane. For quantitative blotting experiments, PVDF membrane should be used since it has better protein retention capacity and is resistant to cracking. After protein has been quantitatively transferred to membrane, the blot can be probed with an appropriate system to assess the level of a specific protein. Quantitative assessment of the amount of a specific protein is performed with an antigen-specific primary antibody and a secondary probe that will visualize the amount of primary antibody binding. There are a number of choices that can be used to assess quantitatively the amount of primary antibody, although perhaps the most reliable quantitative visualization reagent is [^{125}I]-secondary antibody. An alternate system is the use of a biotinylated secondary antibody followed by probing of the blot with [^{125}I]-streptavidin. The relative amount of radioactivity can be assessed using a phosphor imager, or radioactive bands can be detected with X-ray film and used to mark excision of protein bands, which can be counted in a gamma counter. Interestingly, many investigators often utilize a secondary antibody/horseradish-peroxidase conjugate coupled with enhanced chemiluminescence visualization. The signal is captured on X-ray film, and the system is quite sensitive. However, great care should be taken to demonstrate that this system yields a linear response. Thus this technique is best used to detect dramatic changes in protein levels.

Quantitative assessment of protein levels can be accurately determined using an [^{125}I]-secondary antibody system.

3. Summary

The actual level of a cellular protein at any given time-point is determined by the balance between protein synthesis and degradation. The ability to alter the synthesis or degradation of proteins in general, or of a specific protein, by modulators (e.g., growth factors, tumor promoters, receptor ligands) can be assessed in cell culture systems. Unfortunately, turnover rates for proteins cannot be easily assessed in vivo. The rate of protein synthesis or the half-life of a protein can be determined in cultured cells through the use of radioactive amino acids, the most common technique being the pulse-

Rate of protein turnover can be assessed in cultured cells.

chase method. These methods are important in developing a complete picture of the mechanism by which a modulator may alter the level of a given protein, along with the mRNA methods discussed in earlier chapters. Indeed, it is important to note that modulators can affect both mRNA and protein levels. Thus, the possibility of multiple mechanisms needs to be experimentally established prior to developing a complete picture of xenobiotic action.

3

Proteomics

1. Concepts

Development of global scale methods for protein profiling is important to complement existing mRNA analysis techniques, such as DNA microarray analysis. Especially considering that changes in protein levels can occur without any change in an individual protein's mRNA levels. Changes in translational rates, protein stability, protein localization, or posttranslational modifications can all lead to altered protein levels or activity. Only through the global characterization of protein levels can a complete picture of possible changes evoked by a chemical treatment or altered cell phenotype be more fully established. After all, proteins usually are the end point of gene expression.

Proteomics can be defined as the study of structure and expression of proteins through relatively global or high-throughput techniques. Another definition could be the simultaneous analysis of a complex protein mixture to detect differences in individual protein levels. However, this is more difficult to accomplish than the methods available to analyze changes in specific mRNAs. This is due to the fact that proteins are complex molecules made up of a combination of 22 amino acids in unique sequences that fold into an array of secondary and tertiary structures. In addition, each protein can undergo various posttranslational modifications. Thus while each fragment of DNA in the genome behaves virtually the same, proteins can vary widely in their biophysical properties (e.g., membrane bound versus soluble), which leads to a significant problem with examining thousands of proteins at the same time. The hybridization techniques extensively used in DNA/RNA experimentation cannot be used with proteins and no rapid method of global protein analysis is available. The method that allows the most global analysis of protein levels is two-dimensional (2-D) gel electrophoresis. This method separates proteins based on their pI and molecular weight. The combination of these two parameters allows a protein to migrate to a unique position on a 2-D electrophoretic map of crude cellular protein extracts on polyacrylamide gels. However, a limited amount of protein can be applied to each gel (e.g., 100 μg) and not every protein in a total cell lysate resolves into a discreet spot. Also highly basic or

Regulation of Protein Levels and Transcription Factor Function by G. H. Perdew
From: *Regulation of Gene Expression*
By: G. H. Perdew et al. © Humana Press Inc., Totowa, NJ

The problem with 2-D protein gel analysis is that low abundance proteins may not be detected.

Individual protein spots can easily be identified using MS techniques.

acidic proteins (e.g., histones) may not resolve in the pI range used in the first dimension. It is generally assumed that, even after staining the gel with a sensitive stain like silver stain, not every polypeptide will be visualized. Thus, the problem with 2-D gel analysis is that very low abundance proteins, including many regulatory proteins, are difficult to detect by standard 2-D gel analysis. Despite these limitations, 2-D gel analysis can be a powerful method to detect changes in protein expression.

The next step in 2-D gel analysis is the identification of individual protein spots. This can be accomplished through mass spectroscopy (MS), which is a highly sensitive technique coupled with protein databases that examine theoretical trypic peptide fragments for each protein in the database. Prior to these recent developments, protein identification was largely accomplished by isolation of a sufficient amount of an intact protein or a fragment for Edman sequencing, a technique that requires considerably more protein than the current MS approaches require.

Another type of proteomics is "functional proteomics," which can be defined as high-throughput methodology that attempts to detect proteins in the proteome that exhibit a specific functional property. The concept of this methodology is simple; however, producing a microarray with a large number of proteins arrayed is technically challenging, laborious, and expensive. Nevertheless, this methodology holds great promise for globally examining the ability of a protein or small-molecular-weight compound to interact with arrayed proteins.

2. Methods and Approaches

2.1. Global Protein Analysis Techniques

2.1.1. Two-Dimensional Gel Electrophoresis

Historically, the most extensively used method of global protein analysis is the separation of protein samples on individual 2-D gels. This method first

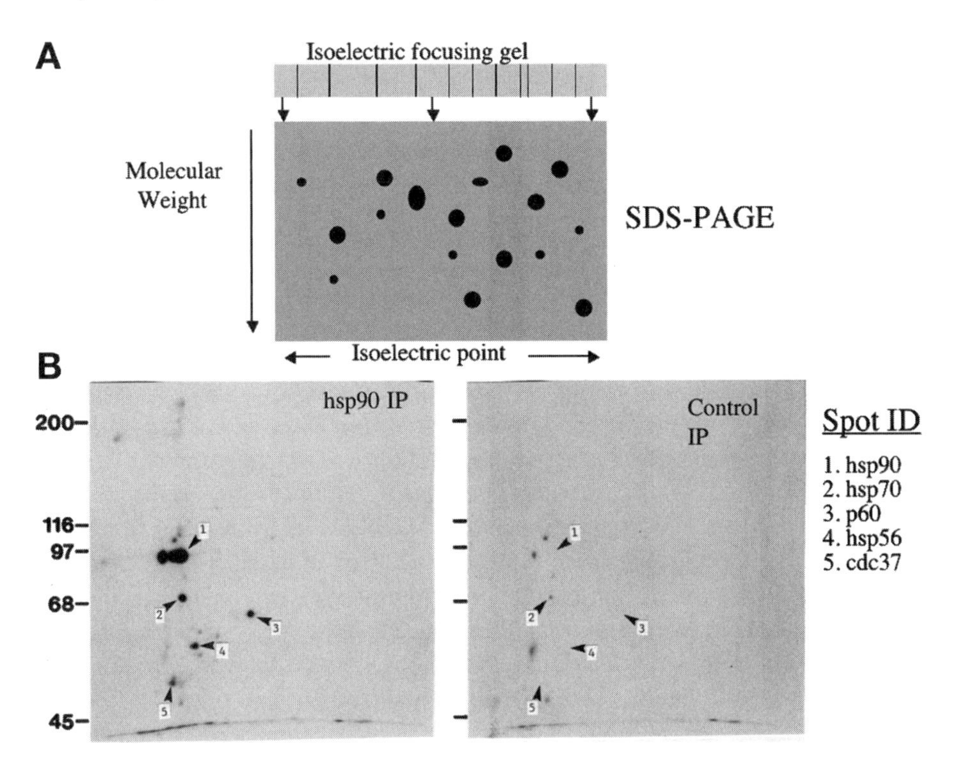

Fig. 3-1. Two-dimensional polyacrylamide gel electrophoresis. (**A**) A schematic representation of the 2-D gel analysis of complex protein mixtures. Proteins are applied to an isoelectric focusing gel and proteins move to their isoelectric point. The gel is placed on a SDS polyacrylamide gel and, upon electrophoresis in the second dimension, proteins migrate as spots. The size of the spot is dependent on the amount of protein and its solubility characteristics. (**B**) 2D PAGE analysis of an hsp90 and a control immunoprecipitation from cytosol obtained from Hepa1 cultured in the presence of [^{35}S]methionine. The identification of the labeled spots is shown and was determined using antibodies and standard blotting techniques.

separates proteins by isoelectric focusing under denaturing conditions, usually in high concentrations of urea and nonionic detergents. Proteins are applied to one end of the isoelectric focusing gel and migrate to their isoelectric point (pI) and focus into a sharp band (Figure 3-1A). The isoelectric focusing gel is then placed on top of a polyacrylamide gel and sealed into place with agarose. In the presence of sodium dodecyl sulfate the proteins migrate as spots on the polyacrylamide gel according to their molecular weight, although posttranslational modifications such as glycosylation can influence their migration independently of their molecular weight.

Traditionally, these gels are usually silver stained, followed by image capture for computer analysis. Silver staining of proteins can be an extremely sensitive method of detection, visualizing as little as 1 ng of protein. More recently, fluorescent-dye staining techniques are being used increasingly. Regions of two separate gels are compared using pattern recognition computer programs. While this method can help map the location of various proteins on a 2-D gel, it is difficult to assess quantitatively relatively small changes in gene expression (e.g., two-fold or less). Nevertheless, this method has been used to map changes in protein levels.

There are several potential technical problems with the separation of proteins on 2-D gels. The immobilized pH gradients utilized can have various pH ranges (e.g., 3–10, 5–7), but the inability to separate proteins of very high or low pI can be a limiting factor with the technique. The incomplete extraction of proteins from whole cells or organelles can lead to nonlinear or variable results. Other potential problems include the failure of some proteins to enter the focusing gel, streaking of proteins upon entry, and the lack of a linear response of the dye or stain. In the second dimension, depending on the percent of acrylamide, high-molecular-weight proteins may not enter properly, while very small proteins often run at the dye front and thus are not properly separated. Despite these limitations, a wide range of proteins separate as discreet spots.

Screening for changes in glycosylation, phosphorylation, or other posttranslational modifications can often be accomplished with 2-D gels if a change in the overall charge of a protein occurs. Many proteins that are posttranslationally modified resolve on a 2-D gel as a series of spots that can all be of the same molecular weight or, in the case of some highly glycosylated proteins, run with varying molecular weights. The ability of 2-D gels to reveal phosphorylation heterogeneity can be used to detect changes in overall phosphorylation status of a transcription factor by allowing the shifts in pI before and after chemical treatment to be assessed. However, it is important to keep in mind that the specific sites that are altered would have to be studied using different procedures. Nevertheless, this is potentially a very useful application of 2-D gel electrophoresis that is often not utilized.

2.1.2. Two-Dimensional Gel Electrophoresis Double-Label Techniques

One of the most useful applications of the standard 2-D electrophoresis protocol is to determine if differential expression of a large number of proteins occurs between two crude protein extracts. This can be accomplished utilizing several distinct methods. One technique developed shortly after the introduction of 2-D gel electrophoresis was the use of two radioisotopes to detect differential protein expression. Briefly, this method involves the incorporation of ^{14}C or ^{3}H amino acids (e.g., leucine) into two populations of cells. One of the cell cultures is treated with the chemical that is being studied and the cell cultures are harvested and combined. Cellular extracts are isolated, the proteins separated by 2-D electrophoresis, and the resulting gels are exposed to X-ray film. Individual protein spots are then excised from the gel and subjected to liquid scintillation analysis. The amount of ^{14}C or ^{3}H radioisotope in each spot is determined and the isotope ratio determined. Assuming that most proteins will not be altered by the chemical treatment, the ratio of most protein will be the same and the average ratio is normalized to one. The normalized ratios are plotted and the ratio of an individual protein spot is considered to be significant if it is greater than two standard deviations from the mean. This method has a number of strengths, including: cells are added together prior to isolation of subcellular fractions (e.g., microsomes), and characterization by 2-D electrophoresis without concern for potential differential losses during the various experimental isolation steps. This method is a powerful approach for detecting differences in the level of individual proteins, although it is largely restricted to cell culture experiments. The usefulness of this technique is well-illustrated in Figure 3-2, where it was determined that cyclopropenoid fatty acids fed to rainbow trout could cause the downregulation of the synthesis of an approx 220-kDa protein identified as acetyl-CoA carboxylase in cultured trout hepatocytes [3]. This study utilized 1-D gel electrophoresis, although 2-D gel electrophoresis would also have been useful to further detect additional individual protein variations according to their pI.

Double isotope system for detecting changes in individual protein expression offers a number of important advantages.

3. Perdew, G.H. et al. Biochim. Biophysics Acta 877 (1986) 9–19.

A second more recently developed approach utilizes the postlabeling of two pools of cellular protein extracts with 1-(5-carboxypentyl)-19-propylindocarbocyanine halide (CY-3) N-hydroxy-succinimidyl ester and 1-(5-carboxypentyl)-19-methylindocarbocyanine halide (CY-5) *N*-hydroxy-succinimidyl ester fluorescent dyes, respectively. The chemically tagged proteins are mixed and separated by 2-D gel electrophoresis and this technique has been termed differential in-gel electrophoresis

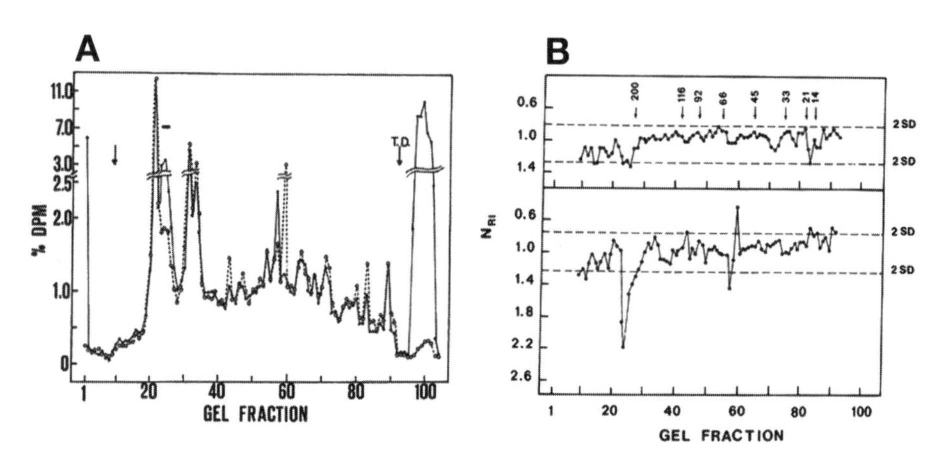

Fig. 3-2. Double isotope labeling of microsomes analyzed by SDS-PAGE. (**A**) Distribution of [^{14}C]leucine (solid line) and [^{3}H]leucine (dotted line) in microsomal proteins from trout hepatocytes from control and cyclopropenoid fatty acid fed trout, respectively. Microsomal proteins were separated using SDS-PAGE and each lane of the resulting gel cut into 2.5 mm slices. The amount of [^{14}C] and [^{3}H] in each gel slice was determined by liquid scintillation counting using a double label program. The arrows indicate the beginning of the separating gel and the bottom of the gel at the point of the tracking dye. The bar between fractions 25–28 marks the region of a significant change in protein synthesis. (**B**) Ratio plot analysis of double isotope labeled SDS-PAGE analysis. Plot of Nri values vs gel fraction from a mixture of ^{14}C-labeled control/^{3}H-labeled control microsomes (upper panel) and ^{14}C-labeled control/^{3}H-labeled cyclopropenoid fatty acid microsomes (lower panel) separated by LDS gradient gel electrophoresis. The numbers at the top of upper panel are molecular mass markers (kDa). The increased Nri value in the range of 200–240 kDa beyond 2 standard deviations indicates a dramatic change in protein synthesis. Adapted from [3].

DIGE analysis using a postlabeling technique that allows the analysis of tissue extracts

4. Friedman, D.B. et al. Proteomics 4 (2004) 793–811.

(DIGE). A system developed by Amersham Biosciences, Inc. can be utilized to image the gels for each fluorescent dye as well as for computer analysis to determine the ratio of the two dyes. Spots that significantly deviate from the mean ratio can be further characterized by mass spectroscopy. This method has the advantage of allowing the characterization of proteins from essentially any source, such as tissue extracts, and has been used successfully to examine differences between normal human intestinal mucosal and colorectal cancer samples. Interestingly a number of proteins were detected at significantly different levels in normal versus cancer samples using this approach [4].

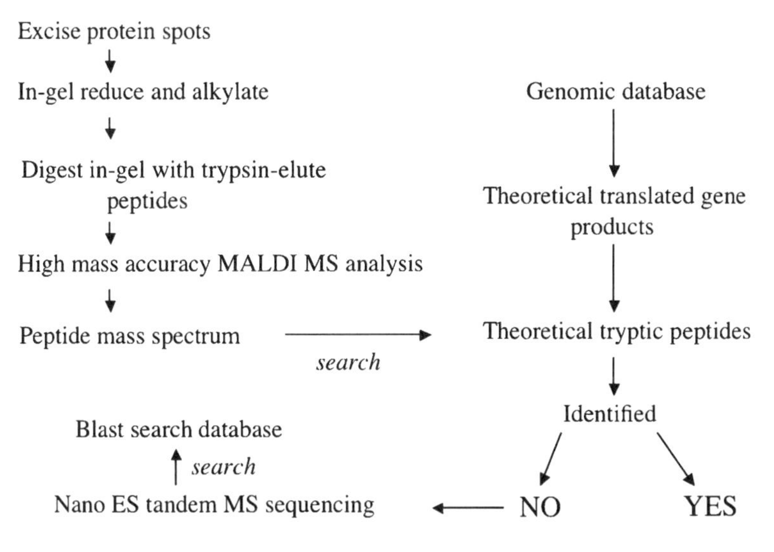

Fig. 3-3. Flow chart depicting the steps used to identify a protein excised from a 2-D gel.

2.1.3. Two-Dimensional Gel Electrophoresis–Mass Spectrometry Systems

After individual protein spots have been identified as being differentially regulated, the identity of the protein needs to be determined. This can be most efficiently accomplished by performing mass spectrometry on peptide fragments, a method entitled "peptide-mass fingerprinting" (PMF). After staining the 2-D gel, gel plugs containing the spot of interest are excised and the protein inside is digested with a proteolytic enzyme such as trypsin. The peptide cleavage products are eluted, then subjected to mass spectrometry using a matrix-assisted laser desorption/ionization (MALDI) time-of-flight (TOF) instrument. The peptides that yield distinct mass spectra are matched against a database with theoretical tryptic peptides. If no possible proteins are identified, additional analysis of the peptide mixture is performed by nanoelectrospray tandem mass spectrometry sequencing to obtain partial peptide sequence information. With two or more short peptide sequences this usually allows unambiguous identification of the protein. The general scheme for protein identification by mass spectrometry is outlined in Figure 3-3.

2.1.4. Isotope Affinity Tag Technique

One non-2-D gel approach is called isotope-coded affinity tag (ICAT). This method involves two reagents with either eight heavy or eight light hydrogen atoms incorporated onto the carbon chain capable of crosslinking to cysteine

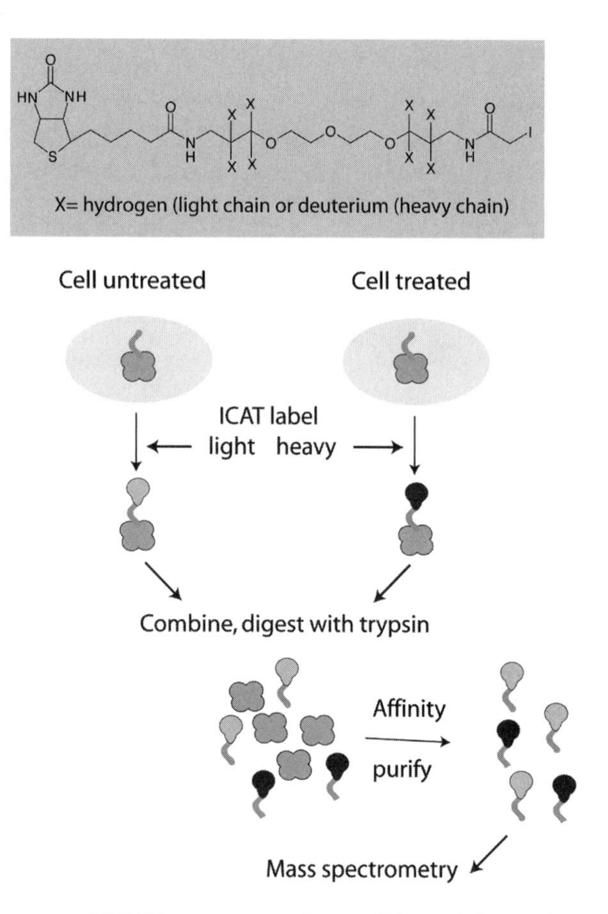

Fig. 3-4. The use of ICAT reagents to detect differential protein expression. Cell extracts from control and treated cells or tissue are labeled with cross-linking reagent that contain either heavy or light hydrogen. This reagent modifies reduced cysteine residues. Samples are combined, digested with trypsin, and affinity chromatography is used to isolate the biotinylated fragments, which are then subjected to mass spectrometry.

residues on proteins and includes a biotin group that allows purification of labeled proteins (Figure 3-4). The two labeled protein extracts are mixed, and proteins are fractionated on an ion-exchange column. Each fraction is digested with trypsin and the affinity-tagged fragments are isolated using a streptavidin affinity column. The isolated fragments are subjected to liquid chromatography–tandem mass spectrometry (LC-MS/MS), and differentially regulated proteins are identified as peaks with altered ratios of heavy- and light-labeled fragments. Differentially regulated fragments are identified by MS/MS.

This method has been successfully used to identify differentially regulated proteins in a number of systems. However, ICAT appears to suffer from some of the same limitations as 2-D gel analysis, such as the level of sensitivity for the detection of low-abundance proteins [5]. A particular disadvantage to the ICAT approach is the inability to detect changes in posttranslational modifications, which often can be detected by 2-D gel electrophoresis. Thus, neither ICAT nor 2-D gel electrophoresis provide comprehensive coverage on a proteome-wide basis. Nevertheless, the ICAT technique has been successfully used by a number of investigators to address a variety of hypotheses.

5. Patton, W.F. et al. Curr. Opin. Biotechnol. 13 (2002) 321–328.

Neither ICAT nor 2-D gel electrophoresis provide comprehensive coverage of expressed proteins.

2.1.5. Antibody Microarrays

Another approach to rapid assessment of relative levels of a large number of proteins is the use of antibody microarrays. Highly specific antibodies are immobilized within a spot on a glass slide similar to the approach used with an oligonucelotide microarray. The size of the library is dependent on the availability of excellent antibodies to the thousands of proteins found in a cell. Protein extracts from cell samples in which one sample is the control and the other has undergone a specific treatment are separately labeled with Cy5 or Cy3. The fluorescently labeled proteins are mixed and incubated on the array. After washing the array, the signal from each fluorescent tag is assessed and a ratio for each spot is determined. The data is analyzed for proteins that exhibit a significant deviation from the mean ratio. Arrays with 500 antibodies are currently commercially available and should be useful screening tools for changes in many regulatory proteins (e.g., cell cycle proteins). The quality of an antibody array is highly dependent on the quality and specificity of the antibodies immobilized on the array. The ability to produce antibody arrays with thousands of antibodies is theoretically possible; however, they

Antibody microarrays offer great promise if all of the antibodies utilized are of high specificity and overall quality.

would be quite expensive and would require extensive validation of the antibodies for there to be confidence in the results obtained. Many commercially available antibodies do not exhibit absolute specificity, and if used on an array, they could exhibit unacceptable background binding. Despite this concern, antibody arrays should be of use as a screening tool for alterations in the level of a large number of proteins in cellular extracts. Positive results can then be validated using standard protein blot and immunoblotting procedures.

2.2. Functional Proteomics

2.2.1. Protein Microarrays and Displays

Proteins can be arrayed either by placing them in solution in 96-well plates for biochemical analysis (e.g., enzymatic assay) or by covalent attachment to glass slides. polyHis-tagged proteins can also be bound to nickel-coated slides. The proper level of hydration needs to be maintained during and after the printing of the slides. The same instrumentation used to produce oligonucleotide arrays can be used to produce protein microarrays. A number of experiments have been published demonstrating that proteins can be arrayed in a way that maintains their functionality.

Protein microarrays can be an effective tool to screen for proteins with specific functional activities or binding properties. Perhaps the first global attempt at screening for protein–protein interactions was the expression, isolation, and immobilization of 5,800 yeast open reading frames as GST/polyHis$_6$ fusion proteins [6]. This represented approx 80% of the yeast proteome and the array was screened for calmodulin-binding proteins. The results revealed 39 calmodulin-binding proteins, 33 of these proteins previously not identified to exhibit this property. Screening this array with six different lipids resulted in the identification of 150 phosphatidylinositol-binding proteins, although some of these

6. Zhu, H. et al. Science 293 (2001) 2101–2105.

Table 3-1
Summary of Pros and Cons of Methods to Examine Global Changes in Protin Levels

	Pros	Cons	Quantitation
Simple 2-D gel comparison	• Requires minimal equipment • Analyze almost any protein sample • Directly examine most proteins in sample	• Gel to gel variability • Difficulty comparing two gels • Unlikely to detect low abundance proteins	• Difficult to quantiate changes in protein levels detected
2-D gel double isotope	• Highly sensitive • Detect two-fold changes in protein abundance • Compare two protein populations on one gel	• Requires radioisotope labeling of cells • Does not require special equipment for analysis	• Accurate assessment of relative changes in individual protein levels
2-D gel-DIGE	• Sensitive • Detect two-fold changes in protein abundance • Compare two protein populations on one gel • Analyze most protein sample	• Relatively expensive analysis system • May not detect low abundance proteins	• Accurate assessment of relative changes in individual protein levels
Isotope affinity tag	• Does not require electrophoresis • Compare two protein populations in one sample • Analyze most protein sample	• No information about posttranslational modifications • Requires analysis of samples by mass • Usually need a mass spectrometry core facility	• Accurate assessment of relative changes in individual protein levels
Antibody microarray	• Highly sensitive • Can detect low abundance proteins	• Only detect a limited subset of proteins	• System can only be considered semi-quantitative

same proteins exhibited broad lipid-binding properties. This approach clearly offers great potential for identifying proteins that bind small ligands. It may also be useful to screen for protein–protein interactions. Future application of this approach to higher eukaryotes may also be promising, but to obtain widespread coverage of most of the proteins expressed in mammals will require a considerably larger effort.

3. Summary

Many important examples exist for regulation of protein activity through targeted degradation or enhanced stabilization. There are a number of global approaches that are now available to detect changes in protein levels, all methods offering certain advantages and disadvantages. The most important issue to keep in mind is that all methods appear to suffer from problems with the detection of low-abundance proteins (*see* Table 3-1).

Combining DNA microarray and proteomic approaches could be used as a means to dissect the mechanism(s) of change in protein levels mediated by a chemical treatment. Each approach would address a specific mechanism(s) that leads to changes in protein levels. This perhaps can be best illustrated through a hypothetical example. Let's say an investigator would like to determine the various mechanisms by which specific protein levels are altered after treatment of mice with a bile acid known to activate a specific receptor in the liver. The first approach chosen is the use of DIGE to detect changes in individual protein levels in liver extracts isolated from bile acid of treated and untreated mice. Twenty proteins were determined to have increased more than twofold and several were markedly decreased. This result might be attributable to a number of factors, such as increased or decreased mRNA synthesis, changes in mRNA stabilization, or altered protein stability. The possible mechanism(s) for the results obtained can be narrowed by performing a DNA microarray experiment on liver mRNA extracted from bile acid of treated and untreated mice. A subset of specific mRNAs are shown to be either up- or down-regulated by bile acid treatment. Next, validation of changes in mRNA levels needs to be established with either Northern blot or real-time PCR. The resulting data set can then be compared with the DIGE data, and if both protein and mRNA are increased for a given protein, then an increase in transcription and subsequent translation is probably the most likely mechanism. However, if a specific protein level increases or decreases with no change in mRNA levels, then a posttranslational mechanism would probably be involved. Approaches outlined in Chapter 2 will allow establishment of altered protein turnover while possible mechanisms for this result will be found in Chapters 5 and 6.

4

Determination of Protein–Protein Interactions and the Motifs That Mediate Them

1. Concepts

Determining whether or not a specific protein interacts with another can be accomplished in a number of ways, which can be divided into in vivo or in vitro approaches. The in vivo approaches include yeast two-hybrid, yeast three-hybrid, mammalian two-hybrid, one-hybrid, and FRET analyses. While the most common in vitro approaches are glutathione-*S*-transferase (GST) pull-down assays, co-immunoprecipitation, immune depletion, gel-filtration or sucrose (or glycerol)-density gradient analysis, far-Western blot analysis, and chemical crosslinking. Each one of these assays has strengths and weaknesses and usually a combination of methods can lead a compelling case that a given interaction actually can occur within the cell. Many of these approaches utilize what can be considered transient overexpression of the proteins being studied and this can lead to interactions occurring that do not normally occur under physiologic conditions. Also, upon overexpression a protein may be found in subcellular compartments where they do not normally exist. For example, upon overexpression, a normally cytosolic protein may be found in significant concentrations in the nucleus. Nevertheless, the use of cellular overexpression systems usually yields information that is physiologically relevant. The following section outlines the various approaches that can be taken to document that one protein is capable of interacting with another.

2. Methods and Approaches

2.1. Cellular Systems to Assess Protein–Protein Interactions

There are a number of approaches available to demonstrate that two proteins interact in vitro. However, demonstrating that two proteins interact within a cell is more restricted, with either the yeast or mammalian two-hybrid being the most widely utilized systems. The use of two-hybrid methodology has a number of advantages over the in vitro assays available. For example, the interaction occurs within the cell so the proteins will fold properly. In addition, the

Regulation of Protein Levels and Transcription Factor Function by G. H. Perdew
From: *Regulation of Gene Expression*
By: G. H. Perdew et al. © Humana Press Inc., Totowa, NJ

interaction will occur under physiologically relevant conditions. These assay systems are also quite sensitive and capable of detecting weak interactions that may not be detected with other approaches. The two-hybrid approach is also an effective system to map interaction domains. In addition, quantitative information can be obtained about the strength of a given interaction within a series of truncated or mutated proteins being expressed. Since the initial development of the yeast two-hybrid assay a number of variations in the general theme of this assay have been developed.

The two-hybrid approach is widely used due to its ability to assess protein–protein interactions within a cell.

2.1.1. Yeast Two-Hybrid Assay

The two-hybrid assay system is based on the fact that most transcription factors consist of two distinct domains, the DNA-binding domain (DNA-BD) and the transcriptional activation domain (AD). The most widely used two-hybrid system utilizes the DNA-BD and AD from the bacterial transcription factor GAL4 [7]. The basic yeast *Saccharomyces cerevisiae* two-hybrid system is outlined in Figure 4-1. The yeast two-hybrid assay was initially developed as a tool to screen for unknown interacting proteins by generating a yeast cDNA expression library fused to the GAL4 AD. This technique has been used extensively for discovering novel protein–protein interactions and continues to be the most utilized approach for this purpose. Several companies offer a series of vectors for two-hybrid analysis, based on either the transcription factor GAL4 or LexA. In the yeast two-hybrid system, the known gene encoding the factor of interest is cloned into the "bait" vector, thus allowing expression of a DNA-BD fusion protein that binds to the enhancer elements in the reporter construct. The ability of another protein to bind to the DNA-BD fusion protein is tested by cloning a cDNA of interest into a "prey" vector containing the AD. If the two proteins interact upon coexpression in yeast, this results in the formation of a functional

7. Criekinge, W. and Beyaert, R. Biol Proced. Online 2 (1999) 1–38.

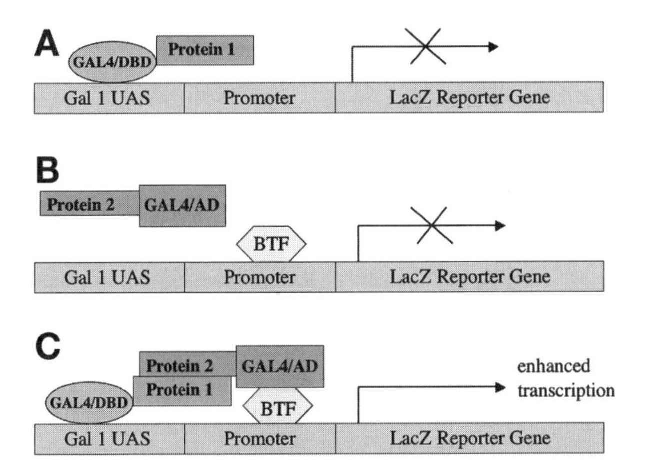

Fig. 4-1. Basic mechanism of the yeast two-hybrid system. **(A)** The GAL4-DBD-hybrid can bind to the GAL1 upstream activating sequence (UAS) site adjacent to the promoter, but fails to activate transcription. **(B)** The GAL4-AD-hybrid protein cannot localize to the GAL1 UAS and does not activate transcription. **(C)** The interaction between the GAL4-AD-hybrid reconstitutes GAL4 function and results in transcription of the LacZ reporter gene.

transcription factor that leads to enhancement of the level of transcription from the reporter vector by bringing together the two domains required to mediate transcription.

Yeast two-hybrid is the most effective means to globally search for protein interaction partners.

When using the yeast two-hybrid system as a screening tool, often the biggest and potentially most time-consuming can be the generation of false positives. These may be generated by interaction of the prey protein with a factor binding to a region of the reporter other than the GAL4 sites, interaction with a factor bound to the TATA box or direct interaction with a DNA sequence near the GAL4 elements, or nonspecific interaction owing to overexpression. Newer two-hybrid systems can greatly reduce the number of false positives attributable to these problems because they have four different reporter sequences stably integrated into the yeast genome. An increase in transcriptional activity with each reporter would eliminate many of the possible mechanisms for false positives (e.g., BD Bioscience Matchmaker™ Two-Hybrid System 3).

One major drawback of using a yeast-based system is that many laboratories are not set up to work with yeast. However, there is an *Escherichia coli*-based system developed using a similar set of vectors.

2.1.2. Mammalian Two-Hybrid Assay

Although yeast is a single-cell eukaryote, it can vary considerably in its regulatory pathways, especially with respect to posttranslational modification (e.g., phosphorylation), when compared to multicellular eukaryotes. Thus, the behavior of a given mammalian protein can be quite different when expressed in yeast. In addition, because of the low level of primary sequence conservation for a wide variety of proteins in yeast as compared to a mammalian organism, the folding of a given protein may differ. Also, yeasts do not express many of the proteins found in higher eukaryotes, which could lead to significant differences in the activity or behavior of a mammalian protein in yeast. For example, on examining the activity of many nuclear receptors that are chaperoned by hsp90/cochaperone complexes, their behavior in yeast often appears to differ from that seen in mammalian cells. This may, at least in part, be attributable to significant differences in the level of amino acid sequence divergence between yeast and mammalian hsp90. Thus, the yeast hsp90 chaperone system's ability to influence mammalian receptor function may be limiting. Another drawback of the yeast system is related to possible differences in subcellular localization of a protein between the two systems. Considering these many possible limitations, protein–protein interactions characterized in yeast are often repeated using the mammalian two-hybrid system to confirm a positive result, as well as for mapping of interaction domains.

2.1.3. One-Hybrid Assay System

This assay system was originally developed to screen a cDNA library for proteins that bind to a specific DNA sequence in yeast. Today a one-hybrid assay is most often performed in mammalian cells to examine the ability of a coexpressed protein to modulate the transcriptional activity of a GAL4/DBD-transcription factor fusion protein in transient transfection experiments. An alteration in transcriptional activity would suggest a possible direct interaction between the two proteins. Note that the GAL4/DBD fusion protein in one-hybrid assays is able to mediate transcriptional activation without expression of an additional protein, in contrast to the two-hybrid system. The use of the two-hybrid system to establish the interaction of one transcription factor with another factor probably would result in high basal activity mediated by the transactivation domain of the bait protein and could make interpretation of the data difficult. In this situation the use of coexpression of the two proteins of interest in efficiently transfected cells followed by immunoprecipitation to establish the presence of both proteins in a complex is an excellent alternate means to establish transcription factor–protein interactions. The one-hybrid system can work well to test the influence of coexpression of one protein on the transcriptional activity of a GAL4/transcription factor fusion protein. As a hypothetical example, an investigator hypothesizes that factor X is capable of binding to transcription factor Y and repressing its activity by blocking coactivator recruitment. This hypothesis can be effectively addressed using a one-hybrid transient transfection experiment in mammalian cells. One potential problem in using the one-hybrid approach is that the manner in which a GAL4-transcription factor fusion protein interacts with a response element may be dissimilar to how the transcription factor interacts with its cognate DNA response element. For example, if a nuclear receptor (NR) heterodimerizes with another receptor prior to binding to its cognate response element, the conformation of the NR could differ from that observed with a GAL4 fusion protein that binds as a monomer to a DNA response element. These differences would probably lead to significant altered recruitment of coactivators. However, upon examining ligand-dependent transcriptional activation mediated by PPARα, this activation is much greater in a GAL4-PPARα fusion protein/GAL4 reporter system compared to a PPARα/RXR-dependent reporter system, yielding increased sensitivity and dynamic range in the assay [8]. Thus, the one-hybrid system can be quite useful in studying specific aspects of receptor function in the absence of its heterodimerization partner.

The one-hybrid assay is an effective system to study the ability of a protein to modulate the transcriptional activity of GAL4 fusion protein.

8. Willson T.M., et al. J. Med. Chem. 43 (2000) 527–550.

2.1.4. Three-Hybrid Assay System

This modification of the two-hybrid system is used to establish the ability of a protein to act as a bridge between two proteins that do not directly interact. For example, the protein p27 is capable of interacting with both cdk7 and MAT1. Whether

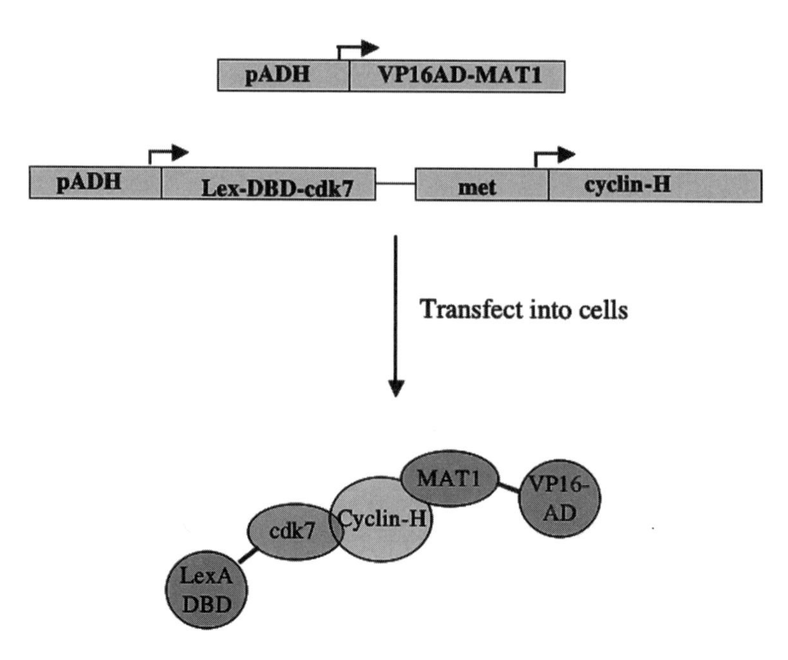

Fig. 4-2. Use of conditional three-hybrid system to examine the ability of cyclin-H to bind to both cdk7 and MAT1.

these proteins are capable of existing in a trimeric complex within a cell can be addressed using a three-hybrid approach. The data outlined in Figure 4-2 revealed that p27 is indeed capable of bringing together LexA/cdk7 and MAT1/VP16, which then leads to transcriptional activation [9]. While the presence of a trimeric complex can be established in vitro using chemical crosslinking or by coimmunoprecipitation, the three-hybrid system is one of the few methods available to establish that this type of interaction actually occurs in a cell.

9. Tirode, F. et. al. J. Biol. Chem. 272 (1997) 22,995–22,999.

2.1.5. Hybrid Screening of Peptide Libraries

One method that defines a minimal protein–protein interacting motif is the expression of a random peptide library fused to a VP16 or GAL4 transactivation domain in a yeast two-hybrid screening assay system. After isolation of positive clones and DNA sequencing, the peptides' amino acid sequences are

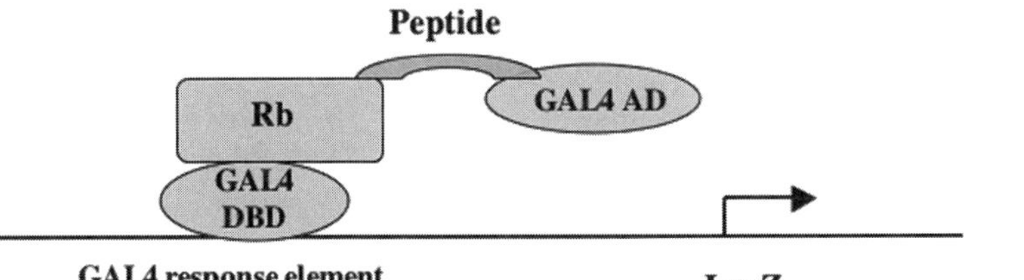

Peptide	Sequence	β-galactosidase activity
LTP	LFCSEEMPSSDDE	53
LTP-Leu	FCSEEMPSSDDE	0.2
P1	YGLWILWCDEEGLDLG	45
P2	NQLLGDVLACYEQEVE	6.9
P3	WTELLFCFEQVYGDPF	6.8
P4	EGGDLGCDESWSEGYT	6.0
P5	CDGLLCTETLL	5.4
P6	GGCPGANLCCFEKSLD	2.5
P7	TTWRERLRCEENGLGV	1.1

Fig. 4-3. Use of hybrid screening of a peptide library to define a motif that binds to RB.

determined and aligned to reveal a consensus sequence that would establish a primary sequence motif. This approach has been used to further define the minimal requirements for the core Rb-binding motif LxCxE. Yang and colleagues observed the core motif sequence in each peptide; however, the flanking sequences appeared to have considerable influence on the affinity of the interaction, which could be effectively assessed by the level of reporter activity, as shown in Figure 4-3 [10].

10. Yang, M. et al. Nucleic Acids Res. 23 (1995) 1152–1156.

2.1.6. Peptide Phage Display

The ability to display peptides on the surface of bacteriophage M13 has lead to the generation of random peptide libraries displayed on phage. These libraries can be screened for phage-containing peptide sequences that bind to a "bait" protein. The random library is constructed by synthesizing random nucleotide sequences of a fixed length and inserting them into a fusion protein vector that upon expression within the bacteriophage leads to display of the fusion protein on the surface of the phage. Perhaps the first use of these phage libraries was the mapping of monoclonal or polyclonal antibody epitope(s). Briefly, this is accomplished by binding an antibody to the bottom of a plastic well of a 96-well plate and placing the phage in the well and allowing phage to bind to the antibody. The well is washed and the phage are eluted and amplified by infecting bacteria. The selected phage is then taken through additional binding/amplification cycles to enrich the pool in favor of binding sequences. The amplified phages are placed through at least three to four rounds of selection, and the library insert DNA sequences for individual clones are determined. Through this process even a rare sequence in the library can be identified. The sequences of the individual clones are aligned to determine the core antibody recognition sequence compared with the sequence of the protein used to generate the antibody. Phage display can also be used to map substrate sites for proteases and kinases. In addition, a binding motif involved in a protein–protein interaction can be determined and defined using phage display techniques.

Phage display is an effective method to define an epitope for a monoclonal antibody.

Both phage display and mammalian two-hybrid can be used together to carefully study protein–protein interaction mediated by a specific motif and the effect of variations in the core sequence, as well as the influence of the flanking sequence on binding affinity. This approach was taken to further characterize the LXXLL motif found in many coactivators (e.g., SRC-1, PBP, RIP140) that interact with transcription factors and lead to enhanced transcription. The McDonnell laboratory has used phage display and a focused peptide library, with the three conserved leucine residues in each otherwise degenerate peptide, to screen for LXXLL motifs that bind with high affinity to the liganded estrogen receptor [11]. Sequence alignment of the phage-selected

11. Chang, C. et al. Mol. Cell. Biol. 19 (1999) 8226–8239.

peptides, as well as LXXLL motifs from coactivators, revealed a preference for a leucine or isoleucine residue adjacent to the first leucine residue. However, this was not a strict requirement. In other amino acid positions no obvious level of amino acid preferences could be detected. These peptides were then cloned into a GAL4 DNA binding domain fusion protein vector for mammalian two-hybrid analysis. This system was used to compare the ability of the various peptides to interact with a series of estrogen-receptor mutants, and the data indicated that the peptides differ in how they interact with the receptor as well as in their affinity. Thus, the combination of phage display and two-hybrid analysis has allowed this group to assess the role of nonconserved amino acid residues to alter LXXLL motif specificity and affinity for steroid receptors. From this example other uses for peptide display in defining core motifs can be envisioned.

2.1.7. Molecular Approaches to Determine Protein–Protein Interactions Within a Transcriptional Complex

One question often asked in the course of studying protein–protein interactions is whether a protein is actually bound directly or indirectly to genomic regulatory-DNA sequences. One particularly useful experimental approach is the ChIP assay. These assays were developed to assess the presence of a protein associated with a specific DNA sequence within the cell. This is accomplished by the immunoprecipitation of crosslinked protein DNA complexes as outlined in Part I, Chapter 4. After resolving the crosslinks, DNA is isolated and RT-PCR is used to detect the presence of specific promoter DNA sequences. Also real-time PCR can be used to quantitate the amount of a specific promoter region in ChIP samples. Perhaps the most important parameter for obtaining excellent results with this technique is the use of a specific high-affinity antibody that has been raised against epitope(s) that are accessible in crosslinked complexes. ChIP assays are now widely used to detect the presence of proteins in transcriptional complexes under physiologically significant conditions. These assays can be performed in tissues after isolation of nuclei and thus can examine protein–DNA complexes that occur in vivo. This technique can be used to detect the presence of any protein that can be crosslinked into large protein DNA complexes; including histones, coactivators, or RNA polymerase. Thus, the ChIP assay is one of the only methods that can provide insight into what type of protein–protein interactions occur within the context of transcriptional complexes found within the nucleus. For example, the ability of Jun dimerization protein 2 (JDP2) to bind to the progesterone receptor on a stably integrated MMTV reporter gene within the nucleus was tested using the ChIP assay [12]. Sheared crosslinked DNA was immunoprecipitated with both a progesterone receptor and a JDP2 antibody to determine the amount of MMTV promoter DNA pres-

12. Wardell, S. E. et al. Mol. Cell. Biol. 22 (2002) 5451–5466.

ent. The presence of MMTV promoter DNA in the JDP2 immunoprecipitation would suggest that JDP2 is bound to the progesterone receptor within the nucleus near the location where the PCR primers amplify the MMTV DNA. The level of resolution obtained is dependent on the size of the sheared DNA, which is usually approx 500–1000 bp.

Re-ChIP assays can determine the presence of two TFs on a single promoter fragment.

Often an investigator would like to address what other factors, or which combination of factors, are recruited to a promoter region (e.g., coactivators). For example, if transcription factor X recruits the coactivator SRC-1 to a specific promoter, does it also recruit the coactivator p300? This can be accomplished by immunoprecipitation of SRC-1 from sheared crosslinked chromatin, followed by elution with SDS, dilution with buffer, and immunoprecipitation with a p300 antibody. This assay has been termed a "Re-ChIP assay" and will determine whether SRC-1 and p300 exist on the same promoter, as outlined in Figure 4-4.

13. Metivier, R. et al. Cell 115 (2003) 751–763.

For an actual example of the utilization of this technique, see the estrogen-receptor transcriptional cycling studies of Metivier et al. [13]. They utilized Re-ChIP assays extensively to determine the presence of various factors associated with a transcriptionally active estrogen receptor complex immunoprecipitated in a ChIP assay that was subjected to Re-ChIP analysis. In this way a profile of the proteins present in oligomeric complexes found on a promoter sequence can be determined.

Yet another potential application for ChIP and Re-ChIP assays is the direct assessment of the proteins found complexed with a specific transcription factor. However, this procedure will require further optimization of the standard ChIP assays to lower nonspecific chromatin binding in the immunoprecipitation step. Once this has been accomplished, the immunoprecipitations can be heated in SDS sample buffer, subjected to SDS-PAGE analysis, and the resulting gel silver stained, to visualize the

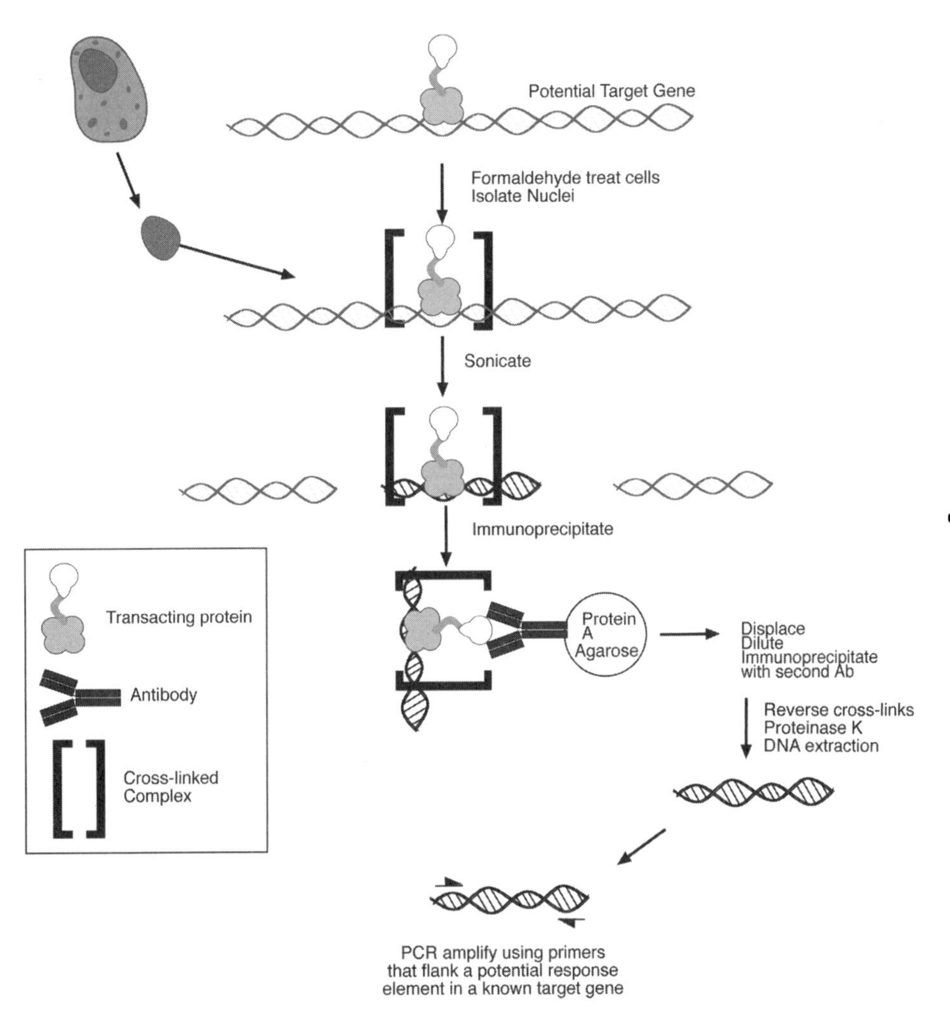

Fig. 4-4. Schematic representation of the steps utilized in a Re-ChIP assay.

various proteins complexed to the immunoprecipitated chromatin fragments. The stained protein bands can be identified by mass spectrometry analysis (*see* Chapter 3). This approach could be utilized to assess the type of regulatory proteins or complexes that are recruited by a transcription factor that was the target of the immunoprecipitation. However, it is important to keep in mind that the average DNA fragment size in a sheared piece of chromatin is approx 500 bp, thus the presence of multiple response elements in a segment of sheared chromatin will potentially complicate the results obtained. Perhaps this can be addressed by further decreasing the size of chromatin fragments used in this type of analysis. Nevertheless, this method offers considerable potential for addressing specific questions about which proteins are recruited to a specific transcription factor bound to chromatin.

2.1.8. Fluorescent Resonance Energy Transfer Analysis

Immunofluorescence microscopy is a common tool used to study subcellular localization of a protein in cultured cells or tissue sections. However, apparent colocalization of two proteins does not necessarily mean that they are actually present in a complex. The highest level of resolution of a light microscope is approx 250 nm, while to demonstrate actual molecular interaction would require 1–10 nm resolution. To circumvent this problem the technique termed fluorescent resonance energy transfer (FRET) has been developed. This technique utilizes two fluorescent fusion proteins wherein one protein, upon excitation, transfers its energy to an adjacent molecule, which then emits at the specific wavelength of the acceptor protein. Usually the donor protein is either blue fluorescent protein (BFP) or cyan blue fluorescent protein (CFP), with CFP exhibiting stronger fluorescence. However, a drawback of using CFP is the possibility that CFP emission spectrum may bleed into the yellow fluorescent protein (YFP) spectrum. YFP is usually the acceptor for either BFP or CFP in FRET analysis.

One potential problem with expression of two fluorescent protein fusion proteins is the possibility that the protein tag will interfere with a potential protein–protein interaction. This problem can be minimized by careful determination of which end of the protein the tag is placed on. An important aspect of this method is the requirement for an inverted fluorescence microscope with the appropriate emission/excitation filters and computer software for data analysis. The better microscopy systems for FRET analysis are quite expensive and are usually found in core microscopy facilities. With the ever-increasing level of sophistication being engineered into new microscope systems, this technique should be more routine in the near future.

2.2. In Vitro Analysis of Protein–Protein Interactions

2.2.1. Density Gradient Centrifugation

One method that can be utilized to assess whether or not a given protein is in a complex or is associated with another protein is glycerol or sucrose density gradient centrifugation. This method involves the use of a density gradient set-up in an ultracentrifuge tube. A protein solution is applied to the top of the gradient followed by sealing of the centrifuge tube. After centrifugation the gradient is fractionated with a gradient fractionator such as an ISCO Density Gradient Fractionator Model 640. The location of the protein of interest can be

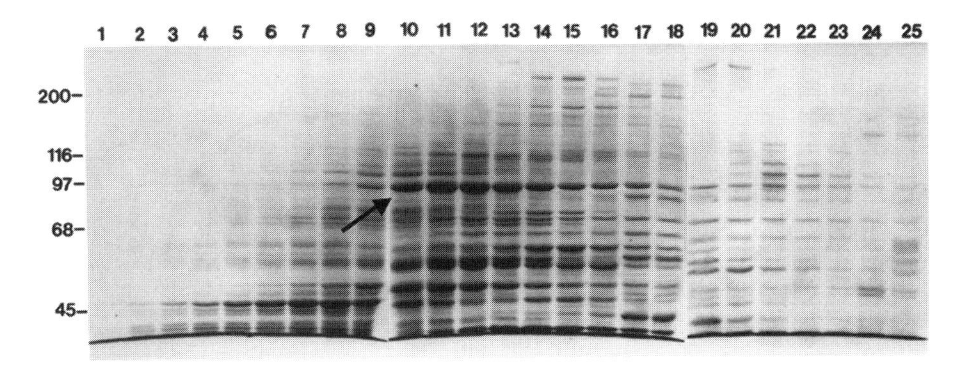

Fig. 4-5. Sucrose density gradient analysis of Hepa 1c1c7 cytosol. Hepa 1c1c7 cytosol (3 mg/mL, 300 (μL) was applied to a 10–30% sucrose gradient. Each fraction was subjected to SDS-PAGE. After electrophoresis the gels were stained with Coomassie blue.

determined by subjecting an aliquot of each fraction to SDS-PAGE and transferring the separated proteins to a membrane. The presence of various proteins can then be assessed with the appropriate primary antibodies, followed by a [^{125}I]-secondary antibody, and the relative concentration of the protein of interest can then be plotted on a graph. While this technique is not high resolution, it does allow one to look at the sedimentation of the entire pool of a given protein under gentle conditions employed by centrifugation to fractionate protein down a density gradient. In addition, several milligrams of protein can be applied to a 5.1 mL density gradient. To illustrate the level of resolution that can be obtained Figure 4-5 reveals the fractionation of Hepa 1c1c7 cytosol on a 10–30% sucrose gradient. Protein was resolved by SDS-PAGE, and the resulting gel stained with Coomassie blue [14]. The arrow in Figure 4-5 identifies the hsp90 band, which fractionates predominantly between fractions 10 and 13. Hsp90 exists as a dimer of 180 kDa, yet note that many low-molecular-weight proteins fractionate in the same region of the sucrose gradient. This illustrates that a large number of cytosolic proteins must exist in multimeric complexes.

Sucrose gradient analysis allows size assessment of the entire protein pool.

14. Perdew, G.H. et al. Exp. Cell Res. 209 (1993) 350–356.

An example of the usefulness of sucrose density gradients is shown in Figure 4-6, which reveals that the Ah receptor exists in the nucleus in two distinct complexes. The Ah receptor was activated and detected by exposing Hepa 1c1c7 cells to the ligand 2-[^{125}I]iodo-7,8-dibromodibenzo-*p*-dioxin, nuclear extracts obtained from these cells were subjected to sucrose density gradient analysis [15]. The presence of iodinated protein complexes was determined by counting each fraction in a gamma counter. Two radioactive peaks were detected which correspond to 180 kDa and 327 kDa AhR complexes (Figure 4-6). Protein standards of known molecular weight are run on individual gradients to generate a relative-molecular-weight standard curve. A time course of ligand treatment of Hepa 1 cells revealed that, between 30 min and 1 h of treatment, the level of 9 S AhR complexes decrease as the amount of 6 S complexes increase. This experiment illustrates the usefulness of sucrose density gradients in examining protein complexes that significantly differ in molecular weight.

15. Perdew, G.H. Arch. Biochem. Biophys. 291 (1991) 284–290.

In order to demonstrate that one protein is associated with another, an antibody can be added to a protein sample prior to sedimentation analysis. The ability of an antibody to shift a protein down can be used to determine whether a potentially complexed protein moves down the gradient with the protein the antibody was raised against. Also, specific fractions can be pooled and immunoprecipitated to look for coimmunoprecipitated proteins. With the combination of immunoprecipitation or chemical cross-linking and sucrose density gradient analysis, a considerable amount of information can be obtained.

Combination of sucrose gradients and immunoprecipitation can be quite useful.

2.2.2. Gel Filtration Chromatography

Gel filtration fractionates protein according to their Stokes' radius, which is dependent on the shape and size of a protein.

Although this method is usually called gel filtration chromatography, it would perhaps more accurately be referred to as permeation or molecular exclusion chromatography. In this type of chro-

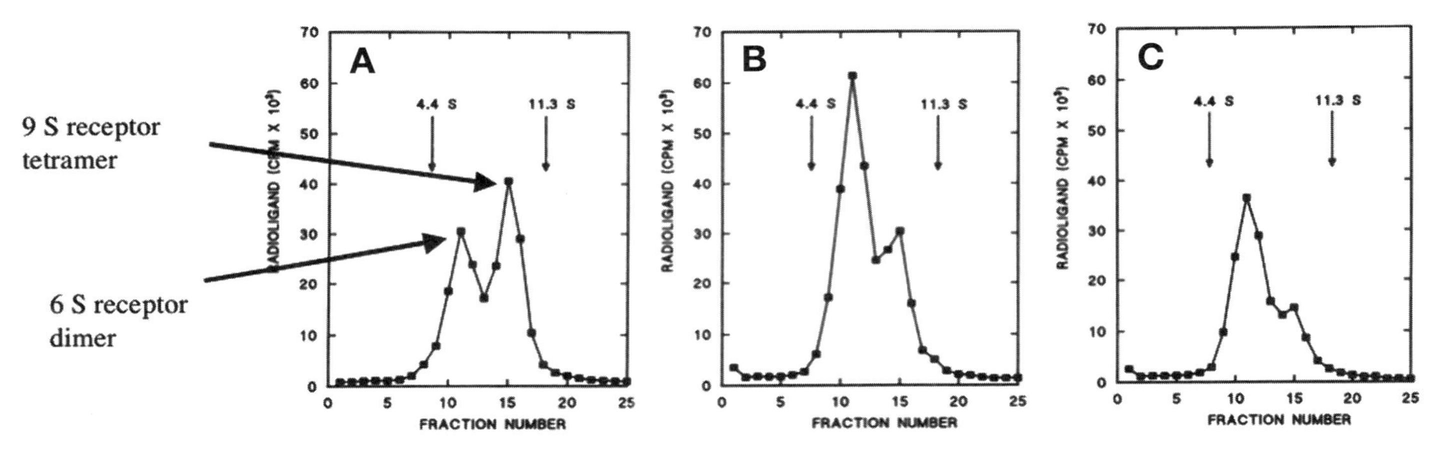

Fig. 4-6. Sucrose density gradient analysis of $I^{125}{}_{2D}$pD-bound Ah receptor complexes present in Hepa 1c1c7 cell nuclear extract. Hepa 1 cells were treated with $[I^{125}]_{2D}$pD for 30 min (**A**), 1 h (**B**), or 2 h (**C**). Cells were harvested, nuclei isolated and extracted with high salt buffer. Extracts were applied to 10–30% sucrose gradient. The amount of radioactivity in each fraction was determined. BSA (4.4 S) and catalase (11.3 S) were analyzed on a separate sucrose gradient as molecular-weight markers. S, sedimentation coefficients.

matography, the stationary phase consists of porous beads with a defined range of pore sizes. This property of the beads yields what is called a fractionation range, which can be defined as the range of molecular weights that can be effectively separated. Proteins that are small enough can fit inside the beads and are thus "included" with the beads, which leads to the protein taking a longer path through the column. Proteins that are too large to fit inside any pores are "excluded" and take the shortest path through the column. Proteins of intermediate size travel through only some of the pores and are thus "partially excluded." A schematic example of the type of data obtained is shown in Figure 4-7. It is important to note that gel filtration cannot be considered a high-resolution technique and generally can only cleanly separate proteins with a molecular weight differing by a factor of 2. However, this level of resolution is adequate to determine whether a protein exists as a monomeric, dimeric, or tetrameric species.

Gel filtration separates proteins based on their Stokes' radius, which is basically the maximum radius of a protein tumbling in solution. Thus, gel filtration can separate proteins of similar size if they have a considerable difference in shape (e.g., ball vs. cylinder). The determination of the molecular weight of a protein can be accomplished by gel filtration if the unknown protein has a similar shape compared to the standards utilized. In general, a series of purified proteins known to be globular are fractionated on the gel filtration column. A plot of the logarithms of the molecular weights of the known proteins versus their elution positions is generated and should yield a linear plot within the fractionation range of the matrix used. For analytical applications, most investigators use high-performance columns that are utilized on a low- or high-pressure liquid chromatography system. These columns are based on agarose, silica, or other materials, but all utilize the same basic mechanism of separation. The range of fractionation obtained is dependent on the type of gel filtration beads and their pore size; several companies offer prepacked high-performance gel filtration columns that can fractionate a variety of protein sizes. Historically, gel filtration chromatography has been used mainly for protein purification and sizing of proteins. However, it can be quite a useful technique to examine the composition of oligomeric complexes.

Protein–protein interactions can be assessed by gel filtration in much the same way as has been discussed using sucrose density gradients. As an example of how this technique can be used to examine oligomeric complexes, Lanz and co-workers used a Superose®6 gel filtration column to fractionate lysate from T47D cells as shown in Figure 4-8 [16]. These investigators wanted to assess whether the coactivator SRC-1 and the RNA coactivator SRA can exist in a complex. The presence of the protein SRC-1 was assessed by protein blot immunochemical analysis, and SRA was detected by RT-PCR of each fraction

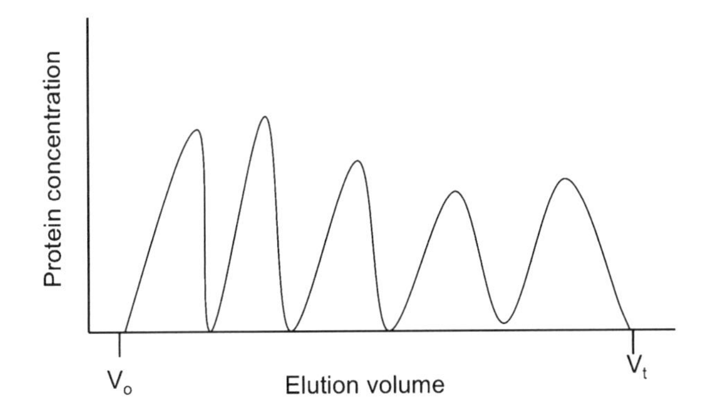

Fig. 4-7. Theoretical separation of nondenatured proteins by high performance gel filtration chromatography. Ideal separation of five proteins, each differs by a factor of 2 in molecular weight. The V_0 and $V0_{t0}$ are defined as void volume and total volume, respectively.

collected. The results indicated that a portion of the SRA pool coelutes with SRC-1. To actually demonstrate that SRA and SRC-1 are in a complex the appropriate fractions were subjected to immunoprecipitation analysis and the results indicated that SRA and SRC-1 are indeed complex. Also note that this technique allows one to see how much SRA is actually in large complexes compared to being either free in solution or in smaller complexes.

16. Lanz, R.B. et al. Cell 97 (1999) 17–27.

2.2.3. Assessment of the Shape of a Protein or Protein Complex

The combination of gel filtration and sucrose density gradient centrifugation can be used to determine a frictional ratio, which is defined by a protein's Stokes' radius and its molecular weight. The determination of a frictional ratio in turn can yield insight into whether the protein is globular or has an elongation shape. The ability to perform this analysis is based on the differing mechanisms of separation by these techniques. The mechanism of separation of proteins by density gradient fractionation is based solely on the molecular weight of each protein [17]. In contrast, gel filtration separates pro-

17. Siegel, L.M. and Monty, K.L. Biochim. Biophys. Acta 112 (1966) 346–362.

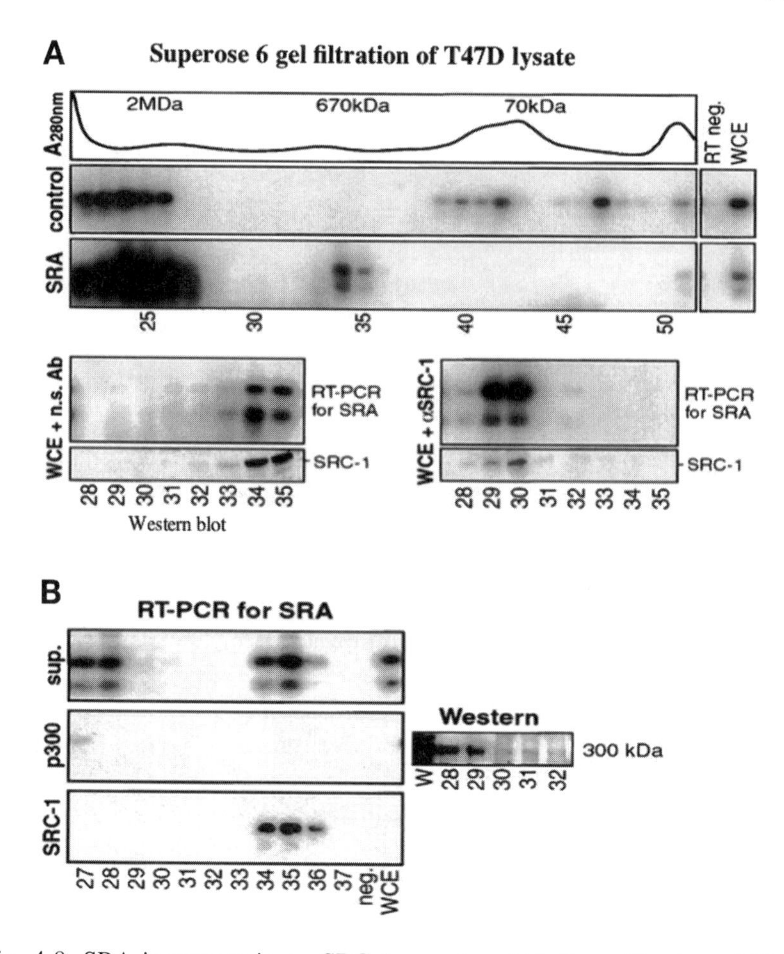

Fig. 4-8. SRA is present in an SRC-1 complex. **(A)** Copurification of SRA and SRC-1 complexes by gel filtration chromatography. Upper panels: T-47D lysates (~400 µg) were fractionated on a Superose 6 column and analyzed for total protein elution (A_{280nm}; top), SRC-1-specific RT-PCR (control), and SRA-specific RT-PCR (SRA). RT-PCR analysis of ~20 µg input whole-cell extract (WCE) in the presence or absence of reverse transcriptase (RT neg.) are shown to the right. Numbers indicate fractions. Elution peaks of molecular size markers are given for mammalian SWI/SNF complex (~2 MDa) and the thyroglobulin (670 kDa); the void volume (4 MDa for globular proteins) was determined at fraction 20 by silver staining (not shown). Lower panels: SRA-specific RT-PCR and parallel immunoblots with SRC-1-specific antibody of fractionated T-47D cells after preincubation of the lysate with either nonspecific antibody (WCE + nonspecific AB) or SRC-1 antibody (WCE + SRC-1 Ab). **(B)** Coimmunoprecipitations of SRA in fractionated cells. T-47D lysates were fractionated as in (A), subsequently immunoprecipitated with antibodies against p300 (middle) and SRC-1 (bottom), and analyzed for SRA by RT-PCR (left panels) or by parallel Western analysis for precipitation of p300 (right). Numbers indicate fractions. Neg., RT-PCR omitting reverse transcriptase; WCE, input lysate; sup., supernatant. (Adapted from [16].)

tein based on both size and shape, also known as a protein's Stokes' radius. The combination of data from these two techniques can yield useful information about whether a protein or protein complex has a globular or elongated shape. This in turn can allow the assessment of how subunits of a complex may be arranged.

2.2.4. Chemical Crosslinking

Chemical crosslinking can determine how many proteins are in a complex, how they are arranged, and which ones are in direct contact with each other.

Results obtained with most of the techniques in this section tell you that one protein is complexed with another or exists in a larger complex with that protein. However, these techniques do not tell you how many proteins are in a given complex or how they are arranged. Through the crosslinking of an oligomeric complex with increasing amounts of a crosslinking reagent, a number of properties of that complex can be determined, including which proteins are bound directly to each other, as well as the actual number of proteins in a given complex. The most commonly used crosslinking reagents are homobifunctional compounds that react with amino groups. A series of hetero- and homobifunctional reagents are available with different distances between the reactive groups. Protein samples are subjected to a crosslinking reagent time-course treatment and samples analyzed using continuous-phosphate SDS-PAGE. The resolved proteins in the gel are transferred to a membrane and the proteins of interest (e.g., receptor, transcription factor) are visualized with an antibody detection system. As an example, Figure 4-9 reveals that crosslinking of the [^{125}I]-photoaffinity ligand–Ah receptor complex results in an increase in dimeric, trimeric, and tetrameric complexes over time [18]. In this experiment a number of crosslinking reagents were tested and dimethylpimelimidate (DMP) was found to yield optimal results.

18. Chen, H-S. and Perdew, G.H. J. Biol. Chem. 269 (1994) 27,554–27,558.

In order to identify which proteins are in each crosslinked complex, the AhR complexes were immunoprecipitated, subjected to SDS gel electrophoresis, and the protein transferred to mem-

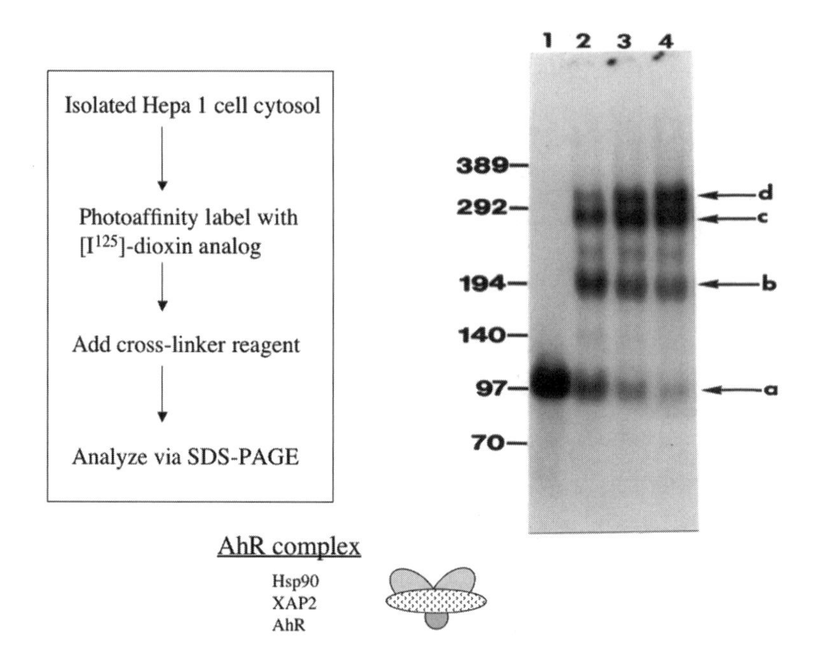

Fig. 4-9. Chemical crosslinking analysis of the cytosolic Ah receptor complex. Continuous phosphate SDS-PAGE analysis of crosslinked cytosolic extracts from $[^{125}I]N_3Br_2DpD$-labeled Hepa 1 cells. Crosslinking was performed for 0 h (lane 1), 1 h (lane 2), 2 h (lane 3), and 3 h (lane 4). Adapted from Perdew G.H., Biophys. Res. Commun. (1992) 182, 55–62.

brane and probed with antibodies. The results reveal that hsp90 is in the dimeric, trimeric, and tetrameric complexes. In addition, the presence of hsp90 in the dimeric complex indicates that hsp90 is in direct contact with the AhR. A model of how these proteins may be arranged can be proposed from these results. Although several questions remain, such as, do both hsp90 molecules bind to the AhR and does the small subunit protein bind directly to the Ah receptor? It turns out that subsequent studies have determined that the 43-kDa protein, now known to be XAP2, is indeed bound to both the AhR and hsp90 [19]. In order to possibly determine the other proteins in these crosslinked complexes, it will be necessary to immunoprecipitate the crosslinked complexes with an antibody to the primary protein

19. Meyer, B.K. et al. Mol. Cell. Biol. 18 (1998) 978–988.

of interest. Because of the possible interference of crosslinking with access to the antibody epitope, the conditions for crosslinking should be carefully optimized. In addition, the antibody should be efficient at immunoprecipitating the protein of interest. However, using chemical crosslinking to define the number of proteins and how they are arranged is not often utilized today. The most probable reason is that this approach requires both careful optimization of each step and the availability of antibodies efficient for immunoprecipitation of crosslinked complexes. The literature describes a number of excellent studies in which protein crosslinking was used in examining the subunit composition of enzyme complexes. For example, the component composition of lactose synthetase complexes was determined by crosslinking these complexes with the homobifunctional crosslinking reagent dimethylpimelimidate [20]. It was established with this technique that the complex exists as a 1:1 complex of α-lactalbumin and galactosyltransferase. Also the galactosyltransferase exists as either a 42- or 48-kDa component. Thus, chemical crosslinking of protein complexes is particularly useful in establishing the number of subunits in a complex and which subunits are in direct contact with each other.

Chemically crosslinked complexes are often difficult to immunoprecipitate.

20. Brew, K. et al. J. Biol. Chem. 250 (1975) 1434–1444.

2.2.5. Choice of Crosslinking Reagent

Carboxy, sulfhydryl, carbohydrate, and amino functional groups can be targeted by crosslinking reagents (*see* Table 4-1). There are both homobifunctional and heterobifunctional crosslinking reagents, with each reagent having an optimal pH requirement. In addition, there are cleavable (e.g., with thiols, periodate) and noncleavable crosslinking reagents. The cleavable reagents can be used to isolate crosslinked complexes and resolve the subunits by chemical treatment for assessment of the molecular weight of each subunit. The other variable in the structure of these reagents is the length

Homobifunctional crosslinking reagents that react with amino groups are most commonly used.

Table 4-1
Crosslinking Reagents Available for Protein Crosslinking Experiments

	Reactive toward				Cleavable by	
Cross-linker[a]	Amines	Sulfhydryls	Reactive nonselective reactive	Carboxyl	Thiols	Periodate
AMAS	X	X				
DMP	X					
DTME		X			X	
DMDB		X				X
EDC	X			X		
LC-SPDP	X	X			X	
APG			X			
BASED			X		X	

[a]For definition of the cross-linker abbreviations visit the Pierce website.

of the compounds. In general, crosslinking reagents of greater length exhibit greater efficiency but may also display undesirable characteristics, such as non-specific crosslinking to protein not actually physically bound to the complex of interest. Which reagent to use in a given situation is usually best determined through a series of pilot experiments utilizing a variety of crosslinking reagents, as well as testing of various concentrations of the crosslinker.

2.2.6. Use of Supershift EMSA to Assess Protein–Protein Interactions

Often after a protein–protein interaction has been established in which one of the proteins is a transcription factor, the question arises whether this interaction can occur while the transcription factor is bound to DNA. An in vitro approach to address this question is the use of supershift EMSA (described in Part I, Chapter 4). As depicted in Figure 4-10, purified protein A binds as a homodimer to a double stranded [^{32}P]-oligonucleotide, the addition of protein B to the gel-shift mixture results in an increased mobility of the complex super-shift. To further illustrate that protein A is bound to protein B, an antibody against protein A or B can be used to further increase the mobility of the protein A/B complex.

This type of experiment is capable of demonstrating that protein can enter into a complex with a transcription factor bound to its response element. An actual published example is the determination of whether the Jun dimerization protein 2 (JDP2) can bind to the progesterone receptor homodimer bound to its response element [12].

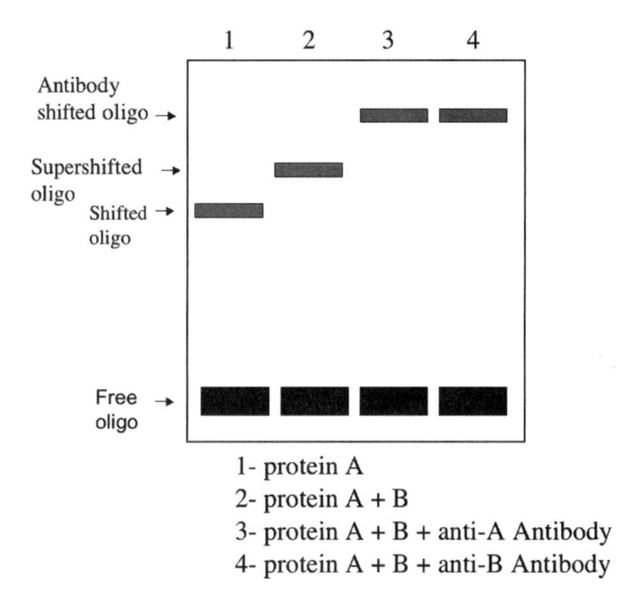

Fig. 4-10. Use of EMSA analysis to examine protein–protein interactions. A schematic representation of a typical result that can be obtained using EMSA analysis.

2.2.7. Immunoprecipitation

The most important aspect in performing immunoprecipitations is the quality of the antibody and the accessibility of a surface epitope in a given protein found in a complex. For example, if you propose to use a monoclonal antibody or a polyclonal antibody made against a synthetic peptide, then one needs to assess whether the epitope is away from the protein–protein interaction domains. However, often this information is not available. Thus polyclonal antibodies made against a large portion of the protein of interest, which ideally contains antibodies to a variety of epitopes, is usually preferred. Therefore, when using a given antibody to assess the presence of coimmunoprecipitated proteins, it is important to keep in mind that an antibody may only immunoprecipitate a subset of the complexes that protein resides in. In addition, only 10–40% of the total amount of protein present in a

12. Wardell, S.E. et al. Mol. Cell. Biol. 22 (2002) 5451–5466.

In general, a high-quality polyclonal antibody with antibodies against multiple epitopes will yield better results in immunoprecipitation experiments.

Often, several antibody preparations against distinct epitopes should be tested to obtain optional IP results.

given amount of lysate will be immunoprecipitated. This would suggest that testing several different antibody preparations will help define the variety of complexes that a given protein resides in. Also, an antibody could preferentially select a subset of a given protein and give the impression that the entire pool of that protein exists, as the results from the immunoprecipitation would suggest. These factors should be kept in mind when interpreting immunoprecipitation results. An excellent example is the protein hsp90, which is capable of complexes with an array of proteins. There are several protein binding sites on hsp90, including its dimerization interface, its cochaperone, p23, and client protein-binding sites. Thus, antibodies against any of these protein-binding sites would preclude the immunoprecipitation of a subset of hsp90 complexes, while an antibody against the dimerization interface would essentially not work in an immunoprecipitation experiment.

2.2.8. Factors Affecting the Specificity of Immunoprecipitation Results

In immunoprecipitations, heavy and light IgG chains can obscure coimmunoprecipitated protein on protein gels.

The majority of immunoprecipitations utilize extracts from unlabeled cell or tissue homogenates and when these immunoprecipitations are subjected to SDS-PAGE the heavy and light chains are prominent protein bands on a gel. The heavy and light chains (e.g., migrate at ~60 and 35 kDa) may obscure immunoprecipitated proteins that comigrate on the gel. Proteins can be visualized on the gel by either Coomassie blue, silver staining, or fluorescent stains. There are several ways to avoid the presence of the IgG heavy and light chains. First, an antibody can be immobilized on agarose-protein G beads (or other IgG affinity beads) and chemically crosslinked to fix the antibody to the protein G. However, the crosslinking can reduce the level of antibody binding activity and thus should be optimized. A second approach is the use of peptide displacement, which can serve two purposes. First, this technique can allow the antibody to remain bound to the resin and a second nonspecific protein bound to the gel is not displaced into the protein sample that is subsequently analyzed by SDS-PAGE. In general, peptide displacement works best with monoclonal antibodies made against a specific peptide sequence such as an eptitope tag like FLAG or HA, although peptide displacement may work with most antibody preparations in which the actual epitope sequence is known. A third means to circumvent the problem of heavy or light chain blocking visualization of immunoprecipitated proteins would be to radiolabel proteins in cultured cells with [^{35}S]methionine or other radioactive amino acids and visualize, using autoradiography, the SDS-PAGE-resolved immunoprecipitations. This would allow detection of protein bands that are immunoprecipitated even when they comigrate with the heavy or light antibody chains.

Another important variable in the overall quality of the immunoprecipitation, as well as the level of background obtained, is what type of resin is used to capture the primary IgG. There are a number of choices, including Protein A-, G-, or L-agarose, or an IgG made in another species that binds to the primary IgG and is crosslinked to agarose. A number of commercial companies make these resins using differing activated resins to immobilize the antibody. Another option is to purchase an activated resin (e.g., Affi-Gel®10, Amino-Link®Coupling Gel) and purified IgG that binds to the species of antibody that is to be immobilized. This allows an investigator to couple the antibody efficiently to resin and thus differently activated resins can be tested for the level of nonspecific binding of proteins. The type of coupling chemistry utilized can have a dramatic effect on the level of nonspecific binding of protein from a crude extract, especially if immunoprecipitations are being performed in low salt or detergent buffer. In general, agarose beads or more rigid crosslinked agarose (e.g., sepharose) are used to couple antibodies or protein A/G for use in immunoprecipitations. However, the porous nature of agarose beads can lead to entrapment of proteins within the structure of the beads. In an effort to lower background, the use of nonporous beads to immobilize antibodies could lead to lower background binding to the beads. However, nonporous bead products tend to have lower binding capacity for antibodies through covalently coupled protein A or G. This problem has been, at least in part, solved by use of very small magnetic beads, which will in effect have a greater surface within a given volume of beads. While, at first, the ability to capture paramagnetic microbeads was found to be more difficult, this problem was recently solved with the development of a superparamagetic MACS® microbead capture system (Miltenyi Biotec). This microbead system allows immunoprecipitations to be performed with shorter incubation times and should be useful in chromatin immunoprecipitations, which often suffer from nonspecific binding problems.

2.2.9. Use of Immunoprecipitation to Identify Complexed Components

Immunoprecipitation of a protein of interest from a cell or tissue extract is a common tool to detect the possible presence of associated proteins. The general approach is to test several antibodies for their ability to efficiently imunoprecipitate the protein being studied. Upon finding an antibody that immunoprecipitates efficiently, immunoprecipitation conditions are optimized to maintain any possible protein–protein interactions and still obtain acceptable background. Usually, a series of incubation and wash strategies are tested to obtain optimal conditions. These immunoprecipitations are subjected to SDS-PAGE analysis, followed by either the staining of the gel or transfer of the proteins to a membrane. Gels are commonly stained with Coomassie blue to visualize relatively abundant proteins, followed by silver staining, which is highly sensitive and can detect proteins that may be present at very low levels (e.g., bound to the primary protein at substoichiometric levels). If, in an immunoprecipitation experiment, an unknown protein coimmunoprecipitates, then a strategy needs to be formulated to identify that protein. The strategies that can be employed will be discussed in the following example. In crosslinking and immunoprecipitation experiments, the Ah receptor was found to exist in a tetrameric complex. There were two possible methods that could have been used to isolate the low-molecular-weight component of the complex. Multiple immunoprecipitations with an anti-Ah receptor antibody could have been performed using cell extracts containing relatively high levels of Ah receptor. Or a second approach, which was actually used in this situation, was to transiently overexpress the FLAG-tagged Ah receptor in a cell line with high-transfection efficiency, followed by isolation of the Ah receptor complex by immunoprecipitation [19]. The latter approach actually was

19. Meyer, B.K. et al. Mol. Cell. Biol. 18 (1998) 978-988.

successfully utilized to identify the low-molecular-weight component of the Ah receptor complex. In either approach the protein would be isolated from multiple immunoprecipitations, displaced with competing peptide, and subjected to SDS-PAGE analysis. The proteins could then be visualized by Coomassie blue staining and the protein band of interest cut out and digested with trypsin while in the gel. The peptide digest could then be subjected to high mass accuracy, matrix-assisted laser desorption ionization–mass spectrometry (MALDI–MS) for identification. Another approach to identify the protein would be to resolve the peptides by C18 microbore HPLC chromatography. Major peptide peaks would then be subjected to microsequencing. There are a number of excellent core university facilities that can perform these analyses on the isolated protein band.

2.2.10. Selection of Antibodies

The selection of an antibody for the study of a given protein is highly dependent on the experiments to be performed. The major experimental approaches that use antibodies are protein blotting, immunoprecipitation, immunohistochemistry, microinjection into cells, and ChIP assays. Perhaps the most common use for antibodies is visualization of a given protein on a protein blot. This procedure does not require that the antibody be highly specific because usually nonspecific or genuine crossreacting antigens are separated on the blot and the molecular weight of the protein of interest is known. In contrast, the other experimental approaches listed above require a highly specific antibody to obtain reliable results.

Both monoclonal and polyclonal antibodies against a specific protein can be very useful reagents. However, a high quality polyclonal antibody is usually a more efficient and sensitive reagent compared with a monoclonal antibody. However, some monoclonal antibodies can be of very high affinity and thus exhibit a high level of sensitivity. The advantage of polyclonal antibodies is particularly dramatic if the antibody preparation is made to the entire protein or a large segment of a protein, which ideally results in a wide spectrum of antibodies to a number of epitopes distributed along a significant portion of the protein. Once this is accomplished the antibody preparation usually is useful for immunoprecipitations, ChIP assays, and immunohistochemistry. The ability of an antibody to recognize more than one epitope is especially useful in ChIP assays, because the antibody needs to recognize a protein that has been crosslinked into a large oligomeric complex. This would suggest that, for many proteins, only a limited number of epitopes would be accessible. However, when one makes an antibody preparation against a large segment of a protein (e.g., 20-kDa fragment), it is possible that this may increase the probability of

Making a polyclonal antibody against an entire protein usually yields antibodies useful for immunoprecipitations.

making antibodies that recognize epitopes in other proteins, while a monoclonal antibody raised against a specific protein recognizes only one epitope and thus should be quite specific. It is important to note that companies, as well as individuals, are making polyclonal peptide antibodies, which are relatively easy to make. However, whether they are efficient for all applications, due to their limited epitope specificity, needs to be tested.

2.2.11. Antibody Production

If one is planning on extensively studying a protein, it is often a good idea to plan on making both polyclonal and monoclonal antibodies, especially if the laboratory has the resources to utilize core or commercial facilities to produce antibodies. The first step in antibody production is the selection of the type of antigen to use. If your goal is primarily to make an antibody for protein blot procedures or routine immunoprecipitation experiments, then the production of peptide polyclonal antibodies against the carboxy- or amino-terminal sequences of the protein being studied is usually adequate. However, if a wide variety of experimental approaches will be utilized, production of antibodies against the entire protein or a segment of the protein will more often yield an antibody preparation of greater utility. This can be accomplished by expression of a bacterial fusion protein (e.g., GST, MBP) or bacterial-tagged protein (e.g., His_6-Tag or T7(TAG®) in *E. coli*. The fusion protein is isolated with the appropriate affinity resin and the isolated protein is cleaved with a specific protease that recognizes a sequence engineered into the vector just preceding the tag sequence. The cleaved protein is isolated by ion-exchange chromatography or another purification method and used as an antigen. Peptides are generally crosslinked to a protein such as BSA or KLH to increase the peptide's antigenicity. An effective approach to couple a peptide is to add a cysteine residue to the amino terminus of the peptide during synthesis and couple the peptide through the sulfhydryl group to a maleimide-activated protein, thus ensuring that the peptide is properly presented during antibody production.

Upon challenge with an antigen, an animal responds by making antibodies to various specific epitopes within a protein sequence. The minimal recognition sequence appears to be approximately six amino acids, and these amino acids can be uniquely recognized by individual antibodies. When using an antigen with greater than 100 amino acids, the threoretical array of antibodies that can be produced is quite large. However, since amino acid sequences vary considerably in their antigenicity, the number of antibodies produced to a protein is actually somewhat restricted. Nevertheless, the serum of an animal after several immunizations contains a large number of individual antibodies being produced by individual B cells and, thus, is considered a polyclonal antibody serum. Polyclonal antibodies are generally made by individual investigators in rabbits, either at a core facility or by a commercial company. This requires milligram amounts of antigen. After several immunizations, the rabbits are bled and antibody can be isolated by affinity chromatography using the antigen immobilized on resin. Usually, several individual rabbits are tested to assess which rabbit is producing the highest antibody titer and specificity. If, for example, antibodies are being made against the entire protein and the level of antibody specificity is not adequate, subpopulations of antibody can be purified using bacterially expressed fragments of the protein. This could lead to production of polyclonal antibodies with different properties that may prove useful for a number of applications.

Sometimes there is a need to make an antibody against a protein that is a member of a closely related family. In this case one has two options available to generate an antibody: one approach involves making an antibody against a synthetic peptide corresponding to a unique sequence of the protein; another approach would be to make monoclonal antibodies against the entire protein as the antigen and select for clones that produce antibodies that are unique to the antigen within its protein family. An overall scheme for monoclonal antibody production is depicted in Figure 4-11 [21].

Monoclonal antibody production is accomplished by immunization of a number of mice (e.g., BALB/cJ) with immunogen as just described for polyclonal antibody production. After several immunizations, serum is obtained from the mice and the presence of specific antibodies can be assessed by either ELISA or protein blotting techniques. The latter technique allows the visualization of not only the specific band of interest, but also the level of nonspecific binding. This allows the selection of an animal for hybridoma production. The spleen is removed and B cells are isolated, followed by fusion with myeloma cells (e.g., SP2/0 or NS-1 cells). A polyethylene glycol solution is used to fuse

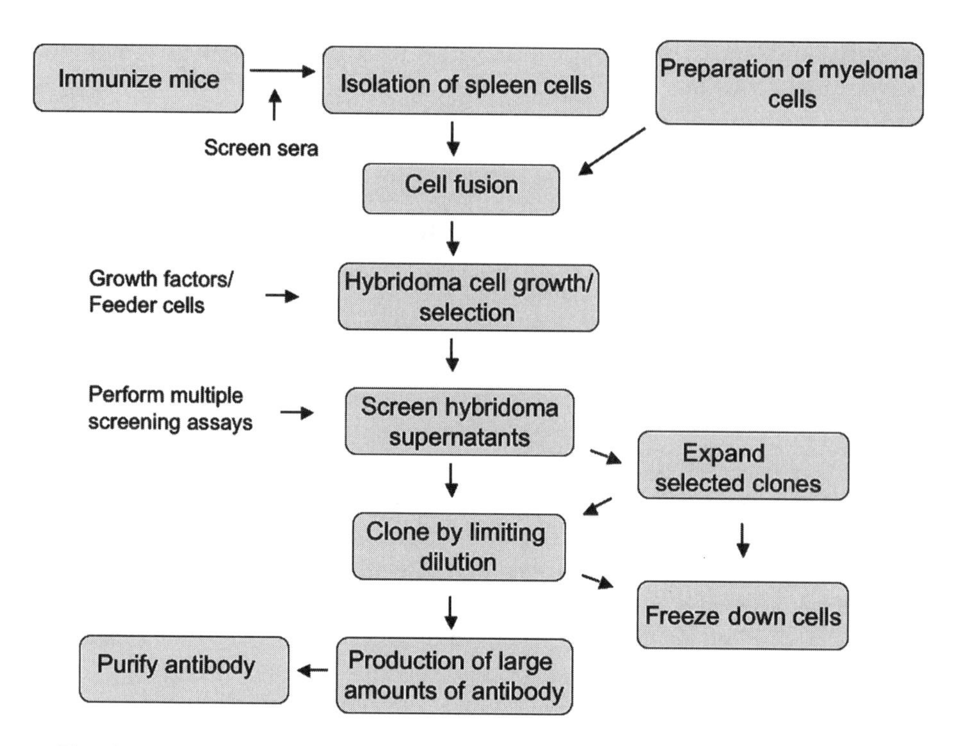

Fig. 4-11. Flow chart depicting the steps used to produce a monoclonal antibody.

21. Harlow, E. and Lane, D. Antibodies: A Laboratory Manual, Cold Spring Harbor Laboratory. 1988, 139–281.

the two cell types together. The fused cells are plated out into 96-well plates and exposed to several selection agents. After 7–14 d only hybrid cells (hybridomas) will grow out as clones. If the cells are plated out properly there should be one or two clones in each well. The hybridomas are screened using ELISA and the positive wells are further screened by an additional assay. It might seem that one could assume that after a number of clones are generated that one of the monoclonal antibodies will turn out to be useful for each assay. This is often a critical aspect of obtaining useful hybridomas. If, for example, there are 40 positive clones, the additional assays should narrow down which clones will be grown out and characterized. The secondary assays should reflect the intended uses for the antibodies, such as for protein blotting analysis or immunoprecipitation. In fact, the use of both

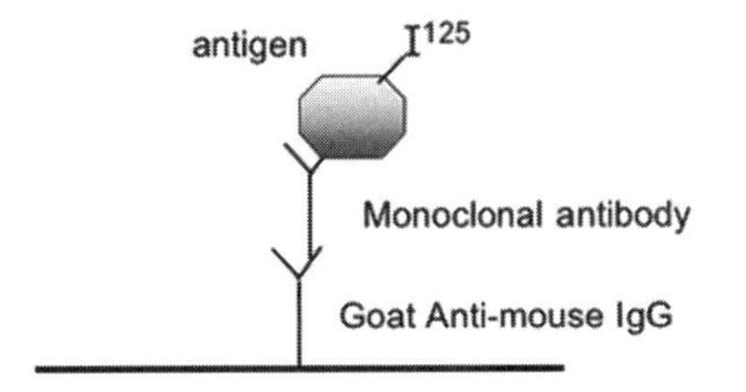

Fig. 4-12. An example of a screening assay for identification of monoclonal antibodies that bind to non-denatured antigen. The goat anti-mouse IgG is bound to plastic in a high-throughput assay system.

of these assays may be warranted to effectively narrow down which clones to expand and subsequently subclone. Perhaps one of the most demanding uses for a monoclonal antibody is in ChIP assays, which require a high-affinity antibody that can immunoprecipitate the antigen crosslinked into a DNA oligomeric protein complex under stringent conditions. If this assay is an important goal of the antibody production project, then this assay should be performed quickly in order to determine whether the goal(s) of the project has been met. Another example of designing an appropriate assay for antibody selection is if one wants to make a monoclonal that binds to a receptor normally complexed with other proteins. In this case, the design of an assay to screen for antibodies that bind to the nondenatured complex could be the key to finding the antibody of interest. Figure 4-12 gives an example of how that assay could be designed.

Screening assays should reflect the intended uses for antibodies, such as for protein blot analysis, ChIP assays, or immunoprecipitation.

Once useful monoclonal antibodies have been obtained, large amounts of antibody can easily be produced in cell culture. One production method utilizes a special flask with a small incubation chamber housed below the main chamber that holds a large volume of cell culture medium where the cells can grow to very high concentrations and produce antibody at approx 1mg/mL of medium. Alternately, cells can be injected intraperitoneally into pristine-primed mice in which ascites fluid accumulates over 10–21 d and the fluid is tapped from the mice. With either of these preparations a large amount of antibody can be purified to homogeneity. Purified antibody can be directly coupled to activated agarose to make an affinity resin for immunoprecipitation experiments. If the epitope recognition sequence for a monoclonal antibody is known, often peptide displacement from the antibody affinity resin can be performed. One such displacement system that is commercially available is the FLAG peptide tag/antibody system.

In summary, by designing the appropriate screening procedures, monoclonal antibodies can be isolated for a variety of applications.

2.2.12. Mapping an Antibody Recognition Site

Determination of the epitope binding sequence for a monoclonal antibody allows the use of peptide displacement from immunoprecipitations.

Often monoclonal antibodies are made to a fusion protein or an entire protein, thus the actual epitope recognition peptide sequence is not known. Determination of the actual recognition sequence can be helpful in understanding whether the antibody would be capable of binding to a protein when certain domains may be masked through protein–protein interactions. Also, once the epitope has been mapped, a peptide can be synthesized and used in peptide-blocking control experiments and for displacement of the antigen from immunoprecipitations. In addition, peptide polyclonal antibody could be made to this sequence, which may also be quite useful. The determination of an antibody's epitope can be accomplished by screening a phage-based random peptide display library, as described in Section 2.1.6. A second approach utilizes a protein domain mapping system such as the Novatope® system developed by Novagen. The basic scheme of this system is to take the cDNA corresponding to the protein of interest and perform a DNase shotgun cleavage, the addition of an adenosine residue (dA tailing). These cDNA fragments are cloned into a bacterial-expression fusion-protein vector and the resulting mixture of vectors is transformed into competent bacteria. The bacteria are plated out and

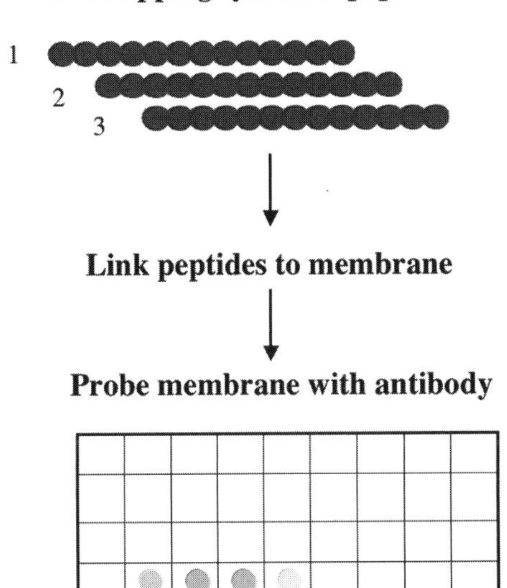

Fig. 4-13. Schematic representation of the use of peptide arrays to map an antibody epitope.

nitrocellulose colony lifts are screened for clones that express a peptide sequence that binds the antibodies. An individual bacterial colony that expressed the peptide is isolated and grown overnight. Plasmid preparations are then isolated from the bacterial culture and the cDNA insert is sequenced. The sequence of several positive clones can help firmly define the minimal antibody recognition epitope.

Another approach to mapping an epitope is the use of a peptide array. The array is produced by taking the amino acid sequence of the protein of interest and synthesizing a series of 13 amino acid peptides, which overlap by 11 amino acids. This series of peptides are immobilized on a cellulose membrane, and the peptides that bind the monoclonal antibody can be assessed (Figure 4-13). This type of an array can be used to precisely map the amino acids that compose an epitope. This methodology has been reviewed by Reineke [22].

22. Reineke, U. et al. Curr. Opin. Biotechnol. 12 (2001) 59–64.

2.2.13. Immune Depletion

Immunoprecipitations can reveal that two proteins exist in a complex, but this experimental approach is generally not able to tell what percentage of a protein is present. This is attributable to the fact that, in a typical immunoprecipitation, only a fraction of the protein of interest is immunoprecipitated, thus it is impossible to make any quantitative assessment. In order to establish what percentage of a protein is complexed with another protein, an immune-depletion experiment can be utilized. Basically, the idea is to set up an immunoprecipitation with a relatively large amount of antibody prebound to a resin, such as protein G sepharose. After a 1-h incubation only about 40% of the antigen is immunoprecipitated, thus after the resin has been spun down, the supernatant is transferred to another aliquot of antibody prebound to resin. A parallel immunoprecipitation using a nonspecific IgG antibody is used as a control to assess the relative level of immune depletion. After a 1- to 2-h incubation, the soluble fraction is subjected to SDS-PAGE and protein blot analysis. The relative amount of the protein immunoprecipitated, and any potential coimmunoprecipitated protein depleted from the sample, can be assessed relative to a control antibody immune-depleted sample using an iodinated secondary antibody. If a single immunoprecipitation fails to completely deplete antigen, a second immunoprecipitation can be performed. With this approach the amount of one protein complexed with another can be accurately assessed on protein blots.

2.2.14. Using Fusion Proteins and Tag Sequences for Immunoprecipitations

As pointed out in the preceding sections, the production of antibodies to a given protein does not always lead to a useful reagent for immunoprecipitation of protein complexes. One approach that is used extensively is the tagging of the N- or C-terminus to a short amino acid sequence, using molecular biology techniques [23]. Placement of the tag

23. Jarvik, J.W. and Telmer, C.A. Annu. Rev. Genet. 32 (1998) 601–618.

Table 4-2
Summary of Commercially Available Peptide Epitope-Tagging Systems

Name	Tag sequence	Antibodies	Vectors	Source
Flag	DYKDDDK	M1, M2, M5 mAbs	+	Sigma
HA	YPYDVPDYA	mAb 12CA5	–	Babco Boehringer Mannheim
c-myc	EQKLISEEDL	mAb 9E10	+	Boehringer Mannheim Invitrogen
T7.Tag	MASMTGGQQMG	T7.TAG	+	Novagen mAb
SV-G	YTDIEMNRLGK	mAb P5D4	–	Boehringer Mannheim
HSV	QPELAPEDPED	HSV.Tag mAb	+	Novagen
V5	GLPIPNPLLGLDST	mAb V-5	+	Invitrogen

sequence on the terminus of the protein that is furthest from functional domains should result in a protein that exhibits the activity of its nontagged form. Many vectors are available with a number of tags (*see* Table 4-2). Also, multiple tags added to a protein can greatly increase the sensitivity of detection of that protein or increase immunoprecipitation efficiency. Several vectors are commercially available that have two or three copies of the tag sequence (e.g., FLAG or HA).

In a number of applications the presence of the sequence tag on the protein proves to be particularly useful. For example, if site-directed mutagenesis is used to mutate a protein–protein interaction motif or a phosphorylation site, a tagged protein can be expressed and immunoprecipitated from cells in the presence of endogenously expressed analogous protein, which will not interfere with the interpretation of the results. Also, tagged proteins can be used in the generation of transgenic animals, the presence of the tag allowing the introduced gene product to be uniquely detected in the presence of its endogenous analog. Yet another useful aspect of using a tag sequence to study a protein is that, when there are no available antibodies against that protein, the addition of a tag sequence to the cDNA corresponding to that protein allows immunoprecipitation experiments to be performed using extracts from transfected cells. A fourth application of a tagged protein is the generation of two different tagged versions of a given protein; these proteins could then be used to assess whether a protein can dimerize. Each protein can be transiently coexpressed in cells and immunoprecipitated with an antibody against one tag, then subjected to SDS-PAGE analysis followed by transfer of protein to a membrane, and then the blot can be probed with an antibody to the second tag. The presence of the first tag in the immunoprecipitation would indicate that this protein is capable of dimerizing. In addition, a mutant form of a protein could carry one of the tags and the "wild-type" protein the second tag, allowing a functional comparison to be made, such as one to test the influence of a point mutation on dimerization. Thus, the use of tag sequences can allow the design of experiments that would be difficult to perform using other techniques.

2.2.15. Overexpression of Proteins for Protein–Protein Interaction Studies

The addition of a tag sequence to a given protein will allow immunoprecipitations to be performed more quickly than the time needed to make a useful antibody.

Bacterial expression of tagged fusion proteins is the most often used means to obtain large amounts of a purified protein.

There are a number of cell-based systems that have been developed to overexpress proteins, including; *E. coli,* yeast, or *Drosophila* cell lines, baculovirus/Sf9 insect cells, and mammalian cells (e.g., COS 1 cells). Usually the protein has a tag sequence or is a fusion protein (e.g., FLAG, HA, GST, His_6, maltose-binding protein, calmodulin-binding protein, and so forth), and this allows for the rapid purification of the protein of interest. Each system has its advantages and disadvantages. By far the most utilized system is bacterial expression, which often results in a high level expression of the protein of interest. Usually, the protein is expressed as a tagged protein or as a fusion protein. The most utilized tag is the six-Histidine tag, which is efficiently purified using a nickel-containing agarose column. The most utilized fusion proteins are glutathione-*S*-transferase, maltose-binding protein, thioredoxin, His_6, or cellulose-binding domain. Affinity resins are available commerically to purify each fusion protein listed. Usually in the fusion-protein vector, a protease cleavage site is placed between the protein tag and the polylinker region where the cDNA for the protein to be expressed is inserted, although for protein–protein interaction studies the fusion protein can be utilized without removal of the tag. However, the bacterial expression system can suffer from a number of limitations. Perhaps the most commonly encountered problem is that a high level of expression leads to solubility problems, which can lead to the majority of the expressed protein being found in inclusion bodies. When this happens the protein in the inclusion bodies can be solubilized in denaturing buffers that include high levels of urea or other chaotropic agents, followed by renaturation of the solubilized protein by dialyzing the protein solution against a series of buffers that incrementally decreases the urea concentration until there is no urea in the buffer. However, proper conformation of the protein will need to be determined by a protein function assay. Often the solubility of proteins in *E. coli* can be enhanced if special commercially available strains of *E. coli* are used. For example, strains are available that stably express tRNAs rarely used in *E. coli,* or stains can contain mutations in both the thioredoxin reductase and glutathione reductase genes, which greatly enhances disulfide bond formation. In addition, growing the *E. coli* at lower temperatures (e.g., 27°C) decreases the level of protein expression, which can lead to increased solubility of the expressed protein. Another possible limitation of bacterial systems is the lack of posttranslational modifications such as phosphorylation or glycosylation. These fusion or tagged proteins can be immobilized on the appropriate affinity resin and incubated with a protein to test for binding to the resin.

Bacterial expression systems are inexpensive systems that can yield large amounts of purified protein. However, there are often problems with protein solubility, as well as obtaining the appropriate folding.

Another system that is commonly utilized for expression of mammalian proteins is the baculovirus-mediated expression of a protein in insect cells (e.g., Sf9 cells). This method is probably the most utilized eukaryotic-based system for overexpression of a protein. There are several advantages of this system, including: (1) posttranslational modification of proteins should occur, (2) proper folding and S–S bond formation, (3) the protein should retain the appropriate subcellular localization, and (4) the protein exhibits greater solubility relative to bacterial expression systems. Some disadvantages are: the procedure is time consuming, relatively

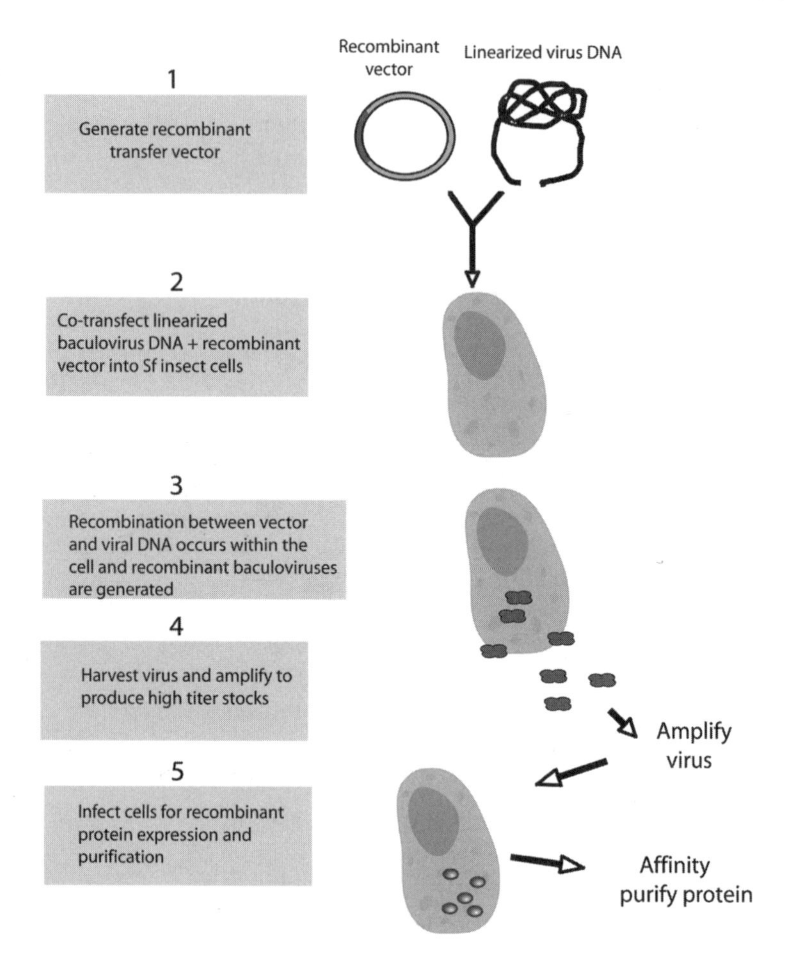

Fig. 4-14. Scheme depicting the major steps in the production of protein using the baculovirus system.

expensive, and requires more manipulations to obtain the protein of interest. An outline of a general production and purification scheme for the baculovirus system is depicted in Figure 4-14.

Overexpression of tagged proteins in COS 1 or 293T mammalian cell lines can be an effective system to obtain a significant amount of protein for protein–protein interaction studies or for monoclonal antibody production. Both of these cell lines express T-antigen, which allows many mammalian expression plasmids to replicate within the cell. This can lead to a significant level of protein expression. In addition, very high transfection efficiencies can be achieved in these cell lines, further enhancing the overall level of protein expression that can be obtained. This system has the advantage that the proteins expressed are usually folded properly and have the appropriate posttranslational modifications. However, the amount of protein obtained by this system is usually less than can be obtained with a bacterial expression system, especially for proteins that poorly translate. Nevertheless, mammalian cell culture can be very useful and most laboratories studying gene expression in mammals can easily isolate adequate amounts of proteins using standard mammalian cell culture techniques.

Yet another technique that may be increasingly utilized in the future is an in vitro bacterial expression system that can achieve high-level expression through the use of continuous exchange cell-free technology (CECF), a technology marketed as Rapid Translation System RTS 500 (Roche Applied Science). Basically, a coupled transcription/translation bacterial lysate system is used to produce protein using a cDNA-expression construct as a template. This mixture is in contact with a semipermeable membrane, and a solution that supplies depleted components is placed on the other side of the membrane. Advantages of this system include: it yields relatively large amounts of protein (up to 5 mg/mL), it allows the addition of factors to the system, such as chaperone proteins or radiolabeled precursors, and it usually results in greater solubility of the protein being produced. The disadvantage of this system is that it is fairly expensive.

2.2.16. Pull-Down Assay

Probably the most utilized fusion protein system for protein "pull-down assays" is glutathione-*S*-transferase (GST), which can be expressed in both mammalian and bacterial expression systems. GST fusion proteins can be efficiently purified using glutathione agarose and eluted with glutathione. These preparations can be dialyzed to remove the glutathione and used in pull-down assays. Alternatively, the GST fusion protein from cellular extracts can be prebound to glutathione agarose, washed, and used in a pull-down assay. Either way, the GST fusion protein bound to glutathione agarose is incubated with a crude extract or a purified protein. The resin is washed and subjected to SDS-PAGE analysis and the protein transferred to a membrane. Antibodies are used to detect the presence of the GST fusion protein and the protein that is being brought down by the fusion protein. GST fusion proteins can also be expressed in cells, such as in COS 1 cells, and the GST fusion proteins isolated using glutathione resin can be isolated from cell lysate. The presence of coisolated proteins can then be assessed on protein blots after SDS-PAGE. Other fusion protein systems also can be effectively utilized in a similar manner, as described previously, including maltose binding protein or His_6 tag (Section 2.2.15.).

A common approach in demonstrating a protein–protein interaction is the use of a coupled in vitro transcription/translation rabbit reticulocyte system.

This system utilizes a protein expression vector containing either a T7, SP6, or T3 bacterial promoter. Usually one protein is in vitro translated in the presence of [^{35}S]methionine, and the second protein is expressed as a bacterial fusion protein and used to pull down the in vitro translated protein. However, another way to set up an in vitro translation assay system for protein–protein interaction assays is to translate both proteins separately in vitro, mix and immunoprecipitate one of the proteins, and determine if the other protein is coimmunoprecipitated using SDS-PAGE analysis. One interesting alteration that can be utilized in this system is the in vitro translation of one of the proteins in the presence of biotinylated-lysine tRNA. After mixing the two translated proteins, the biotinylated protein can be efficiently captured with streptavidin–agarose.

2.2.17. Mapping of Protein–Protein Interaction Motifs

Mapping of protein–protein interaction domains or motifs can be accomplished a number of ways. With a set of deletional constructs utilized in a yeast or mammalian two-hybrid system, the required domain can be effectively mapped. The actual presence of a motif can then be examined by site-directed mutagenesis/alanine screening mutagenesis that walks through the mapped domain. Upon determination of critical amino acid residues, additional studies can determine whether a primary sequence motif exists. The mapping of amino acid residues required for a protein–protein interaction is quite useful in dissecting its role in a signaling pathway (*see* Chapter 6).

2.2.18. Far-Western Blotting

Yet another method that examines protein–protein interactions is referred to as far-Western blotting, because it examines the ability of proteins bound to membrane to bind to a protein in solution. This method can be used in place of GST pull-down assays or immunoprecipitation, at least when one or both proteins can be overexpressed. It has been used to detect how many proteins in a complex extract interact with a specific protein, and it can be used as an assay to purify interacting proteins. Protein extracts are subjected to SDS-PAGE and the protein is transferred to nitrocellulose membrane. After blocking the membrane by incubation in an excess of a protein solution, such as bovine serum albumin, the protein of the membrane is partially refolded by exposure to a series of guanidine-HCl solutions with decreasing ionic strength. The protein that is being used to search for binding partners on the membrane is usually expressed in bacteria. This protein can be a fusion protein (e.g., glutathione-*S*-transferase), which allows rapid purification. Also these fusion proteins can be designed to contain a cAMP-dependent protein kinase site, which allows ^{32}P incorporation into the protein using an in vitro kinase reaction (Amersham Biosciences). This method depends on the proper refolding of the protein–protein recognition sequence of the protein bound to the membrane. Another important point to consider is that after specific molecular weight bands are detected an investigator will need to identify the interacting proteins. This can be accomplished by fractionation of the protein mixture in order to have the protein of interest appear as a purified individual protein band. This will allow the identification of the protein band by mass spectrometry techniques. An alternate approach is to guess from the molecular weight the possible identity of the protein and use antibodies to probe for the protein of interest.

One interesting application has been the utilization of far-Western blots in a search for protein kinase C (PKC) substrates or binding proteins. A number of interacting proteins have been identified, such as 14-3-3 [24]. Studies have also been performed to look for PKC substrates, which are considered a weaker interaction. Thus, washing a blot after incubation with PKC would result in disruption of the complex, and to solve this problem the investigators exposed the membranes to a crosslinking reagent to fix the interactions prior to washing of the membrane. The presence of PKC was determined using an anti-PKC antibody.

24. Van Der Hoeven, P.C. et al. Biochem. J. 345 (2000) 297–306.

2.2.19. Determination of Protein–Protein Equilibrium Binding Constants

Surface plasmon resonance (SPR) is a technology pioneered by Biacore and allows real-time monitoring of biomolecular interactions as well as measurement of the kinetics and affinity of an interaction. The technique can also effectively measure weak interactions as well as factors that can modulate that process. At least in theory, SPR could be used to measure any molecular interaction, with the measure of protein–DNA, ligand–protein, and protein–protein interactions being of particular interest. Interactions are measured on a glass surface of a chip coated with gold, and this surface is coated with a hydrophilic dextran layer to which molecules are chemically linked. SPR measures changes in mass in the aqueous layer adjacent to the sensor

Surface plasmon resonance can be used to measure a wide range of interactions such as protein–DNA or ligand–protein.

chip by measuring changes in refractive index. A computer attached to the biosensor instrumentation monitors changes in refractive index and plots the data on a sensorgram. This method can study a molecule's interaction with a protein without labeling the molecule, and an interaction can be studied in a variety of solution conditions. Protein–protein interactions can be studied to determine the affinity of the interaction or to test the arrangement of the subunits in a protein complex. This is one of the few techniques capable of measuring small-molecular-weight compounds binding to proteins. For example, the ability and specificity of transition metal oxyanion binding to hsp90 can be effectively measured using SPR technology [25].

25. Soti, C. et al. Eur. J. Biochem. 255 (1998) 611–617.

3. Summary

With the sequencing of the human and mouse genome completed, the ability to characterize various aspects of protein regulation through protein–protein interactions takes on increasing importance for understanding the regulation of or alteration in cellular function. After all, most functional and regulatory activities within the cell are mediated by proteins. Usually in any given study, one or more of the in vitro and cell culture approaches outlined in this chapter are taken to firmly establish that two proteins are capable of interacting. However, it is important to keep in mind that many cell- and in vitro-based approaches only infer that two proteins can interact, because in a complex system one or more factors could be bridging between the two proteins of interest. Definitive proof of direct binding can be obtained through with chemical crosslinking, or in vitro protein interaction studies using purified components. After a protein–protein interaction event is determined, often the regulation of this interaction will be examined. These interactions can involve changes in post-translational modifications such as phosphorylation or glycosylation, which will be the subject of the next chapter.

Usually, to demonstrate that a protein–protein interaction occurs in a cell, several approaches should be taken.

5

Posttranslational Modifications

1. Concepts

Posttranslational modifications commonly occur on proteins within a cell and lead to changes in stability, subcellular localization, enzymatic activities, and other protein activities mediated through protein–protein interactions. Because of the large number of posttranslational modifications that occur, this chapter will focus on the modifications that are most frequently encountered when examining regulation of gene expression through transcription factor (TF) analysis. These modifications occur in the majority of proteins involved in transcriptional regulation, including transcription factors, coactivators, repressors, general transcription factors, histones, and RNA polymerases. Protein phosphorylation is probably the most studied posttranslational modification. Interestingly, phosphorylation has been demonstrated to influence a wide range of transcription factor activities, such as transactivation potential, DNA binding, half-life, subcellular localization, dimerization, or heterodimerization, and cofactor or ligand binding. Transcription factors can be phosphorylated on serine, threonine, and to a lesser extent on tyrosine residues. The mammalian cell contains a large number of protein kinases, the specific recognition motif for some being known, while others remain to be determined. For example, the recognition motif for casein kinase II is S/T-X-X-E. The overall level of protein phosphorylation is regulated by the balance between the level of protein-kinase versus protein-phosphatase activity in the cell. The half-life of the average phosphate on a specific amino acid residue is quite short, in the range of minutes to a few hours; thus, the existing dynamic equilibrium can be rapidly altered upon activation of a signaling pathway that leads to activation of a protein kinase. This would result in an increase in the percentage of a target protein carrying a specific phosphorylated site. Another important concept in examining the role of specific phosphorylation events is the ability of phosphorylation of one site to act as a protein kinase recognition site, which leads to phosphorylation by another kinase. Protein kinase GSK-3 requires the recogni-

Regulation of Protein Levels and Transcription Factor Function by G. H. Perdew
From: *Regulation of Gene Expression*
By: G. H. Perdew et al. © Humana Press Inc., Totowa, NJ

A "hierachial" protein kinase requires phosphorylation of one residue to generate a recognition site for another protein kinase.

tion motif S-X-X-S(P), where S(P) is phosphoserine, an example of a kinase that has been termed a "hierarchial" protein kinase.

Although phosphorylation has historically received the most attention as a means of directly regulating transcription factor activity, other modifications, such as O-GlcNAc, acetylation, and methylation, have also been shown recently to be of considerable importance. In fact, the addition of O-GlcNAc to serine or threonine residues, which are often also sites of phosphorylation, have been described for several important regulatory proteins (e.g., SP1, c-myc, ERβ). In general there appears to be reciprocity between O-GlcNAc and O-phosphorylation, where phosphorylation may increase a protein activity and glycosylation decreases its function [26]. Using a monoclonal antibody that recognizes O-GlcNAc amino acid residues, a large number of proteins in a cell have been shown to carry this type of posttranslational modification. Perhaps one of the most interesting examples of how O-GlcNAc modifications may regulate protein function is the ability of O-GlcNAc to modify c-myc on Thr^{58}, a known mutational hot spot in lymphomas. Phosphorylation of Thr^{58} in the transactivation domain of c-myc would appear to modulate cellular transformation potential. Thus, this would suggest that Thr^{58} in c-myc can exist in three different forms, with each form exhibiting a distinct effect on transactivation potential. An important aspect of this observation is that mutation of a potential phosphorylation site to study its role in regulating protein function may be complicated by the disruption of O-GlcNAc modification as well. Another example of O-GlcNAc-mediated regulation is SP1, a ubiquitious transcription factor that can be extensively modified by O-GlcNAc. Interestingly, increased levels of O-GlcNAc have been observed in diabetes and may play a role in decreased responsiveness of adipocytes to insulin. Increased modification of SP1 by O-GlcNAc has been observed

26. Wells, L. et al. (2001) Science 291, 2376–2378.

Many of the mapped O-GlcNAc sites can also be phosphorylated.

when cells are treated with insulin or when cultured with elevated levels of O-GlcNAc. Thus, the posttranslational addition of O-GlcNAc on serine or threonine residues on various proteins appears to play a role in transcription, cancer, and other disease states such as diabetes. Development of methodology to efficiently characterize and map the precise location of these modifications needs to be further developed before we can fully determine the role of O-GlcNAc in modulating gene expression.

A field that is rapidly expanding is the regulation of transcription factors by ubiquitin, small ubiquitin-like modifier (SUMO), and related polypeptide modifiers. Polyubiquitination of a protein with four or more ubiquitin molecules results in a higher affinity for the proteasome, leading to rapid protein turnover. Ubiquitin is a highly conserved polypeptide of approx 76 amino acids in length that can covalently link to a lysine residue on a target protein through a series of steps (Figure 5-1), which can result in protein turnover or regulation of its activity.

Ubiquitin–proteasome system appears to regulate the level of a wide range of transcription factors, especially ones with a short half-life.

In the first step, a ubiquitin-activating enzyme termed E1 forms a thioester linkage with the COOH glycine of ubiquitin. In the second step, the ubiquitin is transferred from E1 to one of more than 20 known E2 ubiquitin-conjugating enzymes through a thioester linkage. The third step is accomplished by E3 ubiquitin ligases, which are responsible for specific recognition of substrate and the transfer of ubiquitin to a lysine residue in the substrate protein. E3 ligases preferentially interact with certain E2 ligases for transfer of ubiquitin. The next step usually involves ubiquitin chain elongation by E4 ubiquitin ligases on substrate proteins. Examples of the steady-state level of a given transcription factor being regulated by E3 ubiquitin ligases include p53 by Mdm2 and HIF1α by VHL E3, among others [27]. Also, there are ubiquitin isopeptidases that can determine the fate of ubiquitinated proteins by cleaving ubiquitin from target proteins. This step

27. Conaway, R.C. et al. Science 296 (2002) 1254–1258.

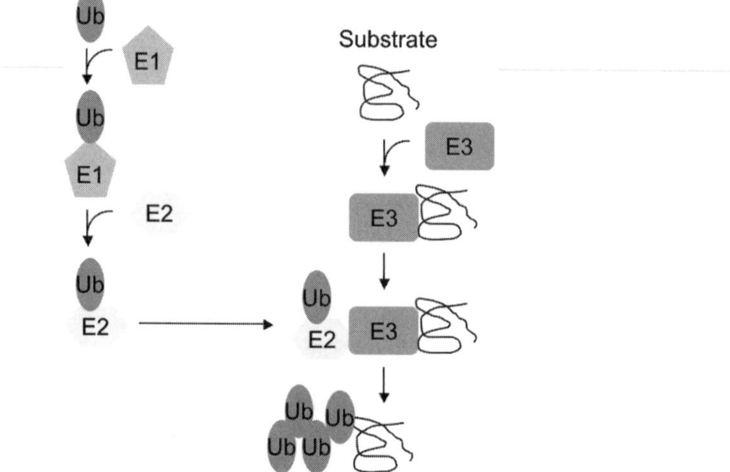

Fig. 5-1. Schematic diagram representing the process of protein ubiquitination that leads to proteasome-mediated degradation.

may also be a possible target of altered regulation. Isopeptidase activity is also found at the lid of the proteasome. In addition to this classical role for ubiquitination of transcription factors, several studies have now indicated that ubiquitination of certain transactivation domains may actually be required for efficient transactivation potential. Also, the nuclear factor kappa B1 (NF-κB1), p105, and p100 precursors are processed into active subunits of NF-κB by ubiquitylation and cleavage by the proteasome. Thus the role of ubiquitination in controlling the activity of many TF appears to be quite complex. Proteins can be monoubiquitinated or multiubiquitinated, this does not necessarily lead to immediate proteolytic turnover. The addition of multiple ubiquitin molecules to form linear or branched ubiquitin chains at one or more lysine residues in the target protein is referred to as polyubiquitination and this leads to the ubiquitinated protein being targeted to proteasomes either in the cytoplasm or the nucleus. The process of polyubiquitination can occur through the addition of ubiquitin to several different lysine residues in the ubiquitin molecule. The type of oligomerization that occurs can dictate the functional result of the polyubiquitination, such as affecting protein activity, causing immediate proteolytic turnover, or cellular trafficking. Interestingly, some proteins actually are degraded by the proteasome in the absence of ubiquitination. Finally, it is important to keep in mind that, while the ubiquitin/proteasomal protein degradation system is an important mechanism of turnover for many proteins involved in transcription, other protein turnover mechanisms exist, such as calpains and lysosomal protein degradation. The lysosomal proteolytic system appears to be especially important for turnover of long lived proteins.

Sumoylation is a posttranslational modification of proteins that utilizes a series of enzymatic steps similar to the ubiquitination process. The sumoylation process results in the transfer of the polypeptide SUMO to a protein. SUMO is a 101-amino acid protein that exhibits only 18% sequence homology with ubiquitin, although they do exhibit close 3-D structural similarities. However, the functional consequence of sumoylation is usually quite different from ubiquitination [28]. The first step in the sumoylation process is the activation of SUMO by a heterodimeric E1 enzyme SAE1-SAE2 that uses ATP to adenylate the C-terminal Gly residue of SUMO, resulting in a thioester bond between the substrate and enzyme complex. Through a transesterification reaction SUMO is transferred from SAE to the E2 SUMO-conjugating enzyme Ubc9. The Ubc9-SUMO conjugate then catalyses formation of a covalent linkage with the β-amino group of a Lys residue in the target protein. In contrast to ubiquitination, the Ubc9-SUMO complex appears to recognize a specific sequence, ψKxE (where ψ represents L,V, I or F). Interestingly, SUMO modification appears to occur primarily in the nucleus, in contrast to ubiquitination, which occurs in both the cytoplasm and the nucleus. The effect of sumoylation on protein function is diverse. For example, SUMO modification of IκBα leads to protein stabilization and thus inhibition of NF-κB mediated transcription, while sumoylation of p53 leads to increased transcriptional activity. Thus, the functional significance of sumoylation needs to be assessed for each protein of interest. The experimental approaches to study sumoylation are similar to ubiquitination and will not be discussed in this chapter.

28. Gill, G. Genes Dev. 18 (2004) 2046–2059.

In contrast to ubiquitination, sumoylation leads to protein stabilization.

Acetylation of histones is a key step in the regulation of transcriptional activity in the nucleus. The role of transcriptional enhancer protein (e.g., SP-1, estrogen receptor) binding to DNA sites upstream from the transcriptional start site appears to be to recruit factors that enhance histone and basal factor posttranslational modifications (e.g., acetylation, methylation, phosphorylation, and ubiquitinylation). These modifications lead to the remodeling of chromatin (loosening of chromatin structure), which allows RNA polymerase to initiate transcription. Through their transactivation domain, TFs recruit coactivator complexes where they either contain histone acetyltransferase (HAT), or the coactivator acts as a bridging factor, bringing a HAT into the transcriptional complex. This leads to histone remodeling and allows basal transcriptional factors to bind near the transcriptional start site. Certain transcription factors can also exhibit transcriptional repression through the recruitment of histone deacetylases (HDAC). A good example is the thyroid hormone receptor (TR) that exists heterodimerized with RXR tethered to DNA in the absence of a TR ligand. This receptor complex actively recruits corepressor complexes that exhibit HDAC activity. Upon ligand binding to the TR the corepressor complex is released and coactivators are recruited, leading to enhanced gene transcription.

The acetylation of transcription factors (e.g., p53, E2F1) and other nonhistone proteins (e.g., α-tubulin) on lysine residues is a posttranslational modification of emerging importance, especially in regulating gene transcription. The list of proteins that can be acetylated is growing, but the functional consequence of protein acetylation can vary considerably. For example, acetylation of the transcription factor NF-E4 leads to increased protein half-life by inhibiting ubiquitin-mediated degradation. Acetylation of the androgen receptor leads to enhanced transcriptional activity, mediated by an increase in coactivator recruitment potential. While acetylation of the HMGI(Y) transcription factor occurs within the DNA-binding domain, this results in inhibition of binding to its recognition element.

2. Methods and Approaches

2.1. Phosphorylation

2.1.1. Determination of Phosphorylation Sites

Determination of phosphorylation sites is most commonly performed using two basic approaches. The first approach involves the use of 2-D mapping of phosphopeptides generated from a [^{32}P]-modified phosphoprotein [29]. The second approach utilizes mass spectrometry to identify isolated phosphopeptides from the protein of interest. The first approach requires the use of cell culture to incorporate [^{32}P]orthophosphate into phosphoproteins. Cells are cultured in medium without phosphate and [^{32}P]orthophosphate is added for a

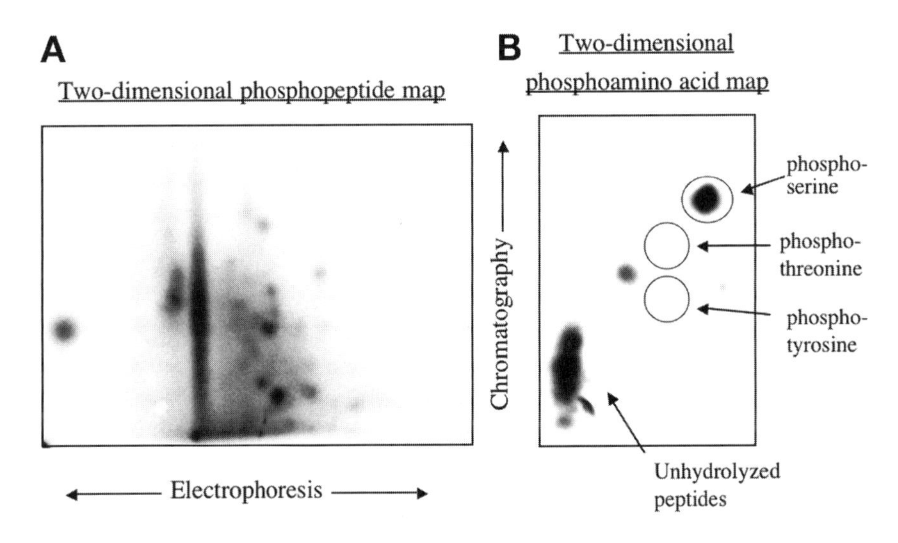

Fig. 5-2. Examples of 2-D mapping of phosphopeptides or phosphoamino acids. Cells were cultured in the presence of [^{32}P]orthophosphate, harvested, lyzed, and then followed by immunoprecipitation of the protein of interest. Protein is resolved by SDS-PAGE and subsequently transferred to nitrocellulose membrane. **(A)** The isolated protein on the membrane is digested with trypsin and the peptide are spotted on a thin-layer plate and subjected to electrophoresis. The plate is then dried, turned 90°, and subjected to standard thin-layer chromatography. **(B)** The isolated protein bound to membrane can also be subjected to acid hydrolysis to yield an amino acid mixture, which can then be resolved by 2-D electrophoresis on thin-layer plates.

short labeling period. Cells are isolated, lyzed, and the protein of interest is immunoprecipitated under stringent incubation and washing conditions. The immunoprecipitation is subjected to SDS-PAGE and the proteins transferred to nitrocellulose membrane. Often the first step in determining the sites that are phosphorylated involves 2-D phosphoamino acid analysis. Briefly, this technique subjects the protein to acid hydrolysis, and the amino acids are then analyzed by application to a thin-layer cellulose plate. The three major phosphopeptides (serine, threonine, and tyrosine) are separated by electrophoresis in the first dimension, followed by ascending thin layer chromatography as depicted in Figure 5-2.

An overall scheme for phosphopeptide mapping is shown in Figure 5-3, however, there are many

29. Boyle, W.J. et al. Methods Enzymol. 201 (1991) 110–149.

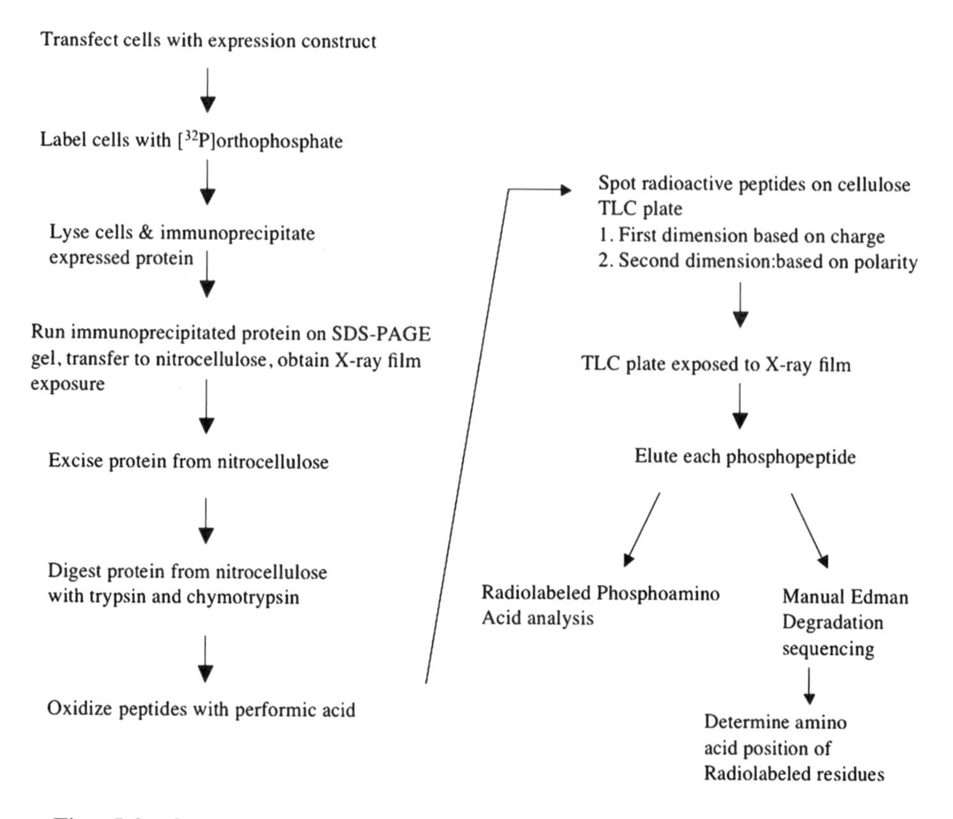

Fig. 5-3. Overall scheme to identify specific phosphorylation sites utilizing [^{32}P]orthophosphate incorporation into cultured cells.

variations in this scheme that have been successfully used. The first step of phosphopeptide analysis involves the excision of the radioactive protein bound to nitrocellulose and digestion with trypsin or other protease. The peptides are then isolated, oxidized, and applied to a cellulose thin-layer plate. The plate is subjected to electrophoresis and the charged peptides move towards the positive or negative electrode. After electrophoresis the plate is dried, turned 90°, and placed in a thin-layer chromatography tank. After solvent reaches the top of the plate, the plate is dried, and the phosphopeptides are visualized by autoradiography using sensitive imaging film. An example of an actual phosphopeptide map is shown in Figure 5-2. If there is sufficient radioactivity the phosphopeptide can be eluted from the plate matrix and manual Edman degradation sequencing can be performed. The radioactive amino acid position from the end terminus can be assessed by the cycle of Edman degradation that releases a radioactive amino acid. A list of possible peptides formed after trypsin digest can be generated using a computer program and peptides containing the amino acid residue previously determined by amino acid analysis to be phosphorylated (e.g., serine). From this list, peptides that have, for example, a serine in the amino acid residue position determined by sequencing would allow tentative identification of the peptides that may be phosphorylated.

An example of the successful application of this technique can be found in the study by Levine et. al., in which a phosphorylation site on the transcription factor ARNT was determined [30]. Conformation of the site(s) mapped can be assessed by using site-directed mutagenesis of the coding sequence to change the putative phosphorylation site to an alanine codon. This mutated mammalian expression vector can then be transfected in cells with high transfection efficiency (e.g., COS 1 and 293T cells) followed by incubation of transfected cells with [^{32}P]orthophosphate. The [^{32}P]protein of interest would then be immunoprecipitated and analyzed as outlined above. Phosphopeptide maps are generated for wild-type versus mutant proteins, and the loss of a phosphopeptide spot(s) on the mutant map suggests that a phosphorylation site has been mutated. Yet another approach to prove that a specific serine residue is phosphorylated employs site-directed mutagenesis to change the serine codon to a threonine codon. This mutant is subjected to both phosphopeptide analysis and phosphoamino acid analysis to detect phosphothreonine. If phosphothreonine is detected this approach can further confirm the identity of a given phosphorylation site. However, not every kinase will recognize a serine to threonine change within the context of a given phosphorylation site. Further confirmation of specific phosphorylation sites can be assessed by mass spectrometry and/or phosphopeptide antibody use. Both approaches will be described later in this chapter.

30. Levine, S.L. and Perdew, G.H. Mol. Pharmacol. 59 (2001) 557–566.

Site-directed mutagenesis must be used to demonstrate that a specific amino acid is phosphorylated.

Another method of [^{32}P]phosphopeptide analysis involves subjecting the phosphopeptides to high-performance liquid chromatography (HPLC) using an analytical C-18 reverse-phase column. [^{32}P]phosphopeptides that resolve as single peptide peaks can be analyzed by mass spectrometry to determine the size of the peptide, or, if a sufficient amount of the peptide is obtained, the actual amino acid sequence can be obtained using an automatic protein sequencer. Also during this procedure the amino acid residue that is phosphorylated on specific phosphopeptides can be determined by 2-D phosphopeptide analysis, as discussed above. This approach has been successful in mapping phosphorylation sites for the progesterone receptor [31]. Radioactive peptides were identified and purified by HPLC, further characterized by phosphoamino acid analysis, manual Edman degradation, and additional proteolytic cleavage digests. Comparing this information with the theoretical trypsin digest peptide list can then narrow down the possible peptides to consider.

31. Knotts, T.A. et al. J. Biol. Chem. 276 (2001) 8475–8483.

Mass spectrometry-based methodology is becoming the method of choice to map phosphorylation sites.

A distinct approach that may soon become the predominant method for mapping phosphorylation sites is the use of mass spectrometric analysis of phosphopeptides, a method that does not require the use of radioactivity. Initial attempts examined the mass of tryptic peptides by mass spectrometry and compared the result to a computer-generated theoretical list of tryptic peptides. An additional list is also generated with the mass of these peptide-containing phosphate groups added to a serine or threonine residue. If the appropriate phosphopeptide masses are obtained, tanden mass spectrometry (MS–MS) fragmentation analysis is performed to confirm the identity of the putative phosphopeptide. An additional means to confirm the identity of phosphopeptides is through digestion of tryptic peptides with a protein phosphatase (e.g., alkaline or potato acid phosphatase) compared with a control sample by mass spectrometry. The disappearance of

Only phosphorylation sites with a relatively high level of modification can be effectively mapped by direct mass spectrometry approaches.

a parent ion for a given phosphopeptide and the presence of the nonphosphorylated form of the peptide would tentatively identify a phosphorylation site. One particular problem with this method is that in practice only phosphorylation sites of fairly high stoichiometry will be observed, and this is often not the case for many phosphorylation sites. To circumvent this problem the phosphorylated peptides need to be isolated from the total mixture of peptides after protease digestion. Recently, a method to isolate phosphopeptide has been described which utilizes Gallium (III) affinity chromatography [32]. Previously, immobilized Fe(III) was demonstrated to bind to phosphopeptides, and this was followed by the recognition that other transition metals can bind phosphopeptides. A comparison of various transition metals has revealed that Ga(III) exhibited the most selective properties in binding phosphopeptides under acidic conditions. In addition, phosphopeptides can be efficiently eluted from Ga(III) affinity resin under basic conditions, desalted, and subjected to MALDI–TOF mass analysis. However, depending on the protein, acidic peptides can bind to the Ga(III) affinity resin, leading to the presence of ions in the spectra that the investigator must differentiate from the phosphopeptides. Phosphopeptide-isolation kits are now commercially available for use prior to mass spectrometry analysis. One particularly promising method to block the binding of acidic peptides is to methylate, or otherwise modify, carboxylic acid groups, which will allow only phosphopeptides to bind to Ga(III). However, these methods are still under development. Another step that can be incorporated after metal ion isolation of phosphopeptides involves splitting the sample into two aliquots and digesting one with alkaline phosphatase. This can be accomplished by using 96-well plates with streptavidin bound to the plastic and biotinylated alkaline phosphatase immobilized on the plate. This method of digestion of the phosphopeptide should minimize contamination of the sam-

32. Posewitz, M.C. and Tempst, P. Anal. Chem. 71 (1999) 2883–2892

The use of iron or gallium affinity chromatography allows detection of phosphorylated peptides even if a peptide exhibits a low level of phosphorylation.

ple prior to mass spectrometric analysis. In the near future these methods will allow efficient routine analysis of the phosphorylated sites on a given protein, although this technique generally requires the investigator to interact with a mass spectrometry facility that has a MALDI-TOF system optimized for peptide analysis. Instrumentation designed for this type of analysis can potentially obtain femtomole sensitivity, which can be critical for investigating low-abundance proteins, such as many transcription factors.

A combination of the two basic approaches just described can also be used to efficiently map phosphorylation sites. First, [^{32}P]phosphopeptides are mixed with unlabeled phosphopeptides generated from an isolated trpysinized protein and are then subjected to 2-D phosphopeptide mapping. The specific radioactive phosphopeptide spots are scraped from the plate, the phosphopeptides eluted, and then subjected to mass spectrometry analysis. The mass spectrometry data is compared with the computer-generated theoretical trypsin digest peptide list.

2.1.1.1. Use of Computer Programs to Identify Putative Phosphorylation Sites

A number of molecular biology programs and websites can predict possible phosphorylation sites within a protein by specific kinases. However, they only search for highly characterized protein kinase sites for which a specific motif has been identified. Table 5-1 lists many of the more-characterized motifs that are phosphorylated by protein kinases for which a consensus motif could be derived by aligning a number of experimentally determined protein phosphorylation sites.

While the information in Table 5-1 can be useful, whether these sites are actually phosphorylated can only be experimentally determined. This is probably attributable to the fact that the overall three-dimensional conformation of these potential kinase sites is an important aspect of whether a given protein kinase can recognize these sites. This information can be used to mutate specific amino acid residues to determine whether a possible phosphorylation site is important to the function of a protein being studied. Ultimately, however, this type of approach will still require demonstration that a given site under study is actually phosphorylated within tissue or cells.

2.1.1.2. Use of Site-Directed Mutagenesis to Characterize the Functional Significance of Specific Phosphorylation Sites

After the specific phosphorylation sites have been mapped, often the next question is whether or not these sites affect a protein's function. Using site-directed mutagenesis, each amino acid that can be phosphorylated is changed into either an alanine or glutamic acid (or aspartic acid) residue. Mutation to an

Table 5-1
Summary of Consensus Protein Kinase Phosphorylation Site Motifs

Kinase	Consensus sequence[a]
cAMP-dependent protein kinase	RSX
	RRXS
	RXXS
	RKXS
Casein kinase I (CKI)	SpXXS/T
Casein kinase II (CKII)	S/TXXE/D
Glycogen synthase kinase 3 (GSK-3)	SXXXS(p)
p34cdc2	S/TPXR/K
Calmodulin-dependent protein kinase II (CaMKII)	RXXS/T
	RXXS/TV
cGMP-dependent protein kinase	R/KXS/T
	R/KXXS/T
Mitogen-activated protein kinase	PXS/T
	XXS/TP
p70s6 kinase	K/RXRXXS/T
Phosphorylase kinase	K/RXXSV/I
Protein kinase A	RXS/T
	RXXS/T
Protein kinase C	S/TXR/K
Protein kinase G	$(R/K)_{2\text{-}3}$XS/T

[a]The bold amino acid residues are the potentially phosphorylated residue.

alanine residue should mimic a nonphosphorylated serine or threonine, while a glutamic or aspartic acid residue mimics a phosphoamino acid, although this latter substitution does not always mimic phosphorylation. Expression of these constructs in cell culture transfection experiments will allow the characterization of a protein as having the potential to be phosphorylated, having no potential for phosphorylation, or mimicking phosphorylation by a glutamic acid substitution. It is important to keep in mind that the wild-type protein may exhibit low stoichiometry of phosphorylation at the site of interest, which could lead to less of a change in activity with a phosphorylation site mutant. In order to circumvent this problem, activation of the protein kinase that phosphorylates this residue could lead to higher levels of phosphorylation and thus to a better assessment of the importance of the site being studied.

If the factor being studied is a TF, or a protein that modulates a transcriptional event, then it can be effectively characterized using cell culture transient transfection assays with a reporter construct. Other properties of the protein phosphorylation site mutant that can also be readily studied include subcellular localization, protein half-life, and the ability to interact with other proteins. Once the functional importance of a phosphorylation site is established, the site can be further studied to determine when and in what cell type this phosphorylation event occurs and which protein kinase is capable of modifying this site.

2.1.1.3. Phosphopeptide-Specific Antibodies

Phosphopeptide-specific antibodies allow the assessment of protein phosphorylation events.

After specific phosphorylation sites of functional importance are identified, often investigators are interested in asking how these sites are regulated within a tissue or a cell. For example, are specific phosphorylation sites regulated during the cell cycle or by exposure to growth factors? Unfortunately, there are only a few approaches available to address these questions. Cells could be incubated with [^{32}P]orthophosphate, undergo the treatment of interest, and then the protein could be isolated and a phosphopeptide map generated. Alterations in the relative amount of a specific radiolabeled phosphopeptide can then be observed. However, this method is only semiquantitative and can only be accomplished in cell culture. Clearly, the preferred method is the production of highly specific antiphosphopeptide antibody [33]. Production of an antibody against a specific phosphorylation site is usually performed by immunizing rabbits with a phosphopeptide. After several injections of phosphopeptide conjugated to an antigenic protein in adjuvant, the rabbits are bled, and antibodies are partially purified by ammonium sulfate precipitation to remove the majority of the other serum proteins. The antibody preparation is then subjected to affinity chromatography using the phosphopeptide immobilized on an activated gel (e.g., Affi-Gel®). Antibody is eluted with a low-pH buffer, neutralized, and passed through a peptide column containing the same amino acid sequence but lacking the phosphoamino acid residue. The non-absorbed antibody is tested for its ability to bind only to the phosphorylated

33. Sun, T. et al. Biopolymers 60 (2001) 61–75.

form of the protein of interest. To ensure the antibody will only react with the phosphorylated epitope, antibody preparations can be mixed with nonphosphorylated peptide prior to use. With this reagent, a detailed analysis of a specific phosphorylation site(s) in a protein can be studied in tissue extracts or cell culture experiments. Thus, the temporal relationship of multiple phosphorylation sites for a given protein can be studied after the production of antibodies against a series of phosphorylation sites. This can be accomplished using either a Western blotting approach or a bead-based multiplex assay system. An ever-increasing number of phosphopeptide-specific antibodies against a wide array of proteins are becoming commercially available.

Use of phosphopeptide antibodies is an effective technique to study phosphorylation events that occur in vivo.

2.1.1.4. Identification of Protein Kinases That Phosphorylate Specific Phosphorylation Sites

After mapping phosphorylation sites and determining that the established site(s) are functionally significant, often the next step is to determine which protein kinases are responsible for the direct phosphorylation of these sites. This can be accomplished through the use of either an in vitro protein kinase or in-gel protein kinase assay. In vitro kinase assays use either a purified bacterial protein or a synthetic peptide. Peptides can be designed to contain amino acid residues flanking the potentially phosphorylated amino acid, as a substrate for a kinase assay. When using a bacterial protein as a substrate, the amino acid residue of interest is mutated and the mutated protein expressed to serve as a reference to ensure that any observed phosphorylation results from phosphorylation of the amino acid residue under study. These proteins are usually produced in a bacterial or baculovirus expression system. In the in vitro kinase assay the purified kinases are incubated with the substrate protein or peptide along with [^{32}P]ATP, and samples are analyzed by SDS-PAGE or a filter binding step.

In vitro or in-gel kinase assays can effectively determine the protein kinases responsible for phosphorylating a specific site.

The in-gel protein kinase assays utilize a peptide or protein crosslinked into the matrix of a polyacrylamide gel. This technique works particularly well with synthetic peptides, and phosphopeptides can also be synthesized as important controls in these experiments [34,35]. Cell or tissue extracts are subjected to SDS-PAGE through the peptide/polyacrylamide gels. After electrophoresis the proteins in the gel are fully denatured and then renatured through a series of washes. After renaturation, the gel is treated essentially like an in vitro kinase reaction and incubated with buffered [^{32}P]-ATP. After fixing the gel and extensive washing, the gel is dried and the presence of radioactive protein bands assessed. The presence of a kinase(s) that can phosphorylate a peptide is seen as specific-molecular-weight bands. A schematic representation resulting from an in-gel protein kinase gel is shown in Figure 5-4.

34. Kameshita, I. and Fujisawa, H. Anal. Biochem. 237 (1996) 198–203.

35. Kameshita, I. et al. J. Biochem. 126 (1999) 991–995.

One strength of in-gel kinase assays is the ability to test for how many protein kinases may be responsible for phosphorylation of a specific amino acid sequence. However this could be limited by the ability of various protein kinases to renature and exhibit activity under the experimental conditions employed. An actual example of the use of both techniques can be seen in the work of Dull et al., in which they demonstrated that casein kinase II (CKII) is the primary kinase that phosphorylates serine residue 43 in the protein XAP2 [36].

36. Dull A.B. et al. Arch. Biochem. Biophys. 406 (2002) 209–221.

2.2. O-Glycosylation

The transfer of β-N-acetylglucosamine from UDP-GlcNAc to serine or threonine residues occurs with a variety of cellular proteins, including transcription factors (e.g., c-myc, SP1). Methodology has been developed to identify proteins that are modified by O-GlcNAc residues, to determine which specific amino acid residues are modified, and to examine the functional consequence of the glycosylation event. There are several approaches to demonstrating that a protein is glycosylated with O-linked GlcNAc. Historically, the most common method utilizes a lectin such as wheat germ agglu-

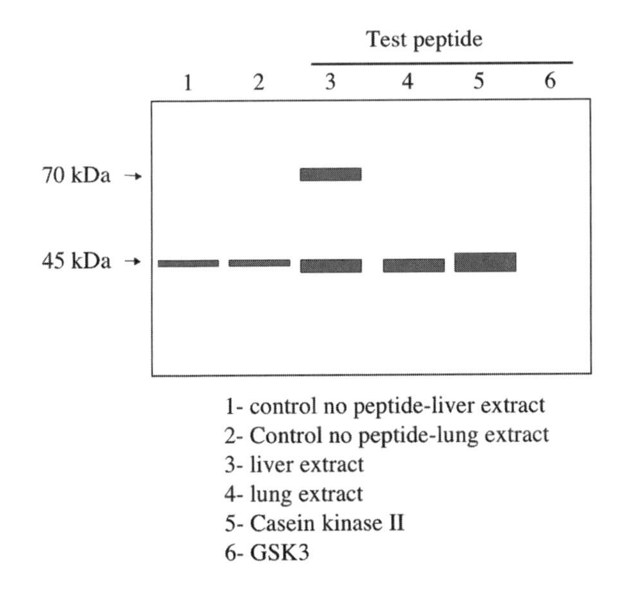

Fig. 5-4. In-gel kinase assay. A schematic representation of a typical result that can be obtained using in-gel kinase assays. The data presented suggests that the test peptide is phosphorylated by casein kinase II in both liver and lung extracts. Note that a low level of autophosphorylation of casein kinase can be seen in lane 1 and 2. An unknown kinase, detected in lung extracts, with a molecular weight of 70 kDa can phosphorylate the test peptide and also does not appear to undergo autophosphorylation.

tinin (WGA), which binds to GlcNAc residues. This lectin can be utilized in several different approaches, including protein–protein interaction mobility-shift assay (PIMSA), WGA-agarose chromatography, or WGA-agarose pull-down assays [37]. A second distinct approach is the use of β-1,4-galactosyltranferase to transfer [^{3}H]galactose to an O-GlcNAc residue in the partially denatured protein under study. This is followed by chromatography on RCA lectin–agarose and demonstration that the protein of interest is capable of binding. More recently, a monoclonal antibody has become commercially available that can bind to O-linked GlcNAc residues in many proteins. This antibody and standard protein blotting techniques could be used to initially assess whether a protein under study carries this type of modification. However, whether this antibody will recognize all or most GlcNAc-modified proteins remains to be assessed.

37. Chou, T-Y., et al. Proc. Natl. Acad. Sci. USA 92 (1995) 4417–4421.

Once established that a protein is covalently modified by O-linked GlcNAc residues, usually the next step is mapping of the actual amino acid residues that are modified. The modified domain of a protein can be mapped by expression of fragments of that protein in a cell line and determining which region of the protein is capable of either binding to WGA or being modified by β-1,4-galactosyltranferase. A site-directed mutagenesis approach, followed by expression experiments in cells, can then be used to further determine the amino acid residues that are modified. Another approach used to map specific glycosylation sites is proteolytic digestion of a [^{3}H]galactosidase-labeled protein, followed by isolation of radiolabeled peptide and mass spectrometric analysis. This approach is similar to the methodology utilized to map phosphorylation sites discussed earlier.

Determining the functional consequence of glycosylation of a particular protein can be a difficult goal, especially if a mapped site of glycosylation is also capable of being phosphorylated. Also, there is no way to easily mimic a glycosylated residue when expressing a protein in cell culture experiments, in contrast to the characterization of phosphorylated residues (e.g., replace a serine with a glutamic acid residue). However, the development of O-GlcNAcase inhibitors such as O-(2-acetamido-2-deoxy-D-glucopyranosylidene (PUGNAc) should allow examination of how enhanced glycosylation levels in a cell affects function of a protein relative to a mutant form that can no longer be glycosylated [38]. Clearly, additional experimental approaches are needed to examine the influence of O-GlcNAc modification on protein function.

Determining the functional role of a specific O-GlcNAc modification is complicated by the fact that often these sites are also phosphorylated.

38. Haltiwanger, R.S. et al. J. Biol. Chem. 273 (1998) 3611–3617.

2.3. Ubiquitination

Assessment of whether a specific protein is a substrate for the ubiquitin–proteasomal degradation pathway can be divided into two basic approaches.

The first utilizes a cell culture-based protocol and the second uses an in vitro reconstitution system.

2.3.1. Determination That a Specific Protein Is Proteolyzed Through the Ubiquitin–Proteasomal Degradation Pathway

This goal is most often initially addressed by demonstrating that a protein can be ubiquitinated. Polyubiquitinated proteins in lysates from cultured cells or tissues are usually not detected, because the proteins are rapidly degraded by proteasomes. Thus, in order to study whether a protein of interest is ubiquitinated, it is necessary to treat cells with proteosome inhibitors to allow ubiquitinated proteins to accumulate. This means it can be difficult to determine whether a specific protein is degraded in vivo through the ubiquitin–proteasomal degradation pathway. Lactacystin and MG132 are the most commonly used proteasome inhibitors in cell culture experiments. However, lactacystin is probably the best available inhibitor because it is cell permeable, has high potency, and is an irreversible inhibitor of proteasomes. To ensure that proteolysis of ubiquitinated proteins does not occur during the production of lysate, ALLnL, an inexpensive peptide aldehyde proteasome inhibitor, should be included in the homogenizing buffer. In addition, the possible removal of ubiquitin by a de-ubiquitinating enzyme can be blocked with a thiol-blocking reagent such as *N*-ethylmaleimide at 10 m*M*. There are no specific inhibitors of these enzymes that can be used in this application. Usually the protein of interest is immunoprecipitated and the presence of ubiquitin bound to the protein being studied is assessed.

Although polyubiquitinated proteins can be subjected to SDS-PAGE analysis, it is important to note that they can be quite large; thus, their transfer to nitrocellulose membrane may take much longer than normal. It is recommended that these protein transfers be extended for a higher number of volt-hours than that normally utilized. After the protein is transferred to membrane, a number of antibodies available from individual investigators or commercial sources are capable of detecting ubiquitinated proteins on Western blots. However, the ability of some of these antibody preparations to effectively recognize ubiquitin requires that the proteins bound to membrane be fully denatured to reveal the appropriate epitope(s). This can be accomplished by autoclaving the blot or boiling them in water [39]. Alternatively, the proteins on the blot can also be denatured by treatment with 6 *M* guanidine HCl. Upon visualization of the ubiquitinated protein, a smear up to the high-molecular-weight range is often seen. In theory, one might expect to see a ladder representing mono-, di- and tri-ubiquitinated species, but, because the target protein can be both polyubiquitinated on one site and ubiquitinated on multiple lysine residues, this leads to the appearance of a smear of the protein of interest.

39. Mimnaugh, E.G. et al. Electrophoresis 20 (1999) 418–428.

2.3.2. Use of Cell-Transfection Experiments to Determine If a Protein Can Be Degraded by the Ubiquitin–Proteasomal Degradation Pathway

Coexpression of HA-tagged ubiquitin is an effective means to test the ability of a protein to be ubiquitinated.

An effective approach to determining whether a specific protein is ubiquitinated is the co-transfection of cells (e.g., COS-1, 293T) with an expression vector expressing the protein being studied and an HA or His_6-tagged ubiquitin expression construct. Cells are treated with a proteasome inhibitor to allow the accumulation of polyubiquitinated proteins. Ubiquitinated proteins can then be isolated from cell lysates with the appropriate affinity resin, subjected to SDS-PAGE, and transferred to a membrane. Then the blot can be probed for the presence of the protein of interest. Also, the specific protein being studied can be immunoprecipitated and, after transfer of SDS-PAGE-resolved protein to membrane, the presence of ubiquitin can be assessed with antibodies against the tag sequence. All of the methods discussed so far rely on the use of proteasome inhibitors, which can have a variety of indirect effects on the protein or pathway being studied. Another approach involves the use of a ubiquitin construct that is mutated to code for arginine at lysine residues that are normally sites for polyubiquitin formation. Coexpression of this construct with a construct that expresses the protein of interest can be utilized to test whether the protein being studied is mono-ubuquitinated [40]. The mono-ubiquitinated protein should be relatively stable and easily detected on a protein blot.

40. Li, M. et al. Science 302 (2003) 1972–1975.

2.3.3. In Vitro Reconstitution System Used to Examine Proteins That Mediate Ubiquitination of a Substrate Protein

After it has been established that a protein's degradation is regulated by the ubiquitin–proteasome system, the question often asked is which E2 and E3 ligases are involved in mediating ubiqitination. An in vitro ubiquitylation assay is composed of

E1, E2, and E3 ligases, ubiquitin, ATP, $MgCl_2$, and a substrate protein. Purified ubiquitin and E1 and E2 ubiquitin ligases are commercially available, while the E3 ligases generally are purified in individual laboratories after overexpression in *E. coli* or in a baculovirus expression system as fusion proteins. Ubiquitin, ubiquitin-HA, ubiquitin-His_6, or ubiquitin-GST are used, with the latter two peptides containing a tag that can be easily used to isolate ubiquitinated proteins on affinity resin. An interesting example is the identification of C-terminal hsp-interacting protein (CHIP) as an E3 ligase that mediates degradation of proteins through their mutual interaction with hsp90 [41].

41. Demand, J. et al. Curr. Biol. 11 (2001) 1569–1577.

2.3.4. Mapping of Ubiquitination Sites on a Protein

Often investigators are interested in mapping the specific lysine residue(s) that are the key site(s) of ubiquitination. In addition, whether ubiquitination modulates the function of a specific domain, such as the transactivation domain is often examined. Usually the first step of such an investigation would be to perform a deletional analysis to determine the domain that is responsible for proteolytic turnover. This can be best accomplished by determining the half-life of a series of truncations of the protein being studied. For determination of a protein's half-life see Chapter 2 in this Part II. These studies can then be followed by mutagenesis of specific lysine residues to alanine and determining their effect on protein half-life. An example is the mapping of a single lysine residue that targets MyoD for rapid degradation [42]. In general, the mapping of the lysine residues that are the key targets for ubiquitination is difficult, because there is no motif that has been identified and often more than one lysine in a protein may be modified. This can result in a difficult analysis if two or more lysine residues together represent the key targets of ubiquitination. An additional potential complication in interpreting results

42. Batonnet, S. et al. J. Biol. Chem. 279 (2004) 5413–5420.

43. Gronroos, E. et al. Mol. Cell. 10 (2002) 483–493.

is that the lysine residues can be a targeted for both acetylation and ubiquitination, as has been demonstrated for the transcription factor Smad7 [43].

2.4. Acetylation

Acetylation of histones and many transcription factors is an important posttranslational modification that is often studied in the context of gene transcription. The methods used to characterize the role that acetylation plays in mediating transcription factor activity are similar to methods described for other posttranslational modifications. Initial studies focus on establishing that the protein being studied is actually acetylated. This can be accomplished in cell extracts by immunoprecipitating a protein, followed by SDS-PAGE and transfer of protein to membrane. The blot is then probed with an anti-acetylated lysine antibody and visualized using standard techniques. Similarly, the anti-acetylated lysine antibody can be used to immunoprecipitate protein from a cellular extract and the presence of a protein assessed using immunochemical techniques. To increase the level of protein acetylation, cells are often treated with such histone deacetylase inhibitors as trichostatin A or sodium butyrate prior to the isolation of cellular extracts. Another approach that may be used to demonstrate acetylation of a specific protein involves incubating the cells with [^{3}H]acetate, immunoprecipitating the protein of interest, and detecting the protein by autoradiography after SDS-PAGE. Collectively, these methods can establish whether a protein is acetylated.

An in vitro assay is commonly employed to demonstrate that a given protein is usually acetylated directly by a specific histone acetyl transferase. A purified bacterially expressed fusion protein of interest can be incubated with a purified histone acetyl transferase, such as p300 or pCAF, in the presence of [C^{14}]acetyl CoA. Often the next step in the characterization of protein acetylation is mapping the actual sites of acetylation, and this can be accomplished effectively by site-directed mutagenesis/deletional analysis and incorporation of [^{3}H]acetyl in the test protein. This is done by transient expression of mutant forms of the test protein in cell culture, coupled with incubation of cells with [^{3}H]acetate and subsequent analysis of the isolated protein. A second approach would utilize an in vitro assay with a histone acetyl transferase to test the mutant protein's ability to be acetylated. Site-directed mutagenesis can be used to map the exact residue that is modified. A complementary approach would utilize a synthetic peptide corresponding to the sequence of interest and a series of peptides containing a mutation(s) of lysine residue(s) to glutamine. After specific sites are mapped, point mutations can be introduced to change the acetylated lysine to a glutamine residue, or another amino acid, in the test protein mammalian expression protein vector. These constructs would be tested in an activity assay. For example, Fu and colleagues mapped an acetylated residue(s) in the androgen receptor to a motif rich in lysine residues, similar to previously characterized sequences found in other acetylated proteins (e.g., p53, ACTR). Using cell culture-based reporter assays, they determined that transcriptional activity of the androgen receptor was enhanced by blocking acetylation [44]. Additional studies also demonstrated that the probable mechanism for this observation was that the deacetylated form of the androgen receptor bound a corepressor and the acetylated receptor bound the coactivator p300.

44. Fu, M. et al. Mol Cell. Biol. 23 (2003) 8563–8575.

3. Summary

The mammalian cell has evolved a sophisticated series of posttranslational protein modifications to regulate TF function. The basic reason for this mode of regulation is the speed at which changes in TF activity can occur within the cell. For example, the half-life of a phosphorylation site on a TF is generally quite short, usually in the range of 5–10 min. The cell invests a considerable amount of energy to maintain this dynamic equilibrium between the activity of protein kinases and phosphatase. This energy expenditure allows rapid regulatory changes in protein activity to take place through changes in the activity of protein kinases/phosphatases. Rapid changes in protein kinase activities usually occur through phosphorylation cascades. The ability to markedly change a TF's activity in 5–10 min within the cell generally could not be achieved at the level of transcription because of the relatively long protein half-life, usually measured in hours. Thus, protein posttranslational modifications allow rapid changes in cellular gene expression to occur without changing the actual level of various regulatory proteins in the cell.

Potential for competition between posttranslational modifications is another important aspect of cellular regulation of protein activity. For example, there can be direct competition between phosphorylation and O-glycosylation for certain serine residues, or competition between acetylation and ubiquitination for specific lysine residues. In addition, there is the potential for one posttranslational modification to sterically hinder the addition of another. These observations would suggest that the development of a complete picture of posttranslational modifications that regulate TF function could be quite complex. Clearly, development of efficient methodologies that allow the identification of several types of modifications will be necessary to fully delineate the role of posttranslational modifications in the regulation of the activity of a specific factor.

Previous chapters have examined the approaches available for studying regulation of protein function through protein–protein interaction and posttranslational modification. The next chapter will detail the molecular and cellular approaches available for examining the role of a given protein in signaling pathways and how it may contribute to overall cellular homeostasis.

6

Dissection of Signaling Pathways

1. Concepts

Basic biological scientists are often concerned with understanding the cascade of events that lead to specific biological end points, such as how cells are driven to enter cell division and the various proteins that control this process, or how the response to hypoxic conditions is regulated. Often this involves examining the ability of proteins to undergo multiple protein–protein interactions and subsequently determining which specific complexation events are key to a given cellular response. One of the most utilized approaches to dissect the key protein–protein interaction events utilizes the introduction of point mutation(s) into a cDNA that can express a mutated protein, this can lead to the disruption of a specific protein–protein interaction. An example can be found in the studies by Lamarche et al., who examined the ability of Rac and cdc42 mutants to interact with either Ser/Thr kinase $p65^{PAK}$ or Ser/Thr kinase $p160^{ROCK}$ [45]. By introducing mutations in Rac or cdc42 that disrupted binding of a specific kinase, they were able to demonstrate which kinase cascades have to be activated to obtain a biological response such as G1 cell cycle progression.

The use of expressed proteins containing point mutations to dissect signaling pathways is an effective and often utilized approach.

45. Lamarche, N. et al. Cell 87 (1996) 519–529.

Polypeptide sequences that contain a protein–protein interaction motif can be introduced into cells to disrupt a specific protein–protein interaction. Several methods have been used to introduce these peptide sequences, including direct microin-

Regulation of Protein Levels and Transcription Factor Function by G. H. Perdew
From: *Regulation of Gene Expression*
By: G. H. Perdew et al. © Humana Press Inc., Totowa, NJ

jection into cells, adjoining the peptide sequence to a fusion protein using standard DNA construct techniques and expressing them in cells, or production of peptides that include both the motif of interest and a penetrating peptide sequence. Microinjection of peptides is perhaps the most difficult technique technically and requires expensive equipment as well as considerable experience. Nevertheless, this is a powerful approach to disrupt specific protein–protein interactions within a cell, thus allowing an assessment of the functional significance of that interaction.

Perhaps one of the most important approaches to determine the role of a specific protein in a signaling pathway is through disruption of its expression. However, it is important to keep in mind that a lack of expression of a given protein can cause unintended effects that can complicate the results. For example, disruption of SRC-1 expression can be used to investigate whether or not it participates directly in coactivating a specific transcription factor. However, considering that SRC-1 can exist in oligomeric coactivator complexes may lead to the lack of other proteins being present in these large complexes. Thus, one needs to be careful when interpreting the results obtained in this type of experiment. Disruption of expression of a specific protein can be accomplished both in vivo and in cultured cells by either inducing mRNA degradation or blocking translation. In this chapter we will explore the methods available to disrupt protein–protein interactions or expression of a specific protein in cultured cell lines, or repress protein function, as well as discuss systems for regulating ectopic expression of protein. The use of ectopic expression, while not a true approach to dissect signaling, is nevertheless an extremely common means to characterize the functional role of a protein.

2. Methods and Approaches

2.1. Disruption of Protein–Protein Interactions

2.1.1. Use of Site-Directed Mutagenesis to Disrupt Specific Interactions

Perhaps the most utilized approach to dissect signaling is the generation of a point mutant by site-directed mutagenesis. After specific protein–protein interacting domain(s) have been mapped, more detailed mapping of critical contact residues can be accomplished using alanine-screening mutagenesis. This entails mutation of one or two amino acid residues through a specific sequence until each amino acid residue has been mutated in a series of constructs. If the original mapping studies have narrowed the critical sequence to a large sequence, then additional mapping should be performed, or the nature of the interaction should be defined. For example, whether or not a given protein–protein interaction results from a hydrophobic or ionic interaction can be defined by performing a series of immunoprecipitations. The immunoprecipitated protein complexes are washed with solutions of increasing salt or detergent concentration. This simple assay could greatly facilitate the selection of residues to test in alanine-screening assays. An example of this is given in Figure 6-1, which illustrates the disruption of the interaction of the Ah receptor and the X-associated protein 2 [46]. The results indicate that detergents can effectively strip the X-associated protein 2 from the Ah receptor, thus suggesting that this protein interaction is mediated through hydrophobic interactions.

46. Hollingshead, B.D. et al. J. Biol. Chem. 279 (2004) 45,652–45,661.

While this system can efficiently generate mutants that fail to bind specific partners, the results obtained could be complicated by unintended misfolding effects attributable to the introduced mutations. This possibility could be addressed at least in part through the use of more conserved residue changes and testing whether they still lead to a disruption of a given protein–protein interaction. While this approach will not prove that a given mutation is not causing a more global folding problem, it can reinforce that this possibility is less likely.

2.1.2. Use of Dominant-Negative Protein Expression

One commonly used approach to block the signaling mediated by a protein is to remove or mutate a domain that is involved in protein–protein interactions. These mutants can still participate in binding to certain proteins through their other domains but would be unable to act as a bridge to other factors via the deleted domain. This can be illustrated by examination of two examples, the first being the deletion of the transactivation domain from a transcription factor, which can then be utilized as a dominant negative protein from a transcriptional standpoint. The mutant transcription factor can bind to its DNA response element but is unable to recruit coactivators and thus blocks transcription. A second example is a kinase that is part of a signaling cas-

An important technical note: an effective method to introduce deletions into constructs is loop-out site-directed mutagenesis using commercially available mutagenesis kits [47].

47. Makarova, O. et al. Biotechniques 29 (2000) 970–972.

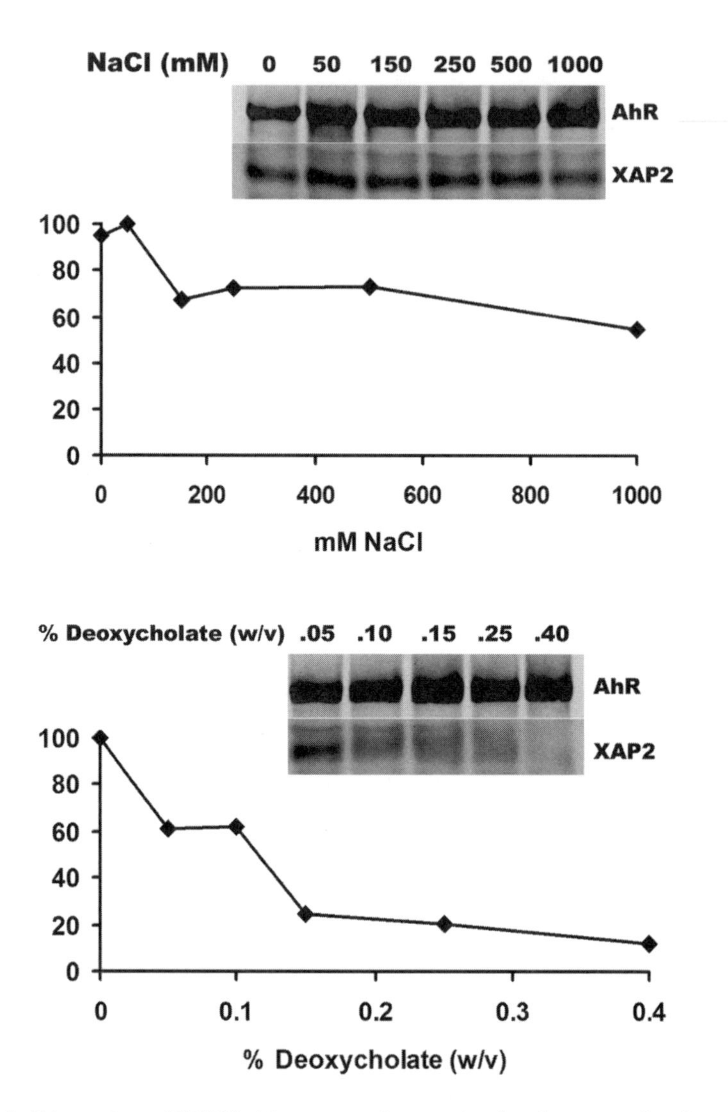

Fig. 6-1. Disruption of XAP2-Ah receptor interaction by detergent. In vitro translation [^{35}S]methionine-labeled Ah receptor and XAP2 were mixed together and incubated on ice for 1 h. Complexes were immunoprecipitated and washed. Complexes were then washed with buffer containing either increasing deoxycholate or NaCl. Samples were subjected to SDS-PAGE and the protein transferred to membrane. The band intensities of AhR and XAP2 were quantified by filmless autoradiographic analysis. (Adapted from [46].)

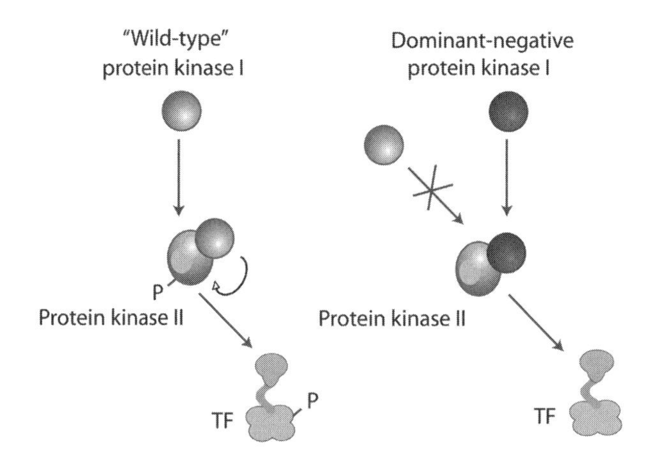

Fig. 6-2. Cartoon depicting the mechanism of dominant-negative protein kinase inhibition of a downstream kinase.

cade. A kinase mutant that has a site-directed mutation in the catalytic domain can bind and be modified by an upstream kinase but is unable to phosphorylate its downstream target proteins, as shown in Figure 6-2.

The stable or transient expression of these dominant negative proteins can be a powerful tool in determining the influence of a signaling cascade on the activity of a factor. For example, a dominant negative mitogen-activated protein (MAP) kinase kinase can be used to repress MAP kinase and test whether this kinase is involved in the regulation of a transcription factor in a transient transfection reporter assay cell culture experiment.

2.1.3. Pharmacologic Disruption of Protein Function

Another popular approach to determine the role of a protein in cellular signaling within a given pathway is the use of small-molecular-weight compounds that are capable of blocking the activity of an enzyme, receptor, or other protein-mediated activity. Literally thousands of compounds have been determined to alter the activity of various proteins, and these compounds can be extremely effective in testing the role of a protein in influencing the activity of proteins downstream in a signaling pathway. Many of these compounds have been characterized as a part of drug discovery involving the screening of libraries of compounds for their ability to block or alter the activity of a protein or pathway. For example, screening assays have been performed to test the ability of a chemical library to inhibit tumor cell growth or inhibit a specific protein kinase. Indeed, pharmaceutical companies are constantly screening for potential therapeutic compounds, such as antiinflammatory chemicals, that block or enhance a biochemical end point in a high-throughput assay system.

A pharmacologic approach allows the efficient screening of a number of already characterized and commercially available inhibitors to a spectrum of signaling molecules (e.g., protein kinases) to be tested in a specific assay system. For example, in order to test the hypothesis that "certain protein kinase(s) are necessary for SP1 to mediate transcriptional activity," a SP1 response element driven reporter vector can be transfected into cells along with the addition of specifically characterized protein kinase inhibitors. This experiment is relatively easy to perform and can yield insight into signaling pathways to examine further. However, this approach should be utilized with caution, often a chemical's ability to influence only one protein is based on limited studies. An example is the compound PD90086, a well-characterized inhibitor of MAP kinase that has been shown to have little effect on a number of other kinases. Studies in the literature have recently shown that this compound is an antagonist for the Ah receptor [48]. This suggests that the level of specificity of a specific inhibitor should be approached with caution and also underscores the importance of approaching a problem using a number of methods. For example, if one wants to modulate MAP kinase activity, this can be accomplished by using more than one type of inhibitor and by transient expression of a dominant negative MAP kinase kinase protein or a MAP kinase phosphatase. If each approach yields a similar answer this would build a case for MAP kinase in the pathway under study. It is important to note that inhibitors of protein activity/function are only available for a limited number of target proteins. Thus, a pharmacologic approach is a powerful tool but should not be used as the sole approach in examining a specific hypothesis.

Any inhibitor may exhibit unintended effects through interaction with proteins not previously identified.

48. Reiners, J.J. et al. Mol. Pharmacol. 53 (1998) 438–445.

2.1.4. Use of Membrane-Permeable Peptides to Deliver Macromolecules Into Cells

Permeable peptides can mediate cargo delivery to virtually every cell in a cell culture experiment.

There are a number of methods to deliver macromolecules to cells, including electroporation, microinjection, antisense, and lipid-mediated protein delivery. Each method has serious drawbacks, such as efficiency of delivery and the percentage of cells that obtain the macromolecule. These delivery systems often are somewhat cytotoxic. In addition, with antisense or the RNAi approaches efficient results are very gene-, species-, and cell line-specific.

A recently developed technique, with a potential for highly efficient delivery of macromolecules, uses a group of peptides or protein transduction domains (PTDs) with the ability to penetrate plasma membranes in an energy-independent manner. The mechanism of transport across the plasma membrane is currently not known and is not dependent on membrane receptors or endocytosis. Interestingly, even at 4°C the permeable peptides are efficiently transferred into the cell. These peptides are able to transport and to deliver DNA or peptide cargo covalently attached to the peptide, and deliver to cells cargo as large as a 130-kDa protein. Also, essentially 100% of cells in culture are capable of transferring the peptide–cargo across the plasma membrane in a concentration-dependent manner. In addition, permeable peptides should efficiently transfer across the plasma membrane of essentially any cell type. Yet another advantage of this method is the ability to design the experiment within a narrow time frame. The three most widely utilized PTDs are from the *Drosophila* homeotic transcription protein antennapedia, the human immunodeficiency virus 1 (HIV-1) transcriptional activator Tat protein, and the herpes simplex virus structural protein VP22 [49]. Short peptide sequences have been identified from two of these proteins that are responsible for their membrane-transversing properties, RQIKIWFQNRRMKWKK, from antennapedia, and YGRKKRRQRRR, found within the Tat protein. All permeable peptides that have been identified to date are highly basic, usually with a large number of arginine residues, which should allow the design of synthetic PTDs in the future.

Permeable peptides can deliver a variety of cargoes.

49. Wadia, J.S. and Dowdy, S.F. Curr. Opin. Biotechnol. 13 (2002) 52–56.

Although the methodology to use permeable peptides to deliver various cargo has been around for a number of years, the technique has seen relatively limited use. This could be attributable to the fact that initial experiments focused on the organic synthesis of the permeable peptide and chemically crosslinking the cargo to this peptide. Although a single large peptide could be synthesized to contain both the permeable peptide sequence and a sequence with a protein binding domain, either of these approaches would be quite expensive. Nevertheless, this approach has been effectively utilized, for example, to block both IKKβ and NFκB essential modulator (NEMO) interaction using a polypeptide containing an antennapedia-permeable peptide sequence and IKKβ NEMO-binding domain within cells, as depicted in Figure 6-3 [50].

50. May, M.J. et al. Science 289 (2000) 1550–1554.

A relatively inexpensive alternative for production and subsequent delivery of proteins into cells is the use of standard molecular biology techniques to produce large amounts of expressed bacterial fusion protein containing the TAT-permeable peptide sequence. This method utilizes a bacterial expression vector arranged to express a His_6-TAT-fusion protein. This vector is transformed into bacteria and grown in culture to allow protein expression. The cells are collected and sonicated in 8 *M* urea and the His-tagged protein is isolated using Ni-chromatography. The isolated protein is further purified on an ion-exchange column, and the urea efficiently removed. The protein is only partly renatured and this mediates more efficient translocation across membranes compared with protein in the properly folded conformation. After translocation the partly denatured proteins are apparently refolded within the cell to the biologically active conformation. This method has been successfully demonstrated for a number of proteins, including $p27^{kip1}$, p16, pRB, and CDK2 dominant-negative [51].

51. Nagahara, H. et al. Nat. Med. 4 (1998) 1449–1452.

Another interesting example is a study in which the investigators proposed to examine the require-

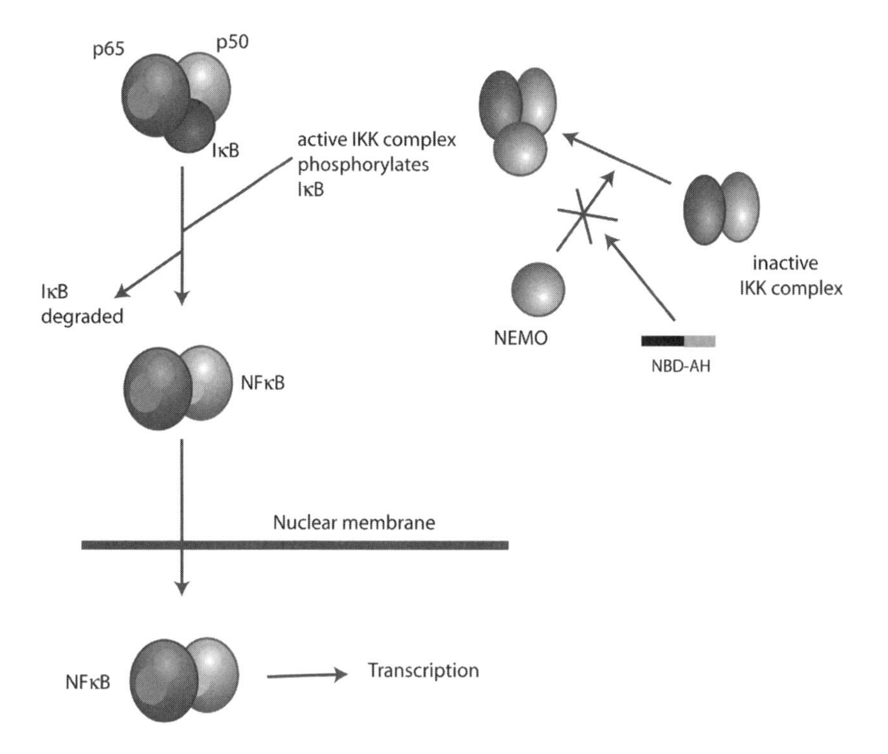

Fig. 6-3. Disruption of NEMO-IKK complex interaction using a permeable peptide containing a NEMO binding domain leads to repression of NFκB mediated transcriptional activity. Treatment of cells with a NEMO binding domain–*antennapedia* homeodomain (NBD–AH) polypeptide leads to sequestration of NEMO preventing its interaction with IKK complexes.

ment of Cdc42 and WASP interaction to mediate SDF-1-induced T-lymphocyte chemotaxis [52]. The investigators produced a His_6-TAT-HA-WASP cdc42/Rac interactive binding (CRIB) domain fusion protein; exposure of CEM cells in culture to this protein led to decreased cell migration. It is important to design a control protein, usually a mutant form of the protein of interest. In this case the CRIB domain was mutated at three key residues involved in WASP/Cdc42 interaction and failed to influence cell migration. In conclusion, the use of protein binding domains linked to permeable peptides can be a powerful approach to disrupt protein–protein interaction in cultured cells.

52. Haddad, E. et al. Blood 97 (2001) 33–38.

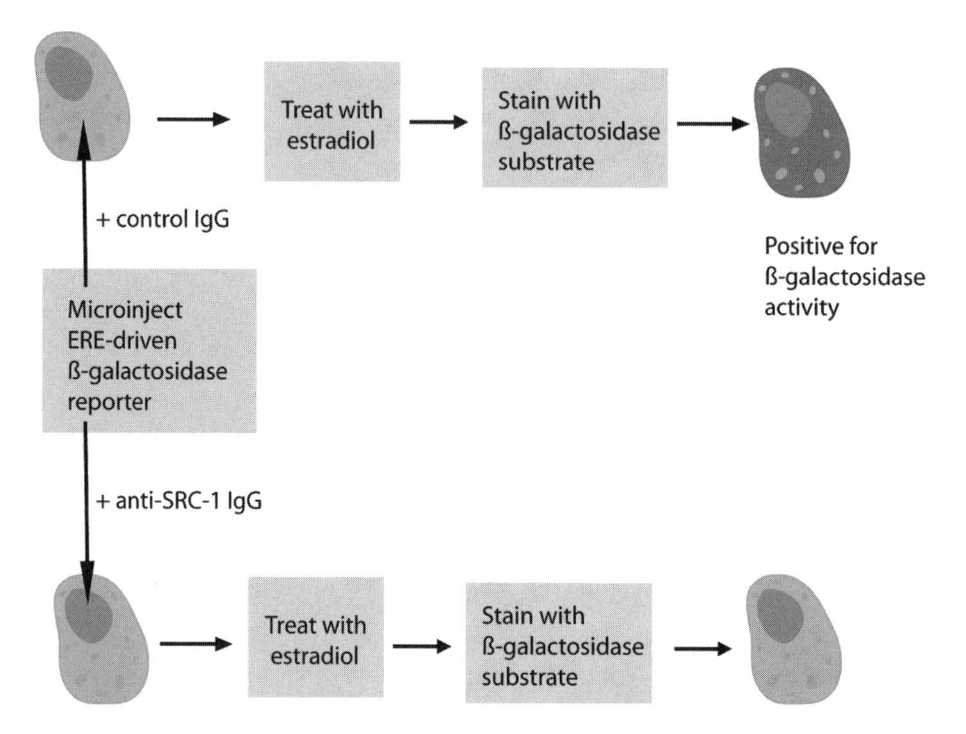

Fig. 6-4. SRC-1 polyclonal antibody blocks liganded estrogen-receptor transcriptional activity.

2.1.5. Microinjection of Antibodies or Peptides Into Cells to Block Protein–Protein Interactions

Microinjection can be a powerful approach to block protein–protein interaction but is technically challenging.

A technically demanding approach to examine the importance of a specific protein–protein interaction within cells is the microinjection of highly specific antibodies directly into cells. This approach requires specialized equipment and a high level of technical expertise. However, this method can be quite useful in addressing specific questions about the role of a protein in the transcriptional activity of a specific gene. A theoretical example is shown in the cartoon in Figure 6-4. The basic experiment involves the microinjection of antibodies along with a reporter vector.

2.1.6. Assessment of the Ability of a Protein to Interact With a Transcriptional Complex Through the Transient Expression of a Fusion Protein Capable of Altering Transcriptional Activity

If one is interested in asking the question does a protein interact with another protein in the context of a transcriptionally active complex, this question can be answered using standard molecular biology techniques to create a fusion protein with transcriptionally repressive or enhancing activity. This can be accomplished by making a fusion protein that has either a histone acetylase or deacetylase activity or a protein that recruits transcriptionally repressive proteins. For example, this approach was used to test whether the glucocorticoid receptor (GR) is capable of binding to Rel A bound to NFκB DNA response elements. To test this possibility they utilized a mammalian construct with the GR and MAD cDNAs fused together [53]. Expression of this fusion protein, if it interacts with a transcription factor bound to its cognate response element, would lead to repression of transcriptional activity. MAD is a transcription factor that heterodimerizes with c-myc and represses transcription. In Figure 6-5 the expression of GR–MAD represses the transcriptional activity of a NFκB response element-driven reporter construct transfected into CV-1 cells, thus demonstrating that, within a cell, the GR binds to NFκB on a promoter sequence.

53. Nissen, R.M. and Yamamoto, K.R. Genes and Dev. 14 (2000) 2314–2329.

2.2. Disruption of Protein Expression

Preventing the expression of a protein is an effective way of determining the actual function of a protein. In the past this has been most effectively accomplished by gene-disruption techniques in mice. However, no similar methodology has been developed that could be utilized in human cells. The two basic techniques that can be used to block expression of a protein both target the stability of a specific mRNA. For a number of years antisense

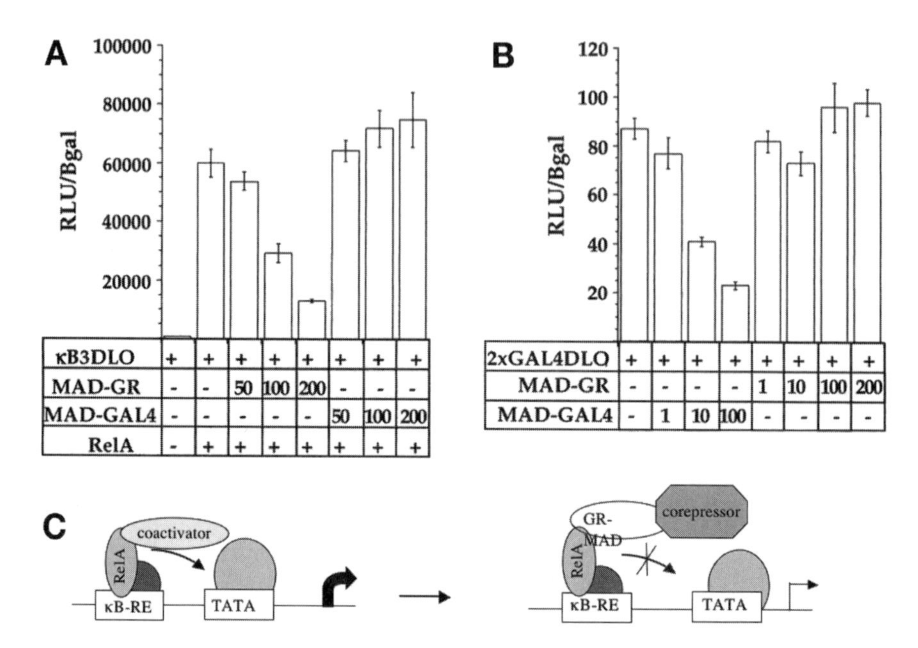

Fig. 6-5. MAD-GR chimera represses RelA transcriptional activity. (**A**) Effect of MAD-GR chimera on RelA activity. CV-1 cells were transiently transfected with a NFκB response element-driven reporter vector alone (κB3DLO), or co-transfected with expression vectors for RelA and increasing amounts of GR-MAD or MAD-GAL4, as indicated. (**B**) Effect of MAD-GAL4 chimera on basal transcription from the GAL4 response element-driven reporter vector (2xGAL4-DLO), or co-transfected with increasing amounts of MAD-GR or MAD-GAL4, as indicated. (**C**) Schematic representation of the influence of MAD-GR on RelA-mediated transcription. (Adapted from [53].)

RNA has been the established technique to either destabilize mRNA or block transcription, and this method, at least initially, appeared to have tremendous potential. However, this methodology has been difficult to establish widely in biological research as there appears to be a number of technical problems impeding its use. In contrast, perhaps one of the most important recent technical advances in biology, is the discovery of RNA interference (RNAi) as a method that can effectively silence expression of specific genes. It is also relatively simple to perform, which should lead to its widespread use, and a number of approaches are available to the research community.

2.2.1. Antisense Oligonucleotides

Antisense oligonucleotides have been extensively studied for the past 25 years and offer great potential to disrupt a variety of diseases, including cancer, viral and bacterial infections, and inflammatory diseases. However, this potential remains to be clinically realized, although one antisense oligonucleotide developed by ISIS Pharmaceuticals named Vitravene has been approved by the FDA to treat CMV retinitis. The mechanism of action of antisense technology is relatively simple; the use of a complementary single-stranded oligonucleotide to a specific mRNA sequence can block translation or lead to degradation of that mRNA. The first antisense oligonucleotides utilized the phosphodiester linkage found in DNA and were relatively short, 13–25 nucleotides. These single-stranded oligonucleotides rapidly degrade in cells owing to intracellular endo- and exonucleases, which limits their usefulness in obtaining an efficient degradation of their target mRNA. This problem led to the development of the first wave of chemically synthesized modified oligonucleotides (e.g., methylphosphonates and phosphorothioates), with phosphorothioate oligonucleotides (PSOs) being the most widely utilized type of antisense used in clinical and research applications. In this class of oligonucleotides, one of the nonbridging oxygen atoms in the phosphodiester bond is replaced by sulfur. PSOs are highly soluble, have excellent antisense activity and are fairly stable within cells. They are efficient at activating RNase H and are readily taken up by cells because they are negatively charged. Numerous reports have demonstrated the successful use of PSOs for research purposes. However, a number of antisense PSOs would need to be designed in order to obtain an efficient down-regulation of a specific mRNA. In addition, the fact that they have a charge can lead to nonspecific effects, such as the potential to bind to a number of heparin-binding molecules like laminin, fibronection, and basic fibroblast growth factor. Thus, PSOs can be designed to successfully down regulate expression of a specific gene, but this can be an expensive undertaking that also requires the careful design of the appropriate controls.

Another type of antisense oligonucleotide analog uses phosphorodiamidate morphino oligomers (PMOs). These oligomers utilize a DNA-like chemistry with a six-membered morpholine ring instead of a deoxyribose sugar, and the charged phosphodiester internucleoside linkage is replaced by an uncharged phosphorodiamidate linkage. The lack of charge appears to reduce nonspecific effects commonly seen with charged oligonucleotides. PMOs are also highly resistant to nucleases and have been demonstrated to sterically block translation, especially when designed to bind to translational start sites. This appears to be their primary mechanism of action as they fail to recruit RNase to a PMO–mRNA complex. This is in contrast to the mechanism of action of charged oligonucleotides such as PSOs, which lead to mRNA turnover by RNase H recruitment. However, the neural charge characteristic of PMOs leads to problems with their efficient uptake into cultured cells. Unlike antisense oligos, which undergo endocytosis, PMOs exhibit essentially no significant uptake. To overcome this problem an efficient delivery system has been developed that hybridizes the PMO to a complementary DNA molecule, which can be further complexed with weakly basic ethoxylated polyethylenimine (EPEI). This complex can efficiently undergo endocytosis and the PMO subsequently released from the endosome into the cytoplasm [54]. While not widely used, this system has significant potential. However, the use of the small interfering (si)RNA system has apparently decreased the need for this antisense approach (next section).

54. Morcos, P.A. Genesis 30 (2001) 94–102.

Perhaps the most exciting potential of PMOs is their use in in vivo models such as rodents and zebrafish and the possibility of a disease therapy for humans. Unlike cultured cells, most tissues appear to be capable of internalizing PMOs, although the mechanism is not clear. The efficiency of PMOs in blocking expression has been demonstrated for the target gene c-myc in rat liver using a partial hepatectomy protocol [55]. Another interesting study performed by the same group of investigators used PMOs designed to bind to nascent mRNA across a splice junction. This was demonstrated to result in the production of an improperly spliced mRNA that expressed a truncated c-myc protein. PMOs are being increasingly used in the zebrafish model to efficiently "knock-down" expression of a wide range of proteins. This system has worked so well that it can be used to test the role of a given protein in development or biological responses to chemical treatments.

55. Arora, V. et al. J. Pharmacol. Exp. Ther. 292 (2000) 921–928.

2.2.2. Small Interfering RNA (siRNA)

RNAi are double stranded RNA molecules that can suppress the expression of complementary genes by causing sequence-specific degradation of mRNA. RNAi has been known for a number of years and has been utilized for repression of specific genes in *C. elegans, Drosophila,* and plants. It has also been established that expression of long (>30 nt) RNAi in mammalian cells leads to activation of an antiviral defense mechanism that induces the production of interferon. This, in turn, leads to nonspecific degradation of RNA transcripts and a shutdown of protein synthesis. However, in 2001 it was established that 21-nt siRNA can mediate suppression of a specific gene in mammalian cells without mediating the nonspecific toxic effects of longer siRNAs [56]. The mechanism of action is believed to occur through the entry of siRNA with RNA-induced silencing complexes (RISCs). The siRNA are then unwound to form activated RISCs, which subsequently associate with the target mRNA and lead to RNA cleavage. The potential application of this technology in silencing genes in mammalian cells is tremendous. With the entire sequence of the human and mouse genomes mapped, it can be envisioned that RNAi could be designed to test the function of every gene expressed within a cell. The ability to actually perform this type of experiment has been confirmed by a study that used RNAi against a number of structural and signaling proteins, where the outcome of the repression should lead to a cellular phenotype [56]. The expected phenotypes were obtained and their results confirmed the efficiency of RNAi in disrupting gene expression.

siRNA is emerging as the method of choice to "knock-down" expression of a specific protein.

56. Elbashir, S.M. et al. Nature 411 (2001) 494–498.

Several methods can generate RNAi for cell culture experiments: chemical synthesis, in vitro transcription, and expression from a plasmid or viral vector. The first step in the production of siRNA against any target mRNA is to find the appropriate sequences to utilize in making an siRNA. Several rules need to be followed to ensure a high success rate. First, use sequences at least 100 bp from the start or termination of translation; second, find a 19-nt stretch that is flanked by AA at the 59 end and TT at the 39 end, and third, the GC content should be more than 30%. This sequence is followed by a loop

A detailed discussion of the use of controls in siRNA experiments can be found at www.stke.org/cgi/content/full/sigtrans;2002/147/pl13.

sequence and the antisense of the initial sequence. The original RNAi used in mammalian cells were synthesized in vitro. The use of an in vitro transcription kit to produce siRNA is relatively inexpensive and easy to prepare. An oligonucleotide with the target sequence is made containing the target sequence and an 8-bp sequence complementary to a T7 primer. The two oligos are annealed and a fill-in with Klenow DNA polymerase is performed. This double-stranded oligo serves as a template for T7 RNA polymerase in an in vitro transcription system. After the siRNA is made, it can be transfected into cells with one of a number of transfection reagents made especially for this purpose, and high transfection efficiency can be obtained, at least in some cell lines. A second method involves the digestion of a long in vitro transcribed double-stranded RNA with RNase III into siRNA. The advantage of this technique is that a number of siRNAs need not be tested, and the disruption of gene expression can be quickly accomplished. However, one possible drawback is the possibility of silencing a related gene as well as the target gene. A third approach to the production of siRNA is chemical synthesis, which is a relatively easy approach for the investigator. However, this method can be expensive considering that usually three to four siRNA need to be made to ensure that optimal gene disruption is obtained. Another benefit of chemically synthesized siRNA is the flexibility it allows to easily scale-up the amount produced. The chemically synthesized siRNAs appear to require a higher concentration of siRNA compared with the in vitro-translated siRNA, but the same level of gene disruption can be achieved. The fourth method utilizes a different approach by expressing siRNA within the cell using an siRNA-expressing plasmid. The available plasmids utilize either the human or mouse U6 promoters or the human H1 promoter. These promoters were chosen because they naturally produce small RNAs in mammalian cells, and they terminate transcription with three to six uridine residues. Double-stranded oligonucleotides are made following the rules described in this section and subcloned into an expression vector. One advantage of this system is that, once the constructs are made, unlimited amounts of vector can be produced for cell-transfection experiments. However, probably the biggest advantage is that stable cell lines can be produced that exhibit long-term suppression of the gene of interest. Also, retroviral systems have been developed and these systems should be particularly useful in cell lines with low transfection efficiency when transfected with vector-based systems.

Regardless of the siRNA system to be utilized, it is important to design controls that will establish the specificity of the gene silencing obtained. A number of controls can be considered, the most obvious being the mutation of two to three bases with the specific siRNA to be tested. Another control is the use of an siRNA that is against an mRNA not expressed in the cell being studied.

2.2.3. Use of Cell Lines Generated From Null Mice

Another method for examining the role of a specific protein in the cell is the generation of primary or established cell lines from animals that have undergone gene disruption. This can be a very effective technique and it is widely used. The advantage of this method, compared with the two methods just described is that the gene is totally disrupted, as opposed to the 90% disruption often obtained with siRNA. Further details about establishment and the experimental use of these cell lines is given in Part III, Chapter 4.

2.3. Regulation of Ectopic Protein Expression

There are numerous viral- and plasmid-based systems to mediate transient or stable expression of proteins in cultured cells. These systems can be used to overexpress a protein in a cell line in order to study the effects of enhanced expression on a given pathway. This is usually accomplished using vectors that have strong enhancer/promoter regions (e.g., CMV, EF-1α, SV-40, Ubc). If the goal is to produce a stably expressing cell line, a variety of selection markers (e.g., neomycin, hygromycin, zeocin) can be used, neomycin being the most utilized. If the goal of the experiments requires a high level of transfection efficiency, such as >80%, then either the production of a stable cell line or a viral system must be utilized. A number of excellent viral infection systems are commercially available, such as lentiviral expression systems, that work well in a variety of primary and established cell lines.

2.3.1. Inducible Protein-Expression System

The tet-on system for regulating the level of ectopic protein expression is a powerful system. However, it requires the use of two vectors to obtain tight control over expression.

A variety of plasmid- and viral-based systems have been developed to regulate transcription from a plasmid for use in transiently or stably transfected eukaryotic cells. Examples of regulated expression systems include the use of the zinc-, tetracycline-, and ecdysone-inducible promoters. Each system has its advantages and disadvantages, and they should be carefully considered prior to embarking on a project. The tetracycline-inducible plasmid-based system is probably the most utilized approach and will be described here in more detail. This system requires the use of two vectors. The first expresses the tetracycline-controlled transcription enhancer (tTs). This vector is stably integrated into the cell line of interest, stable cell lines are established in the presence of the appropriate selection agent, and the presence of tTs confirmed. This line is then transfected with a second vector carrying the gene of interest under control of the TRE promoter. In the presence of the tTs this promoter is repressed, while upon the addition of tetracycline derepression occurs through tetracycline-mediated release of tTs from the promoter. Regulation of ectopic expression of PPARδ in MCF-7 cells is an example of the utilization of the tet-on system [57]. The investigators used a commercially available MCF-7 cell line that stably expresses the tTs protein. The expression of PPAR was accomplished by stably integrating a pTRE/hyg-PPARδ vector. This system allowed these investigators to clearly demonstrate that PPARδ mediates increased the proliferation of human breast and prostate cell lines.

57. Stephen, R.L. et al. Cancer Res. 64 (2004) 3162–3170.

2.3.2. Use of Ribozymes to Control Protein Expression

Another approach to regulate ectopic protein expression is through the control of the degradation of specific mRNA. This can be accomplished through the inclusion of a self-cleaving RNA motif. The proper positioning of these motifs can lead to control of mRNA degradation. Yen and co-workers have optimized the positioning and structure of a schistosome Sm1 ribozyme to effectively mediate rapid turnover of an mRNA [58]. Using a high-throughput assay to screen chemical libraries, they identified toyocamycin as a potent inhibitor of the

58. Yen, L. et al. Nature 431 (2004) 471–476.

ribozyme. Treatment with toyocamycin, a nucleoside analog, resulted in rapid increase in luciferase activity in cells transfected with a luciferase plasmid that expresses a ribozyme containing mRNA. This would suggest that additional compounds in the future could be identified to regulate ribozyme activity and not exhibit toxicity. Another means to block ribozyme cleavage activity is the treatment of cells with a morphilino oligonucleotide that binds directly to the ribozyme sequence. This approach is quite promising as long as efficient delivery of the oligonucleotide into cells is achieved. This system has the advantage over the tet-on system because there is no need to generate a cell line expressing the tet-repressor. Thus, the use of ribozymes designed into standard mammalian expression vectors offers tremendous potential for efficiently regulating the expression of proteins in cells and perhaps in the future in vivo.

3. Summary

A number of different approaches have been described to disrupt protein–protein interactions. Each approach has its strengths and weaknesses. Usually it is best to utilize more than one approach to firmly establish the significance of either a protein in a specific pathway or the role of a specific protein–protein interaction within a cell. The approach taken depends mostly on the availability of reagents, equipment present in the laboratory, and the expertise of the research group. Table 6-1 summarizes the major techniques available along with specific comments as to their strengths or weaknesses.

The focus of this and previous chapters has been largely on the use of cell culture models. It is important to keep in mind that if experiments are not using primary cells (e.g., hepatocytes) but rather are using transformed or tumor cell lines, these cell lines can yield results that may not be observed in normal cells. Even normal or primary cells once they are removed from a tissue and cultured in the presence of serum may no longer exhibit properties observed in the intact tissue. For example, hepatocytes, once placed in cell culture, no longer efficiently express certain cytochrome P-450s such as CYP1A2. Thus, while these cell culture systems are extremely useful they can yield results that may not extrapolate to the in vivo situation. So, once mechanistic information is obtained in cell culture experiments, the next logical step is to design an experiment to test whether the newly discovered mechanism also occurs in an animal model such as mice. The use of mouse models in mechanistic studies is the subject of Part III.

Table 6-1
Summary of Methods to Disrupt Protein–Protein Interactions and Protein Expression

Method	Comments
Disruption of protein–protein interaction	
Point mutagenesis to disrupt protein interactions	Specific point mutations can be introduced into an expression construct and, upon transfection into cells, express a mutated protein that fails to bind to another protein. The activity of the mutated protein or its binding partner can be assessed to determine if the lack of binding to another protein modulates activity.
Dominant negative protein expression	TFs can be made into dominant negative proteins by removal or mutation of the transactivation domain. This mutant can bind to its cognate DNA-binding element and inhibit transcription, in effect blocking the activity of the "wild-type" TF. Another example of a dominant negative protein is a catalytically inactive kinase that binds to another kinase but fails to activate it.
Pharmacologic disruption	Low-molecular-weight compounds are often used in experiments to block the activity of an enzyme (e.g., COX2, protein kinases). An emerging approach to blocking a protein–protein interaction is through the development of low-molecular-weight compounds that are capable of binding to a specific protein domain.
Permeable peptides	This method can be an efficient technique when the protein–protein binding domain has been defined as a short peptide sequence that, when delivered to cells, can bind with high affinity to the protein of interest. The binding protein peptide sequence is synthesized along with a cell-permeable peptide sequence.
Microinjection of antibodies/ peptides	Microinjection of antibodies is an efficient technique to block a protein–protein interaction or the activity of a protein. However, the technique is quite difficult and requires expensive specialized equipment.
Fusion protein/peptide expression	Using standard molecular techniques, a protein-binding domain or peptide sequence can be fused to a protein. Expression of this fusion protein in a cell-transfection experiment should lead to blocking the interaction of the two proteins being studied within the cell.
Disruption of protein expression	
Antisense RNA	This is a technique that has been around for a relatively long time but has not been widely used, due to a number of technical problems. Antisense RNA's mode of action is simple: upon entry into a cell, these RNA analogs bind to specific mRNA and either block translation or lead to RNase H cleavage. There are several types of RNA analogs available that vary in their efficiency and resistance to intracellular degradation. This technique should only be considered a "knock-down" procedure and is unlikely to totally block expression of the protein of interest.
iRNA knock-down of gene expression	This is now the method of choice to "knock-down" expression of a specific protein. There are two basic techniques that can be used to generate the iRNA. The first is the synthesis of iRNA followed by delivery to cells using transfection techniques. The second technique utilizes either a vector that can be transfected into cells or a retroviral vector system to allow production of the iRNA within the cell.
Cell lines generated from transgenic/null mice	The potential problem with the preceding techniques is their inability to *totally* disrupt cellular protein expression. Protein expression can actually be prevented through gene disruption techniques in mice. Cell lines can be generated from a variety of tissues from mice that contain a null allele for a specific protein by immortalization with a *ts*SV-40 virus, or through other immortalization techniques. Also, mouse embryonic fibroblasts can be isolated and grown as primary cultures.

Study Questions

1. You are assigned to determine the role of a newly identified coactivator. One way to gain insight into its function is to determine the type of transcription factors that it can interact with. Describe two methods to determine their identity.
2. What methods will demonstrate that two proteins directly contact each other in a protein complex.
3. Design three distinct assays to screen hybridomas for useful and specific monoclonal antibody production.
4. Describe two methods that can assess whether the entire amount of a protein is complexed with another protein in a cytosolic extract.
5. Explain the differences between polyclonal and monoclonal antibodies with respect to specificity and possible uses.
6. Describe two methods for mapping the binding epitope of a monoclonal antibody.
7. What is the difference in the mechanism of protein separation mediated by gel filtration compared to sucrose density-gradient centrifugation.
8. What can be learned from the use of Re-ChIP assays during the study of the regulation of transcriptional activity by a specific TF.
9. What is the level of resolution of ChIP assays in terms of establishing the location of a protein bound to DNA detected by PCR analysis. Explain your answer.
10. Describe a method or series of methods to define precisely the minimal protein domain that is responsible for a protein–protein interaction.
11. Describe a method that can yield nearly 100% introduction of a protein into cultured cells. List the advantages and disadvantages.
12. Describe the two major approaches to map phosphorylation sites. What are the major advantages and limitations of each approach?
13. A low level of phosphorylation-site stoichiometry results in an inability to map specific phosphorylation sites. How can this overcome?
14. Describe a method to determine the half-life of a protein within a cell.
15. Define and give an example of a dominant negative transcription factor and how it can be used to define its own function.
16. Describe two distinct methods to "knock down" expression of a specific protein in cultured cells. List the advantages and disadvantages.
17. The tet-on transcriptional regulatory system is a powerful means to control ectopic protein expression. What is the major draw back of this system?

References

1. Sumanasekera WK, Tien ES, Davis JW, 2nd, Turpey R, Perdew GH and Vanden Heuvel JP. Heat shock protein-90 (Hsp90) acts as a repressor of peroxisome proliferator-activated receptor-alpha (PPARalpha) and PPARbeta activity. Biochemistry 42(36): 10726-35, 2003.
2. Chen HS, Singh SS and Perdew GH. The Ah receptor is a sensitive target of geldanamycin-induced protein turnover. Arch Biochem Biophys 348(1): 190-8, 1997.
3. Perdew GH, Schaup HW, Becker MM, Williams JL and Selivonchick DP. Alterations in the synthesis of proteins in hepatocytes of rainbow trout fed cyclopropenoid fatty acids. Biochim Biophys Acta 877(1): 9-19, 1986.
4. Friedman DB, Hill S, Keller JW, et al. Proteome analysis of human colon cancer by two-dimensional difference gel electrophoresis and mass spectrometry. Proteomics 4(3): 793-811, 2004.
5. Patton WF, Schulenberg B and Steinberg TH. Two-dimensional gel electrophoresis; better than a poke in the ICAT? Curr Opin Biotechnol 13(4): 321-8, 2002.
6. Zhu H, Bilgin M, Bangham R, et al. Global analysis of protein activities using proteome chips. Science 293(5537): 2101-5, 2001.
7. Van Criekinge W and Beyaert R. Yeast Two-Hybrid: State of the Art. Biol Proced Online 2: 1-38, 1999.
8. Willson TM, Brown PJ, Sternbach DD and Henke BR. The PPARs: from orphan receptors to drug discovery. J Med Chem 43(4): 527-50, 2000.
9. Tirode F, Malaguti C, Romero F, Attar R, Camonis J and Egly JM. A conditionally expressed third partner stabilizes or prevents the formation of a transcriptional activator in a three-hybrid system. J Biol Chem 272(37): 22995-9, 1997.
10. Yang M, Wu Z and Fields S. Protein-peptide interactions analyzed with the yeast two-hybrid system. Nucleic Acids Res 23(7): 1152-6, 1995.
11. Chang C, Norris JD, Gron H, et al. Dissection of the LXXLL nuclear receptor-coactivator interaction motif using combinatorial peptide libraries: discovery of peptide antagonists of estrogen receptors alpha and beta. Mol Cell Biol 19(12): 8226-39, 1999.
12. Wardell SE, Boonyaratanakornkit V, Adelman JS, Aronheim A and Edwards DP. Jun dimerization protein 2 functions as a progesterone receptor N-terminal domain coactivator. Mol Cell Biol 22(15): 5451-66, 2002.
13. Metivier R, Penot G, Hubner MR, et al. Estrogen receptor-alpha directs ordered, cyclical, and combinatorial recruitment of cofactors on a natural target promoter. Cell 115(6): 751-63, 2003.
14. Perdew GH, Hord N, Hollenback CE and Welsh MJ. Localization and characterization of the 86- and 84-kDa heat shock proteins in Hepa 1c1c7 cells. Exp Cell Res 209(2): 350-6, 1993.
15. Perdew GH. Comparison of the nuclear and cytosolic forms of the Ah receptor from Hepa 1c1c7 cells: charge heterogeneity and ATP binding properties. Arch Biochem Biophys 291(2): 284-90, 1991.

16. Lanz RB, McKenna NJ, Onate SA, et al. A steroid receptor coactivator, SRA, functions as an RNA and is present in an SRC-1 complex. Cell 97(1): 17-27, 1999.
17. Siegel LM and Monty KJ. Determination of molecular weights and frictional ratios of proteins in impure systems by use of gel filtration and density gradient centrifugation. Application to crude preparations of sulfite and hydroxylamine reductases. Biochim Biophys Acta 112(2): 346-62, 1966.
18. Chen HS and Perdew GH. Subunit composition of the heteromeric cytosolic aryl hydrocarbon receptor complex. J Biol Chem 269(44): 27554-8, 1994.
19. Meyer BK, Pray-Grant MG, Vanden Heuvel JP and Perdew GH. Hepatitis B virus X-associated protein 2 is a subunit of the unliganded aryl hydrocarbon receptor core complex and exhibits transcriptional enhancer activity. Mol Cell Biol 18(2): 978-88, 1998.
20. Brew K, Shaper JH, Olsen KW, Trayer IP and Hill RL. Cross-linking of the components of lactose synthetase with dimethylpimelimidate. J Biol Chem 250(4): 1434-44, 1975.
21. Harlow E and Lane D. Antibodies: a laboratory manual. Cold Spring Harbor Laboratory, Cold Spring Harbor, NY, 1988.
22. Reineke U, Volkmer-Engert R and Schneider-Mergener J. Applications of peptide arrays prepared by the SPOT-technology. Curr Opin Biotechnol 12(1): 59-64, 2001.
23. Jarvik JW and Telmer CA. Epitope tagging. Annu Rev Genet 32: 601-18, 1998.
24. Van Der Hoeven PC, Van Der Wal JC, Ruurs P, Van Dijk MC and Van Blitterswijk J. 14-3-3 isotypes facilitate coupling of protein kinase C-zeta to Raf-1: negative regulation by 14-3-3 phosphorylation. Biochem J 345 Pt 2: 297-306, 2000.
25. Soti C, Radics L, Yahara I and Csermely P. Interaction of vanadate oligomers and permolybdate with the 90-kDa heat-shock protein, Hsp90. Eur J Biochem 255(3): 611-7, 1998.
26. Wells L, Vosseller K and Hart GW. Glycosylation of nucleocytoplasmic proteins: signal transduction and O-GlcNAc. Science 291(5512): 2376-8, 2001.
27. Conaway RC, Brower CS and Conaway JW. Emerging roles of ubiquitin in transcription regulation. Science 296(5571): 1254-8, 2002.
28. Gill G. SUMO and ubiquitin in the nucleus: different functions, similar mechanisms? Genes Dev 18(17): 2046-59, 2004.
29. Boyle WJ, van der Geer P and Hunter T. Phosphopeptide mapping and phosphoamino acid analysis by two-dimensional separation on thin-layer cellulose plates. Methods Enzymol 201: 110-49, 1991.
30. Levine SL and Perdew GH. Aryl hydrocarbon receptor (AhR)/AhR nuclear translocator (ARNT) activity is unaltered by phosphorylation of a periodicity/ARNT/single-minded (PAS)-region serine residue. Mol Pharmacol 59(3): 557-66, 2001.
31. Knotts TA, Orkiszewski RS, Cook RG, Edwards DP and Weigel NL. Identification of a phosphorylation site in the hinge region of the human progesterone receptor and additional amino-terminal phosphorylation sites. J Biol Chem 276(11): 8475-83, 2001.

32. Posewitz MC and Tempst P. Immobilized gallium(III) affinity chromatography of phosphopeptides. Anal Chem 71(14): 2883-92, 1999.
33. Sun T, Campbell M, Gordon W and Arlinghaus RB. Preparation and application of antibodies to phosphoamino acid sequences. Biopolymers 60(1): 61-75, 2001.
34. Kameshita I and Fujisawa H. Detection of protein kinase activities toward oligopeptides in sodium dodecyl sulfate-polyacrylamide gel. Anal Biochem 237(2): 198-203, 1996.
35. Kameshita I, Taketani S, Ishida A and Fujisawa H. Detection of a variety of Ser/Thr protein kinases using a synthetic peptide with multiple phosphorylation sites. J Biochem (Tokyo) 126(6): 991-5, 1999.
36. Dull AB, Carlson DB, Petrulis JR and Perdew GH. Characterization of the phosphorylation status of the hepatitis B virus X-associated protein 2. Arch Biochem Biophys 406(2): 209-21, 2002.
37. Chou TY, Dang CV and Hart GW. Glycosylation of the c-Myc transactivation domain. Proc Natl Acad Sci U S A 92(10): 4417-21, 1995.
38. Haltiwanger RS, Grove K and Philipsberg GA. Modulation of O-linked N-acetylglucosamine levels on nuclear and cytoplasmic proteins in vivo using the peptide O-GlcNAc-beta-N-acetylglucosaminidase inhibitor O-(2-acetamido-2-deoxy-D-glucopyranosylidene)amino-N-phenylcarbamate. J Biol Chem 273(6): 3611-7, 1998.
39. Mimnaugh EG, Bonvini P and Neckers L. The measurement of ubiquitin and ubiquitinated proteins. Electrophoresis 20(2): 418-28, 1999.
40. Li M, Brooks CL, Wu-Baer F, Chen D, Baer R and Gu W. Mono- versus polyubiquitination: differential control of p53 fate by Mdm2. Science 302(5652): 1972-5, 2003.
41. Demand J, Alberti S, Patterson C and Hohfeld J. Cooperation of a ubiquitin domain protein and an E3 ubiquitin ligase during chaperone/proteasome coupling. Curr Biol 11(20): 1569-77, 2001.
42. Batonnet S, Leibovitch MP, Tintignac L and Leibovitch SA. Critical role for lysine 133 in the nuclear ubiquitin-mediated degradation of MyoD. J Biol Chem 279(7): 5413-20, 2004.
43. Gronroos E, Hellman U, Heldin CH and Ericsson J. Control of Smad7 stability by competition between acetylation and ubiquitination. Mol Cell 10(3): 483-93, 2002.
44. Fu M, Rao M, Wang C, Sakamaki T, et al. Acetylation of androgen receptor enhances coactivator binding and promotes prostate cancer cell growth. Mol Cell Biol 23(23): 8563-75, 2003.
45. Lamarche N, Tapon N, Stowers L, et al. Rac and Cdc42 induce actin polymerization and G1 cell cycle progression independently of p65PAK and the JNK/SAPK MAP kinase cascade. Cell 87(3): 519-29, 1996.
46. Hollingshead BD, Petrulis JR and Perdew GH. The aryl hydrocarbon (Ah) receptor transcriptional regulator hepatitis B virus X-associated protein 2 antagonizes p23 binding to Ah receptor-Hsp90 complexes and is dispensable for receptor function. J Biol Chem 279(44): 45652-61, 2004.

47. Makarova O, Kamberov E and Margolis B. Generation of deletion and point mutations with one primer in a single cloning step. Biotechniques 29(5): 970-2, 2000.
48. Reiners JJ, Jr., Lee JY, Clift RE, Dudley DT and Myrand SP. PD98059 is an equipotent antagonist of the aryl hydrocarbon receptor and inhibitor of mitogen-activated protein kinase kinase. Mol Pharmacol 53(3): 438-45, 1998.
49. Wadia JS and Dowdy SF. Protein transduction technology. Curr Opin Biotechnol 13(1): 52-6, 2002.
50. May MJ, D'Acquisto F, Madge LA, Glockner J, Pober JS and Ghosh S. Selective inhibition of NF-kappaB activation by a peptide that blocks the interaction of NEMO with the IkappaB kinase complex. Science 289(5484): 1550-4, 2000.
51. Nagahara H, Vocero-Akbani AM, Snyder EL, et al. Transduction of full-length TAT fusion proteins into mammalian cells: TAT-p27Kip1 induces cell migration. Nat Med 4(12): 1449-52, 1998.
52. Haddad E, Zugaza JL, Louache F, et al. The interaction between Cdc42 and WASP is required for SDF-1-induced T-lymphocyte chemotaxis. Blood 97(1): 33-8, 2001.
53. Nissen RM and Yamamoto KR. The glucocorticoid receptor inhibits NFkappaB by interfering with serine-2 phosphorylation of the RNA polymerase II carboxy-terminal domain. Genes Dev 14(18): 2314-29, 2000.
54. Morcos PA. Achieving efficient delivery of morpholino oligos in cultured cells. Genesis 30(3): 94-102, 2001.
55. Arora V, Knapp DC, Smith BL, et al. c-Myc antisense limits rat liver regeneration and indicates role for c-Myc in regulating cytochrome P-450 3A activity. J Pharmacol Exp Ther 292(3): 921-8, 2000.
56. Elbashir SM, Harborth J, Lendeckel W, et al. Duplexes of 21-nucleotide RNAs mediate RNA interference in cultured mammalian cells. Nature 411(6836): 494-8, 2001.
57. Stephen RL, Gustafsson MC, Jarvis M, et al. Activation of peroxisome proliferator-activated receptor delta stimulates the proliferation of human breast and prostate cancer cell lines. Cancer Res 64(9): 3162-70, 2004.
58. Yen L, Svendsen J, Lee JS, et al. Exogenous control of mammalian gene expression through modulation of RNA self-cleavage. Nature 431(7007): 471-6, 2004.

PART III

USE OF TRANSGENIC AND KNOCKOUT MICE TO STUDY GENE REGULATION

Jeffrey M. Peters, PhD

Contents

1

OVERVIEW

The application of transgenic and knockout (null) mouse models to the study of how specific proteins regulate gene expression—which in turn modulates cellular homeostasis, toxicology, and carcinogenesis-has become relatively routine in the last decade. Advances made in the use of these model systems are based in large part on classic research undertaken in the 1970s and 1980s, when embryologists were examining the fate of cells within early embryos and attempting to introduce exogenous DNA into mammalian embryos. It was first demonstrated in 1974 that viral DNA could be detected in mice whose blastocele cavity was injected with SV40 DNA during preimplantation development [1]. It was later shown that Moloney murine leukemia virus could be stably incorporated into the germ line of adult mice after viral infection of preimplantation mouse embryos [2]. In 1980, it was reported that pronuclear injection of exogenous DNA could integrate and be detected in somatic tissue in mice [3]. Subsequently, several other research groups reported similar findings using the same methodology of pronuclear injection to demonstrate detection of exogenous DNA in somatic tissue as well as in the germ line [4–7]. Owing in large part to these novel findings, the field of transgenic technology began growing extensively, since it was recognized that introduction of foreign DNA into mice would provide invaluable tools for studying the effects of overexpressed proteins in an intact mammalian model.

The use of transgenic and knockout mouse models to study homeostasis, toxicology, and carcinogenesis has become routine.

1. Jaenisch, R. and Mintz, B. PNAS 71 (1974) 1250–1254.

2. Jaenisch, R. PNAS 73 (1976) 1260–1264.

3. Gordon, J.W. et al. PNAS 77 (1980) 7380–7384.

4. Costantini, F. and Lacy, E. Nature 294 (1981) 92–94.

5. Brinster, R.L. et al. Cell 27 (1981) 223–231.

6. Gordon, J.W. and Ruddle, F.H. Science 214 (1981) 1244–1266.

7. Harbers, K. et al. Nature 293 (1981) 540–542.

Use of Transgenic and Knockout Mice to Study Gene Regulation by J. M. Peters
From: *Regulation of Gene Expression*
By: G. H. Perdew et al. © Humana Press Inc., Totowa, NJ

Developmental biologists were instrumental in establishing transgenic and null mouse model systems through their quest to understand fundamental developmental processes.

While introducing exogenous DNA into mice by pronuclear injection allowed for analysis of overexpressed proteins, a related technology that combines integration of foreign DNA via homologous recombination in embryonic stem (ES) cells and subsequent germ-line transmission from chimeric offspring produced after microinjection of recombinant ES cells into recipient blastocysts was also developed. This approach was greatly facilitated by the discovery in 1981 that cell lines derived from the inner cell mass of mouse preimplantation embryos could be maintained in vitro in an undifferentiated state [8,9]. Subsequent to these findings, the use of ES cells as vectors for introducing specific genetic mutations in mice became a methodology that was quickly utilized by numerous investigators [10]. Initially, the standard approach involved introduction of targeting vectors containing selection cassettes via homologous recombination into ES-cell genomic DNA that enabled researchers not only to disrupt the coding sequence of the gene target, but also to select for the recombinant event. However, some, but not all, of the derived mutant mouse models produced by this method exhibited lethal phenotypes, or phenotypes that were likely due in part to alterations induced by the selection cassette. This disadvantage can now be overcome with the use of conditional deletion approaches that rely on recombinase transgenic mice crossed with mutant mice that contain flanking recombinase sites around the target gene of interest. In this way, specific genes can now be silenced in both tissue and temporally-specific mechanisms in the absence of confounding variables, such as the presence of an artificial selection cassette. While the latter approach is gaining wide acceptance among scientists, this methodology is considerably more complex than the original method of targeted gene disruption but provides a sophisticated experimental tool that can be applied to many areas of science.

8. Evans, M.J. and Kaufman M.H. Nature 292 (1981) 154–156.

9. Martin, G.R., PNAS 78 (1981) 7634–7638.

10. Robertson, E.J. Biol. Repro. 44 (1991) 238–245.

The objective of Part II is to describe and summarize the application of transgenic and null mice for studying the regulation of gene expression in mammalian models. (There are a number of excellent textbooks describing the methodologies required for the production of these mice and the reader is encouraged to examine these for that purpose [11,12].) The first two chapters focus on transgenic mice and null mouse models, and the final chapter will address the use of cells and cell lines derived from these models and how they can be utilized to determine how gene expression is regulated in mammals.

11. Nagy, A., Gertsenstein, M., Vintersten, K. and Behringer, R. Manipulating the Mouse Embryo, 3rd Ed., Cold Spring Harbor Laboratory Press, Cold Spring Harbor, NY, 2003.

12. Pinkert, C. A., ed. Transgenic Animal Technology, 2nd Ed., Academic Press, San Diego, CA, 2002.

2

Production of Transgenic Animals

1. Standard Production

In this chapter, the term "transgenic" mouse will be used for mice generated by microinjection of DNA into pronuclei of embryos. This type of model has numerous applications and takes advantage primarily of expressing, or overexpressing, a gene product of interest in a mouse model (Figure 2-1). In Chapter 3, the term *null* mouse will be used for mice generated by microinjection of recombinant embryonic stem cells into blastocysts. This type of model also has numerous applications and focuses primarily on the phenotype observed in a mouse when a specific gene product is deleted, or silenced (Figure 2-1). The term *conditional null* mouse will be used for mice that are generated by a combination of these approaches. Conditional null mice typically contain both a transgene that is usually used to regulate expression of Cre recombinase and a targeted allele containing flanking *LoxP* sites that enable deletion of the gene product in a tissue- and time-specific manner.

There are a number of approaches that have been used to successfully generate mice that express specific proteins. The methods for expressing transgenes have steadily improved over time. Initially, it was quite common to use DNA vectors for microinjection that simply contained the genomic sequence of interest coupled with the endogenous promoter (Figure 2-2A), or a cDNA of interest coupled with a tissue-specific or inducible promoter: for example, a cDNA construct linked with the albumin promoter for liver-specific expression (Figure 2-2B), or a cDNA construct linked with a metallothionein (MT) promoter (Figure 2-2C) to enable inducible expression in the presence of exogenous zinc. While these tissue-specific and inducible systems offered the advantage of being more selective in terms of expression of the transgene, the disadvantages of these systems included expression in tissues other than that desired, weak expression patterns, expression when not desired, and the requirement of using zinc at a concentration that causes changes in addition to induction of the transgene, all of which could potentially interfere with interpretation.

Use of Transgenic and Knockout Mice to Study Gene Regulation by J. M. Peters
From: *Regulation of Gene Expression*
By: G. H. Perdew et al. © Humana Press Inc., Totowa, NJ

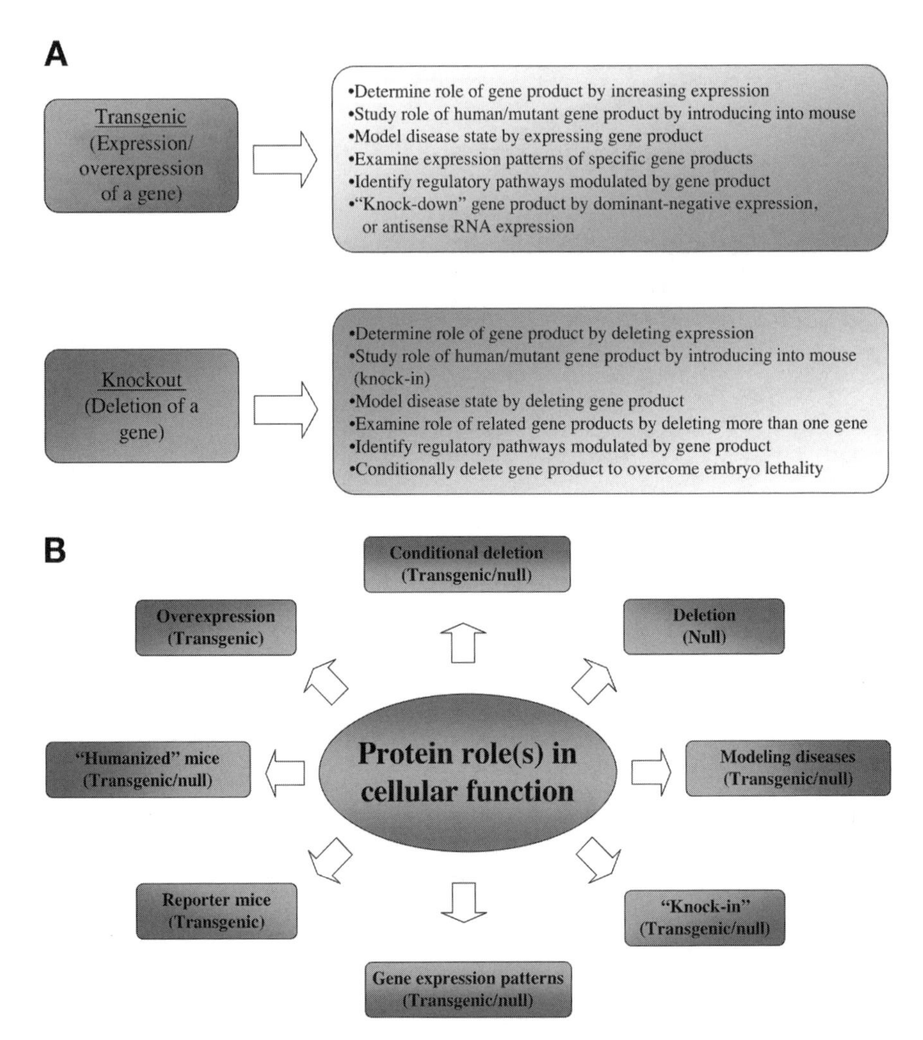

Fig. 2-1. Contrasting the uses of transgenic versus null mouse models. (**A**) Transgenic mice typically overexpress a gene product of interest and have many applications. Similarly, null mouse models have numerous applications that take advantage of specific deletion of a particular gene product. (**B**) The role of a protein in cellular function can be determined using transgenic and null mice. There are significant differences and similarities in these approaches that rely on either overexpression or deletion of a specific protein.

More recently, the use of a cDNA sequence of interest coupled with tissue-specific/inducible promoters has become the norm, because it has the advantage of being able to selectively induce (or repress) expression of the protein of interest [13,14]. Complex promoter systems have been developed, and DNA

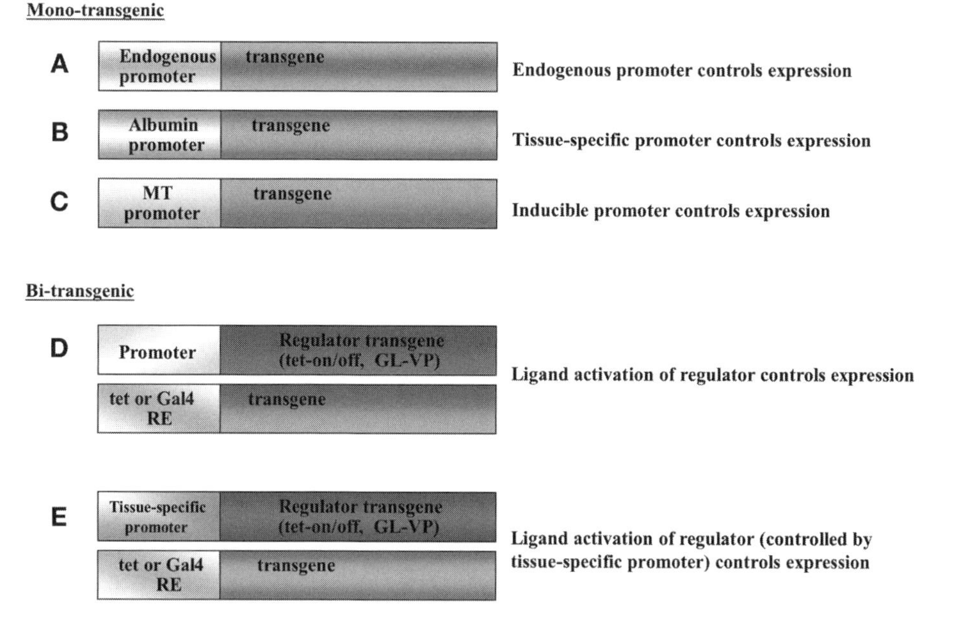

Fig. 2-2. Examples of approaches used to drive expression of transgene. Mono-transgenic expression can be controlled by the endogenous promoter associated with the gene of interest (**A**), by a tissue-specific promoter (**B**), or an inducible promoter (**C**). Bi-transgenic expression can be controlled by a promoter that drives expression of the regulator protein in a non-tissue-specific fashion, which in turn regulates the transgene by inducing expression in response to ligand (e.g., doxycycline) activation (**D**). Bi-transgenic expression can be controlled by a promoter that drives expression of the regulator protein in a tissue-specific fashion, which in turn regulates the transgene by inducing expression in response to ligand activation (**E**).

constructs have also evolved to include bi-directional function that enables investigators to monitor expression patterns by reporter gene expression, in addition to selectively inducing/repressing transgene expression. The use of bi-transgenic mice has become increasingly common and has several advantages compared to the earlier models. The first attempts to generate more selective expression with bi-transgenic mice utilized regulated induction of gene expression with a constitutively active "regulator" transgene consisting of a transcription factor under the control of a tissue-specific promoter, and

13. De Mayo, F.J. and Tsai, S.Y. Trends Endocrinol. Metabol. 12 (2001) 348–353.

14. Lewandowski M. Nat.Rev. Genet. 2 (2001) 743–755.

The ability to selectively induce transgene expression has provided an invaluable approach to study gene regulation in vivo.

The tet on and tet off systems are common tools used to selectively regulate transgenes.

a target transgene under the control of a minimal promoter containing appropriate cis-acting elements. This would then allow for activation of the target transgene by the regulator transgene (Figure 2-2D, Table 2-1). Limitations to this approach included difficulty in designing regulator constructs that did not alter the phenotype of the target tissue, and difficulty in obtaining selective target transgene expression owing to expression in the absence of regulator or differences in cell types within a specific tissue. To improve this system, investigators developed regulator transgenes composed of transcription factors whose activity could be regulated with specific ligands (Figure 2-2E). Three related systems are currently in use, the tetracycline-, mifepristone-, and ecdysone-regulated models.

The tetracycline-regulated system was the first bi-transgenic model developed, and has since been redesigned. The original system relied on a chimeric protein that consists of a modified tetracycline repressor (tetR) sequence fused with the activation domain derived from the HSV thymidine kinase, and termed the tetracycline transactivator (tTA). The tTA acts as a transactivator and, in the presence of tetracycline (more commonly doxycycline), it does not bind to the tetracycline operon so that gene transcription does not occur. With this system (also known as *tet off*), transgene expression can be modulated by continual treatment with doxycycline, and induction of gene expression occurs with withdrawal of doxycycline. To allow for induced transgene expression in the presence of tetracycline, mutations of the tetR were performed, resulting in conformational changes that led to the ability of the transactivator to bind to the tetracycline operon in the presence of doxycycline, thus activating gene transcription. This reverse tet transactivator system (rtTA; also known as *tet on*) offers the advantage of being able to selectively turn on transgene expression upon ligand administration.

Table 2-1
Evolution of DNA Constructs for Transgenic Mice

Transgene regulator	Transgene promoter/ response element	Mono or bi-transgenic	Advantage	Disadvantage
None	Endogenous promoter	Mono	Expression regulated by "normal' factors	Global expression, not selective
None	Inducible (e.g., MT)	Mono	Selective	Prolonged treatment with agent, not tissue specific, toxicity of Zn
None	Tissue-specific (e.g., albumin)	Mono	Tissue specific	Not temporally specific, may not sufficiently drive expression
tTA or rtTA	Tet-responsive	Bi	Ability to determine timing of expression in response to Dox, ability to induce varying levels of expression	Not tissue-specific; effect of Dox?
GL-VP	Gal4	Bi	Ability to determine timing of expression in response to RU486, ability to induce varying levels of expression	Not tissue-specific
EcR	EcR response elements	Bi	Ability to determine timing of expression in response to phyto-ecdysterol, ability to induce varying levels of expression	Not tissue-specific, expensive inducer
Tissue-specific tTA, rtTA, Gl-VP	Tet or Gal4	Bi	Expression is targeted to tissue of interest, ability to determine timing of expression, ability to induce varying levels of expression	Extensive development of constructs and promoters may be required, or dependent on other investigators' development

MT, metallothionein; tTA, tetracycline transactivator; rtTA, reverse tetracycline transactivator; Gl-VP, a mutant ligand binding domain of the progesterone receptor fused with the DNA binding domain of Gal4 and transactivation domain of herpes simplex virus VP16 protein; Tet, tetracycline; Gal4, Gal4 yeast transcription factor; EcR, ecdysone regulated.

Inducible Cre/Flp transgenic mice are essential for conditional null mouse models.

Another commonly used gene-switch regulator system is the mifepristone system, which is based on a modified human progesterone receptor. A mutant ligand-binding domain (LBD) that does not bind to progesterone but binds to the antagonist mifepristone (RU486) is fused to the DNA-binding domain (DBD) of the yeast transcription factor Gal4 and the transactivation domain of the HSV VP16 protein. Upon ligand (RU486) binding, the encoded transcription factor (GL-VP) activates transcription of target transgenes containing Gal4 binding sites. There are several advantages of this system over the tet on/tet off system: the dose of RU486 required to induce target gene expression is very low, RU483 does not compete with any naturally occurring ligand, and, at the doses used to induce transgene expression, RU486 has no other known effects.

The ecdysone gene-switch system is based on the ecdysone receptor (EcR) that is involved in insect development and differentiation. In response to ligand activation, the EcR forms a heterodimer with another gene product and regulates transcription of target genes with appropriate response elements. A number of modifications to the insect receptor system have been made to improve its application for transgenics, including mutations to the DBD of the EcR and corresponding DNA response elements, and improving the transactivation ability by inserting a VP16 activation domain.

To allow for tissue and temporal specificity of transgene activation, fusion of tissue specific promoter units to the regulator transgene have been engineered to enable tissue-specific expression of the regulatory unit (e.g., rtTA, GL-VP) that can only be activated in the target tissue in response to chemical-mediated signaling of the regulator (e.g., doxycycline, RU486). Many of the different gene-switch systems described previously that are used for transgenic mice also have application for conditional mutant mouse models, in particular with respect to expression of Cre recombinase; which will be addressed in the the next section. Additionally, tissue-specific expression of Cre has also been used to activate transgenes in a time- and tissue-specific fashion. This is accomplished by using a DNA construct that has a disrupted promoter of the transgene with *LoxP* sites surrounding this disrupting sequence. Bi-transgenic mice with this type of transgene construct and a gene-switch Cre transgene can then be used effectively to turn on transgene expression in a time- and tissue-specific way—a powerful tool for more selective expression of transgenes.

Production of transgenic mice that express particular proteins of interest is achieved through standard procedures available in many good textbooks. An overview of producing a transgenic mouse line is presented in Figure 2-3. Table 2-1 and Figure 2-2 illustrate a variety of DNA constructs that have been used to produce transgenic mice. Linearized DNA can be used from a number of sources, including plasmids, cosmids, yeast artificial chromosomes (YAC), bacterial artificial chromosomes (BAC), and P1 clones. Purification, linearization, and preparation procedures for DNA will vary greatly depending on the vector used, and an appropriate text should be consulted before these steps are undertaken. Linearization of large DNA fragments can be difficult given the

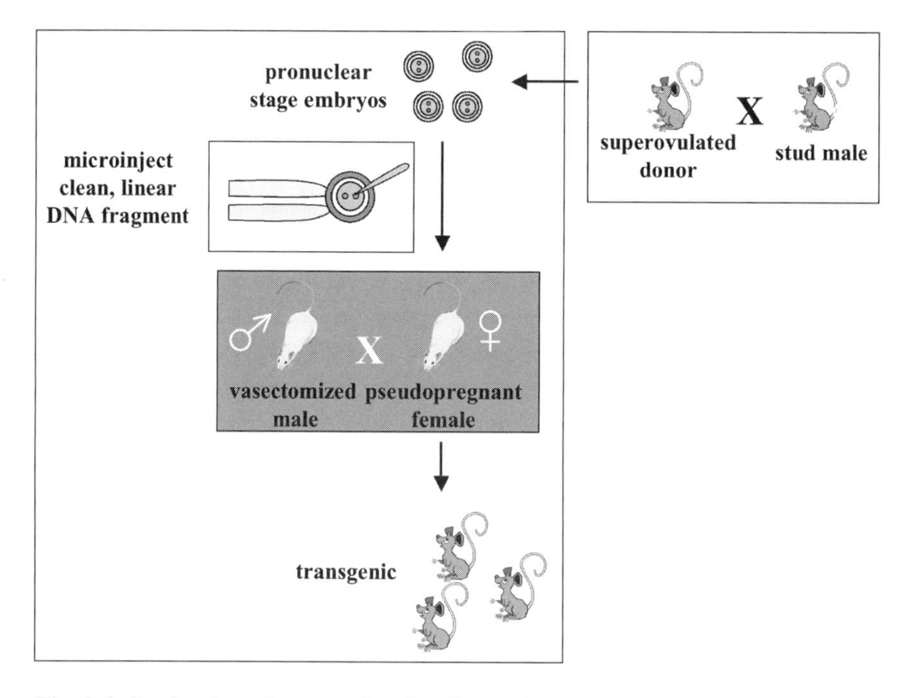

Fig. 2-3. Production of transgenic mice. Pronuclear-stage embryos are microinjected with linearized DNA containing the gene of interest. Microinjected embryos are transferred to pseudopregnant female mice, which give birth to transgenic offspring that contain the gene of interest. The DNA is randomly incorporated into the genome, and subsequent expression is dependent on many factors. Transgene expression can be driven by various tissue and/or inducible promoter systems.

increased likelihood that the sequence may not contain unique restriction sites that can be used for this purpose. Purified DNA fragments containing the sequence of interest are subsequently microinjected into pronuclear-stage embryos, where the DNA will be randomly integrated into chromosomal DNA. There are several different hybrid mouse lines that are typically used to obtain pronuclear-stage embryos for microinjection, but inbred mouse lines can also be used, and, in some instances, are advantageous (e.g., introducing a transgene into a null mouse on a C57BL/6 background). After transfer of microinjected embryos back into pseudopregnant mice, transgenic offspring can be screened from the forthcoming progeny using standard PCR or southern blotting methods. After confirmation of transgene incorporation, the true test in the generation of mice that express the protein of interest is to demonstrate protein expression in the tissue(s) of interest. Depending on the specific gene of interest, this type of screening/confirmation usually involves western blotting, but it can become complicated when confronted with the issues of similar endogenous gene expression. Hemizygous transgenic mice can then be bred to obtain colonies for studies, and it may be wise to breed the hemizygous mice to homozygosity to reduce the need to confirm genotype with each litter. The following sections will focus on the specific applications of this technology to study regulation of gene expression.

2. Applications of Transgenic Models

Since the advent of transgenic technology, numerous transgenic mouse lines have been produced (*see* http://www.jax.org). Mouse models offer the researcher a number of advantages compared to in vitro methods, such as the presence of endogenous promoters targeted by transcription factors, an intact physiological environment of neighboring cells/factors, the ability to express selectively using various promoters and/or inducing agents, and the ability to monitor changes in whole animal models. The disadvantage of mouse models is the difficulty, if not impossibility, of controlling for the number of transgene copies that integrate into the genome. In addition, expression levels of the transgene are not always identical for different regulatory systems, and there is no control for the disruption of other genes at the site of integration and its effects. Overexpression or induced expression of specific transgenes can be used for a number of applications to delineate mechanisms underlying regulation of gene expression, including (but not limited to) (1) identifying roles of transcription factors, (2) determining novel roles of coeffector proteins in physiology, disease, and cancer, (3) determining biological functions of growth factors/cytokines, (4) "humanizing" mouse models, (5) modeling disease states, (6) examining roles of critical regulatory enzymes, (7) determining expression patterns, (8) identifying regulatory pathways using microarray analysis, (9) altering functions by dominant-negative inhibition, and (10) producing "knock-down" transgenics with antisense transgenes. Numerous examples of transgenic mice developed for these purposes have been produced, but a select number have been chosen for illustration purposes. Figure 2-4 generically illustrates how transgenic overexpression can be applied to study how specific proteins of interest modulate physiological responses.

2.1. Transcription Factor Transgenics

While disruption of a gene encoding a transcription factor (loss of function) can lead to the discovery of novel changes in gene regulation, overexpression of a transcription factor or growth factor (gain of function) also provides an alternative and different approach to obtain related information. Among specific proteins that are known to elicit modulation of target gene expression are oncogenes, nuclear hormone receptors, and growth factors. *C-myc* is a proto-oncogene belonging to the basic helix–loop-helix family of transcription factor

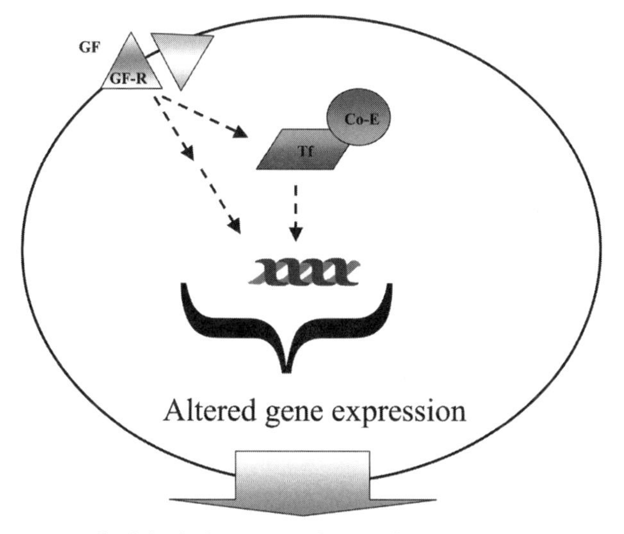

Fig. 2-4. Applications of transgenic mice to study gene regulation. Transgenic mice can be used to study how transcription factors (Tf), coeffector molecules (Co-E), growth factors (GF), growth factor receptors (GF-R) and numerous other gene products regulate cellular homeostasis and/or lead to specific diseases.

with known functions in the transcriptional regulation of genes involved in cell cycle control. A considerable number of studies have examined gene regulation mediated by *c-myc* using transgenic mice overexpressing this proto-oncogene in liver under control by the albumin promoter. Using increased expression of hepatic *c-myc* in mouse models has lead to the discovery of other interacting regulatory pathways that likely lead to carcinogenesis, including those dependent on epidermal growth factor receptor (EGFR) [15] and transforming growth factor-α (TGFα) [16]. These two interacting pathways were determined by examining EGFR expression patterns and function in mice overexpressing hepatic *c-myc*, or by crossing TGFα-transgenic mice with *c-myc* transgenic mice and showing that these mice spontaneously develop liver tumors [16,17]. In addition to identifying novel roles in carcinogenesis, the *c-myc* transgenic mouse has also

15. Woitach, J.T. et al. J. Cell. Biochem. 64 (1997) 651–660.

16. Murakami, H. et al. Can. Res. 53 (1993) 1719–1723.

17. Sanders, S. and Thorgeirsson, S.S. Mol.Carcinog. 28 (2000) 168–173.

been used to demonstrate that this gene product also regulates the expression of genes encoding key regulatory glycolytic enzymes, including glucokinase, L-type pyruvate kinase, and 6-phosphofructo-2-kinase [18]. These examples illustrate how expected, and unexpected, gene regulation mediated by transcription factors can be uncovered using a transgenic mouse model.

18. Riu, E. et al. Biochem. J. 368 (2002) 931–937.

Similar to proto-oncogenes, nuclear hormone receptors can also regulate transcription in a ligand-dependent fashion, and transgenic mouse models have been used to determine how this class of receptors regulate gene expression and alter physiological responses. For example, transgenic mice overexpressing the vitamin D receptor in mature cells of the osteoblastic bone-forming lineage (using a tissue specific promoter to induce expression) exhibit stronger cortical bone and increased trabecular bone volume in addition to reduced bone resorption; presumably through modulation of receptor activity by 1,25-dihydroxyvitamin D and subsequent changes in target gene expression [19]. Transgenic mice expressing estrogen receptor under the control of the MT promoter have significantly impaired reproductive performance [20], and neonatal exposure to an estrogen receptor ligand (DES) leads to enhanced uterine tumor formation [21], in the absence of increased expression of known estrogen receptor target genes such as lactoferrin. Combined, these two examples show how targeted expression of a nuclear hormone receptor can lead to modulation of physiological responses likely mediated by the respective receptor, and indicate that ligand activation does not necessarily lead to modulation of target gene expression. Obviously, these types of transgenic mouse models and others overexpressing nuclear hormone receptors offer an alternative strategy, in contrast to null mouse models, to use to confirm or identify novel target genes when coupled with microarray technology.

19. Gardiner, E.M. et al. FASEB J. 14 (2000) 1908–1916.

20. Davis, V.L. et al. Endocrinology 135 (1994) 379–386.

21. Couse, J.F. et al. Mol.Carcinog. 19 (1997) 236–242.

Overexpression of a gene product can reveal novel roles by enhancing activity or function of a particular protein.

The use of transgenic mouse models to study the roles of transcription factors and growth factors is indeed a powerful approach, not without limitations that are related to expression. While there are numerous issues to consider in designing the DNA construct including which promoter and which sequence (genomic vs cDNA) to use, as integration of DNA is random with this method, it may be difficult to control for effects that could result from disruption of neighboring gene products, which in turn may or may not influence expression of the integrated transgene. In addition to this obvious limitation, one also has to consider where the protein of interest will function once expressed. For example, transmembrane receptors must not only be expressed effectively in the tissue(s) of interest, but they must also be processed such that the protein product is inserted into a membrane. This factor could be significantly influenced by numerous variables. Therefore, careful characterization, such as demonstration of cellular localization to the plasma membrane, must be included when initially evaluating the suitability of a transgenic mouse model for a transmembrane receptor, to ensure that increased receptor binding will lead to subsequent subcellular signaling. Obviously, great care must be taken when designing DNA constructs to verify that correct localization signal sequences are inserted correctly. Similar issues should be considered for proteins with unique subcellular distribution (e.g., nuclear localization signals, etc.).

2.2. Transcriptional Coeffector Transgenics

Coactivators and co-repressors are proteins that function in association with ligand-activated nuclear hormone receptors to facilitate remodeling of chromatin via acetylation or deacetylation that subsequently allows or prevents transcription of target genes, respectively. Transgenic mouse models for both coactivators and co-repressors have been described and have provided clues into functional roles for these proteins that, in some cases, extend our understanding of how these cofactors regulate gene expression at the whole-animal level. For example, using a transgene encoding peroxisome proliferator-activated receptor (PPAR-α) coactivator-1 (PGC-1α) under the control of a muscle-specific promoter, it was discovered that this coactivator appears to regulate conversion of type II muscle fibers to type I muscle fibers as evidenced by increased expression of genes characteristic of type I muscle fibers [22]. Thus, by overexpressing a coactivator, a novel role for this protein with known transcriptional function appears to have been identified. Research suggests that PGC-1α may have critical roles in signaling that leads to specific muscle-fiber formation, which could impact diseases including diabetes and obesity. Further, data from the PGC-1α transgenic mouse also demonstrated that the coactivator activates transcription in cooperation with Mef2 proteins and serves as a target for calcineurin signaling. A related transgenic mouse line overexpressing another coactivator, p300, has been used to examine the possibility that the amount of the coactivator was limiting for glucocorticoid receptor/T-cell receptor competitive transcriptional regulation [23]. By driving expression of p300 in T-cells using a tissue-specific promoter, antagonism between glucocorticoid receptor and T-cell receptor can be prevented, suggesting that the amount of coactivator could be limiting in T-cells. Interestingly, p300 transgenic T-cells also exhibited increased proliferation and interleukin (IL)-2 production when stimulated, which could be attributable to a general increase in transcriptional activity upon increased expression of p300.

22. Lin, J. et al. Nature 418 (2002) 797–801.

Coeffector transgenic mice provide a more physiological approach to study how these proteins function on DNA with intact chromatin, in contrast to approaches that evaluate coeffector influence on plasmid DNA.

23. Yu, C-T. et al. Mol. Cell. Biol. 22 (2002) 4556–4566.

2.3. Growth Factors and Cytokine Transgenics

Similar to nuclear hormone receptors and coeffector molecules that function to modulate gene expression, growth factors and cytokines can also lead to significant changes in gene expression after binding to respective receptor and activating second-messenger signaling pathways. By creating transgenic mice that overexpress these factors, which regulate responses including cell proliferation, differentiation, and inflammation, one can identify and/or confirm molecular pathways that regulate various processes ranging from maintenance of cellular homeostasis to activities that lead to disease states. Although treating mice with growth factors and cytokines can provide similar information, overexpression accomplished by transgenic approaches has the added advantage of relatively consistent increased exposure to a particular growth factor. In addition, it enables selective increase in expression in specific tissues and at specific time points.

Mice overexpressing cytokines offer the advantage of enabling the study of regulatory pathways during disease progression without continued exogenous administration, since this is accomplished with transgene expression.

For example, using a transgenic mouse overexpressing tumor necrosis factor alpha (TNFα) in the lung, regulatory cytokines known to be modulated by TNFα were determined to be differentially expressed during early, mid-, and late-stage progression of pulmonary fibrosis [24]. Through this type of analysis, it was determined that ILs (known TNFα-responsive genes) are indeed induced by TNFα but that expression patterns varied as fibrotic lesions progressed from the early to later stages, showing that other factors in addition to TNFα are involved in the gene regulation of IL-1a and IL-6. This kind of analysis would not be readily reproducible using exogenously provided TNFα and illustrates an advantage to transgenic analysis of growth factors. Given its significant role in tumor development and immune function, it is also worth noting that TNFα transgenic mice have been used extensively to delineate how this cytokine functions in the etiology of cancer and immune dysfunction.

24. Sueoka, N. et al. Cytokine 10 (1998) 124–131.

2.4. "Humanizing" Mice

Transgenic mouse technology has also been used to create mice that express human genes, enabling scientists to study human responses through an appropriate model species. Most notably, this approach has been used to create transgenic mice that express xenobiotic metabolizing enzymes and nuclear hormone receptors that regulate xenobiotic metabolizing enzymes. Cytochrome P450 (CYP) consists of a large superfamily of proteins, including CYP1, CYP2, CYP3 and CYP4, that function by metabolizing foreign and endogenous chemicals. Metabolism by CYP can lead to detoxification and/or bioactivation and therefore represent critical proteins that can prevent and/or lead to toxicity and cancer. Since there are notable species differences in CYP expression and function, humanizing mice to allow expression of human CYP provides a useful model to study xenobiotic metabolism. This approach has been successfully used to generate transgenic mice that express (1) human CYP2D6 [25], (2) CYP3A7, which is limited to fetal expression in humans [26], (3) human CYP3A4 in intestine [27], (4) human CYP4B1 in liver [28], (5) human CYP1B1 using an inducible bi-transgenic mouse model [29], and (6) human CYP19 (aromatase) in ovary and adipose tissue [30].

25. Kimura, S. et al. Am. J. Hum. Genet. 45 (1989) 889–904.

26. Li, Y. et al. Arch. Biochem. Biophys. 329 (1996) 235–240.

27. Granvil, C.P. et al. Drug Metab, Disp. 31 (2003) 548–558.

28. Imaoka, S. et al. Biochim. Biophys. Res. Comm. 284 (2001) 757–762.

29. Hwang, D.Y. et al. Arch. Biochem. Biophys. 395 (2001) 32–40.

30. Hinshelwood, M.M. and Mendelson, C.R. J. Ster. Biol. Mol. Biol. 79 (2001) 193–201.

Creating transgenic mice expressing human CYP isoforms is a straightforward approach when the identical isoform in question is not expressed in mice, but some are. In this case, creating transgenic mice using mice that are null for a specific CYP has been successful. For example, a humanized transgenic mouse line has been produced in the CYP1A2-null mouse that expresses the human isoform primarily in the pancreas. These humanized transgenic mice can be used for treatments with xenobiotics and analysis of changes in gene expression, biochemistry, and physiology.

"Humanizing" mice may prove to be instrumental as tools to evaluate drug development.

A humanized mouse model has also been developed that allows for studying CYP induction in response to xenobiotics via activation of a receptor. While the pregnenolone X receptor (PXR) regulates CYP3A in response to ligand activation in mice, the steroid and xenobiotic receptor (SXR) regulates a similar pathway in humans. By creating a transgenic mouse that expressed human SXR in a PXR-null background, a mouse model was established that allowed the study of CYP3A induction and metabolism by compounds that are routinely administered to humans [31]. This mouse model and others using similar approaches provides a useful tool for examining CYP3A induction and metabolism, which contributes a substantial proportion of our knowledge of xenobiotic metabolism in humans.

31. Xie, W. et al. Nature 406 (2000) 435–439.

2.5. Transgenic Models of Disease States

Altered expression of oncogenes and other proteins is known to occur in tissues from individuals with various diseases, including diabetes and cancer. For example, *c-myc* and epidermal growth factor receptor (EGFR) are often expressed at high levels in experimentally-induced and human tumors, in addition to transformed cell lines. Similarly, increased expression of specific proteins is also found in disease states such as diabetes, e.g., regulatory and functional proteins involved in fatty acid catabolism. These observations have led researchers to test hypotheses that increased expression of specific proteins is causally related to cancer and other diseases, and this becomes an obvious application for transgenic mouse technology. By overexpressing a specific protein or combination of proteins, one can determine if this leads to the development of cancer or other diseases and, when it does, one can assay the molecular events that contribute to the etiology of these conditions.

Construction of transgenic mice that exhibit symptoms of human disease has greatly improved our ability to elucidate the etiology of, and develop novel therapies for preventing or treating diseases.

One of the first demonstrations that increased expression of an oncogene will lead to cancer was provided by a group who created a transgenic mouse that expressed *c-myc* in response to glucocorticoids and demonstrated increased tumor incidence in mammary gland, testes immune cells, and other tissues. Using a similar approach, others have shown that restricting *c-myc* expression to the liver with albumin promoter leads to a 50–60% rate of tumor formation in liver of transgenic mice. This incidence can be further increased to 100% by simultaneously increasing liver expression of TNFα, as assessed in compound transgenic mice [16]. This is also one of many examples of how transgenic mice can be crossed with other transgenics to study interactions between gene products. This approach has been applied to many model systems, including crossing transgenic mice with null mouse models (discussed in Chapter 3). Targeted overexpression of simian virus 40 large-tumor antigen in prostate has provided an excellent model for studying cancer in this tissue [32]. Enhanced expression of oncogenic neu/erbB2 in mammary epithelium leads to mammary tumors and metastases [33], and overexpression of amyloid precursor protein driven by neuronal-specific promoter leads to amyloid plaque formation and neurodegenerative symptoms similar to that found with Alzheimer's disease [34]. Another example is diabetes, a disease caused by a number of mechanisms for which there are limited models available to study specific pathways. Producing a transgenic mouse that overexpresses PPARα in cardiac tissue results in a phenotype resembling that found in diabetes [35], a new model for delineating molecular changes that occur in this disease.

All the mouse models that essentially mimic cancer and other disease states provide invaluable tools for determining the molecular changes that lead to the condition under study. Obtaining a useful phenotype establishes its potential as a model. Subsequent biochemical and molecular experiments comparing varying responses between control and transgenic mice (plus or minus treatments) are currently standard in science, in addition to application of primary cells and cells lines obtained from these models (*see* Chapter 4).

16. Murakami, H. et al. Cancer Res. 53 (1993) 1719–1723.
32. Greenberg, N.M. et al. PNAS 92 (1995) 3439–3443.
33. Guy, C.T. et al. PNAS 89 (1992) 10578–10582.
34. Moechars, D. et al. J. Biol. Chem. 274 (1999) 6483–6492.
35. Finck, B.N. et al. J. Clin. Inv. 109 (2002) 121–130.

2.6. Critical Enzyme Transgenics

DNA repair involves numerous proteins, and the regulatory pathways are not completely understood. It is known that alterations in DNA repair can lead to the development of cancer resulting from mutations in DNA encoding important genes being replicated into proliferating cells. Since DNA damage occurs with aging and/or exposure to mutagens, and DNA repair enzymes might be involved, producing a transgenic mouse that overexpresses a DNA repair enzyme could in theory prevent tumor formation associated with aging or mutagen exposure. One enzyme that repairs DNA is O^6-methylguanine-DNA methyltransferase (MGMT), which removes alkyl lesions at the O^6 position of guanine. Transgenic mice driven by a transferrin promoter to overexpress the human isoform of MGMT [36] spontaneously develop hepatocellular carcinoma at a significantly lower level than controls, and chemically-induced liver tumors are also significantly reduced in this transgenic line, compared to controls. Models such as the MGMT transgenic mouse, and other mice expressing DNA repair enzymes, provide a way to analyze how alterations in DNA composition can regulate development, carcinogenesis, and more specifically, modulation of related changes in gene expression.

36. Walter, C.A. et al. Ann. NY Acad. Sci. 928 (2001) 132–140.

Another class of regulatory proteins that have more of a global impact on cellular processes are proteins and enzymes that either enhance or prevent intracellular oxidative stress. For example, superoxide dismutase (SOD) has several isoforms, which catalyze the formation of hydrogen peroxide and oxygen from superoxides generated by various intracellular reactions. Thus, SOD serves to protect cells from oxidative damage and, in turn, modulates the signaling mediated by reactive oxygen species. Transgenic mice overexpressing human Cu/Zn SOD have served in a tremendous amount of research that focused on how this enzyme prevents oxidative damage leading to numerous diseases, including neurodegenerative disorders, diabetes, cancer, and others. By overexpressing SOD in mice, one can identify pathways that are modulated by this enzyme. For example, if increased reactive oxygen species are thought to lead to signaling that causes cancer, one can examine this in a SOD transgenic mouse and compare differences in oxidative damage, changes in gene expression, and ultimately tumor formation between transgenic and control animals.

2.7. Gene Expression Patterns

Transgenic mice have been used to study developmental expression patterns using several different approaches. Myf5 and Mrf4 are two muscle-specific transcription factors that are located in a 14-kb locus that regulates a transcriptional cascade during skeletal muscle development in mammals. Using bacterial artificial chromosome (BAC) clones containing this locus, unique reporter constructs were inserted near each coding sequence for Mrf4 (human alkaline phosphatase) and/or Myf5 (lacZ). Transgenic analysis of mice created with these and other mutant forms of BAC clones provided striking evidence of temporal expression patterns of each transcription factor [37]. This demonstrated that the Mrf4 reporter construct did not interfere with expression of the lacZ-Myf5 indicating that the Mrf4 sequence did not contain regulatory elements required by Myf5. This is also a good example of a transgenic mouse model produced using the endogenous promoter for molecular analysis rather than an inducible promoter system. This type of system could also be applied to study other gene products for use not only in developmental analysis but also in response to physiological effectors, including dietary factors, drugs, toxicants, and so forth. Another approach used to study relative expression patterns has been facilitated with the use of trans-

37. Carvajal, J.J. et al. Development 128 (2001) 1857–1868.

Reporter mice can be used to elucidate the target tissues/cells influenced by drugs/chemicals that activate specific regulatory pathways.

genic mice created with reporter constructs fused with specific promoters. Several excellent examples of this type of model are the AP-1 [38], nuclear factor (NFκB) [39], and nuclear factor of activated T cells (NFAT)-luciferase [38] reporter transgenic mice. By fusing the respective promoter to a luciferase construct, luciferase activity of cells/tissues can be measured and used as an indicator of activation or inhibition of these pathways in response to various treatments, including growth factors, cytokines, drugs, etc. More recently, bidirectional DNA transgenes have been in use whereby activation of the transgene by a regulator not only induces expression of the targeted transcript of interest but also increases reporter activity (e.g., β-gal activity, luciferase, etc.) allowing for both expression patterns in addition to functional analysis of the transgene.

38. Rincon, M and Flavell, R.A. Mol. Cell. Biol. 16 (1996) 1074–1084.

39. Millet, I. et al. J. Biol. Chem. 275 (2000) 15,114–15,121.

2.8. Identification of Regulatory Pathways Using Microarrays

Transgenic mouse models can be applied to microarray technology (described in Section 1) to tentatively identify putative target genes or genes that are regulated by the protein of interest. By examining differential gene regulation between wild-type and transgenic mouse tissue, one can identify novel roles for ligands, receptors, enzymes, or other functional proteins that have been overexpressed. For example, demonstrating increased or decreased expression of a particular gene product or regulatory pathway in wild-type tissue, as compared to transgenic mouse tissue, can provide clues as to the role of the protein of interest. This type of application can be used for all of the examples provided in this chapter.

2.9. Dominant-Negative Transgenics

Expression of a dominant-negative molecule is an alternative approach to null mouse models for inhibiting specific pathways and can be accomplished with transgenic mice. In fact, this approach has been used a great deal for delineating complex signal transduction pathways in immune cells. Because of their extensive application in studying immune cell function, a number of critical features of dominant-negative effectors have been described and should be considered before transgenic analysis. Determining the relative ratio of dominant-negative to native molecules to effectively inhibit protein function is essential and may be limited due to the effectiveness of the promoter system(s) used to drive expression of the dominant-negative. Dominant-negative molecules must be able to specifically sequester a key upstream regulator or downstream substrate. One common problem encountered with putative dominant-negative constructs is that the specificity for inhibiting a specific function is difficult to obtain. For example, some catalytically inactive Akt (PKB) or protein kinase C molecules are ineffective for specific inhibition since they can inhibit other related isoforms. One of the most extensively studied signaling pathway in T-cells using dominant-negative transgenic mice is the guanine-nucleotide-binding protein Ras. This protein is activated in antigen-receptor or cytokine-stimulated T-cells and is a central regulator for mediating antigen-receptor function in pre-T-cells. A dominant-negative form of Ras (N17Ras) was shown to function by sequestering guanine–nucleotide-exchange factors required for Ras activation, thereby selectively blocking activation of all Ras isoforms but not other GTPases. Using a tissue-specific transgenic mouse expressing this dominant-negative form of Ras in the thymus, investigators showed that N17Ras prevents TCRαβ-induced positive selection of single-positive T-cells, but didn't prevent negative selection [40]. Further support for the specificity of this effect was provided by analysis of a null mouse model for the Ras-exchange protein GRP, confirming results from the dominant-negative transgenic mouse model. Applications of transgenic mice expressing dominant-negative mol-

Dominant-negative transgenics are commonly used to study immune function, but have other applications as well.

40. Swan, K.A., et al. EMBO J. 14 (1995) 276–285.

ecules is a complementary and alternative approach to null mouse models to study loss of gene product function and is not limited to studying signaling pathways in immune cells. For example, using cardiac cell-specific transgenic expression of a dominant-negative cyclic-AMP (cAMP) response element binding protein (CREB), a novel model to study dilated cardiomyopathy was established [41]. Using this model, investigators have also shown that the adenylate cyclase activity of cardiac cells in response to stimulation is diminished in transgenic mice expressing the cardiac-specific dominant-negative CREB molecule [42]. Therefore, this transgenic mouse line provides an excellent model to study signaling pathways and the molecular events orchestrated by cAMP stimulation, and this mouse line could be used to other tissues, if the transgenic model is otherwise appropriate. Additionally, transgenic mouse models expressing dominant-negative molecules coupled with tissue-specific inducible promoters have been produced, allowing for analyses in specific tissues of interest with this approach.

41. Fentzke, R.C. et al. J. Clin. Invest. 101 (1998) 2415–2426.

42. Eckhart, A.D., et al. J. Mol. Cell. Cardiol. 34 (2002) 669–677.

2.10. "Knock-Down" Transgenics Using Antisense Approach

Similar to the dominant-negative transgenics approach to inhibiting gene product function is the use of transgenes encoding antisense deoxynucleotides. Using a neurofilament gene promoter to drive expression of an antisense oligodeoxynucleotide against the type II glucocorticoid receptor in a transgenic mouse line, mRNA encoding this receptor can be reduced by approx 50–70% in the hypothalamus and parietal cortex, with a significant reduction in effect being observed in other tissue such as liver [43]. While expression of the glucocorticoid receptor was not completely eliminated in this model, increased fat deposition and weight gain suggest that this reduction has a functional consequence. Similarly, a partial reduction in mineralocorticoid receptor in cardiomyocytes using a tet-off,

Overexpression of an antisense molecule in a transgenic model is an alternative to making a knockout.

43. Pepin, M.C. et al. Nature 355 (1992) 725–728.

tissue-specific expression of an antisense construct is reported to lead to a phenotype of dilated cardiomyopathy associated with interstitial fibrosis; all of which can be reversed by the administration of doxycycline, which results in diminished expression of the antisense transgene [44]. A related system was recently described to down-regulate and determine the function of the pancreas duodenal homeobox 1 *(PDX1)* gene with a tissue-specific tet-on antisense transgenic mouse line [45]. A transgenic mouse was produced from a DNA construct that contained both the tet-on regulator subunit and the tet-responsive transgene, thereby eliminating the need to cross two transgenic mouse lines. By selectively treating mice with doxycycline and increasing expression of the antisense RNA against *PDX1* in the pancreas, glucose tolerance is impaired, which correlates with significantly reduced expression of glucose transporter 2 and glucokinase, suggesting that *PDX1* is involved in the regulation of these gene products. Although antisense technology has been used effectively in many plant systems, its application in mammalian models has been hampered by nonspecific features associated with this approach. However, recent research has provided an improved method that takes advantage of small interfering RNAs (siRNAs) that appear to reduce more efficiently the expression of target mRNAs. Interestingly, a transgenic mouse model designed to constitutively express an antisense siRNA against green fluorescent protein (GFP) has been described [46]. By crossing this transgenic mouse with another transgenic reporter mouse line that constitutively expresses GFP, these investigators were able to demonstrate that GFP expression could be significantly reduced in many tissues. Additionally, a transgenic rat line was produced, crossed with GFP-transgenic rats, and used to produce similar results found in mice. This suggests that the use of transgenes encoding siRNAs

44. Beggah, A.T. et al. PNAS 99 (2002) 7160–7165.

45. Lottmann, H. et al. J. Mol. Med. 79 (2001) 321–328.

Transgenic mice expressing siRNA is another alternative for a knockout model and could be used for making "knockdown" rat models.

46. Hasuwa, H. et al. FEBS Lett. 532 (2002) 227–230.

may be of great value in creating "knockout" rats that can be manipulated for transgenic production but whose embryonic stem cells are of no use for null rat models. With the use of antisense transgenic mouse lines, investigators have an alternative to produce knock-down/knockout models. This approach offers the advantage of decreasing the length of time required to produce some of the more complex null and conditional null mouse models that will be described in Chapter 3.

3

KNOCKOUT ANIMALS

1. Standard Production

There are both "conventional" and "conditional" approaches to generating null mice, with the former approach being more common until the application of Cre/*LoxP* and Flp/*FRT* systems were developed. Both of these approaches are similar, but conditional methods take advantage of transgenic mice that express recombinase. There are several very good text books describing the construction of null mice, which should be consulted for more in depth coverage of the advantages and disadvantages of this method [11,47]. The overall strategy is depicted in Figure 3-1. In order to generate a conventional null mouse, a DNA targeting vector must be designed using isogenic genomic DNA. Construction of complex targeting vectors has recently been greatly improved. Since the mouse genome is now sequenced, investigators can identify BAC clones containing the gene of interest without screening a genomic library. Additionally, a recently developed method, termed "recombineering," allows for rapid construction in less than one month of targeting vectors containing selection cassettes and *LoxP*/*FRT* sites [48]. A positive selection cassette (e.g., neo-resistance minigene) is inserted into the coding sequence of the gene of interest, and the encoded gene product will, in theory, be rendered nonfunctional (Figure 3-1). Additionally, a negative selection cassette (e.g., HSV-TK) is also subcloned into the targeting vector to increase the efficiency of

11. Nagy, A., Gertsenstein, M., Vintersten, K. and Behringer, R. Manipulating the Mouse Embryo, 3rd Ed., Cold Spring Harbor Laboratory Press, Cold Spring Harbor, NY, 2003.

47. Joyner, A. L., ed. Gene Targeting: A Practical Approach, 2nd Ed. Oxford University Press, New York, NY, 2001.

48. Copeland, N.G. et al. Nat. Rev. Genet. 2 (2001) 769–779.

Use of Transgenic and Knockout Mice to Study Gene Regulation by J. M. Peters
From: *Regulation of Gene Expression*
By: G. H. Perdew et al. © Humana Press Inc., Totowa, NJ

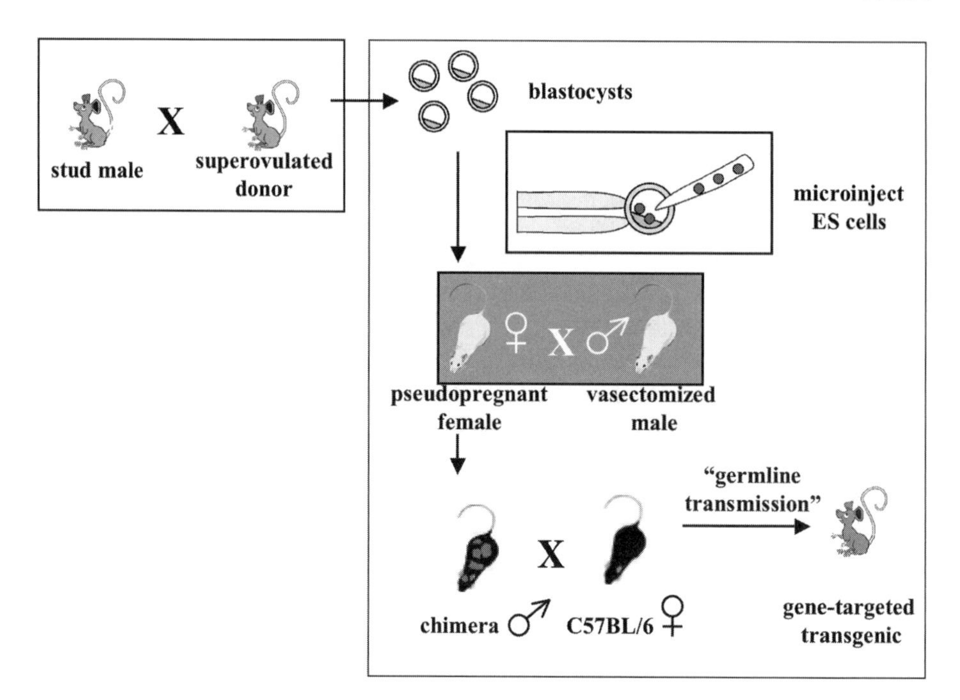

Fig. 3-1. Production of null mice by microinjection of (embryonic stem) ES cells into blastocysts. Recombinant ES cells containing the targeted mutation(s) of interest are microinjected into blastocyst-stage preimplanataion embryos and transferred to pseudopregnant female mice. Chimeric offspring are born and mated with wild-type mice to obtain heterozygous mice. Subsequent breeding allows for generation of homozygous wild-type and null alleles for the gene of interest. This system can also be coupled with transgenic Cre mice to enable conditional deletion of a gene of interest.

isolating recombinant embryonic stem (ES) cells. The completed targeting vector is electroporated into ES cells, where homologous recombination will occur with varying levels of efficiency, and heterozygous ES cells can be obtained. Heterozygous ES cells with the targeted mutation are then microinjected into recipient blastocysts, where they will integrate with the inner cell mass and become part of the chimeric offspring. When the targeted allele integrates into the germ line of the male chimera, it can be transmitted into subsequent offspring and is termed "germ-line" transmission. Homozygous offspring are obtained by breeding heterozygous offspring from the mating of chimera with wild-type mice. It is best to use congenic mice for subsequent studies and there are good textbooks available describing methods for obtaining these lines. The production of a conventional null mouse requires approx 6 mo to 2 yr to produce. One can utilize mice with targeted mutations for numerous applications that will be the focus of the remainder of this chapter.

Before describing applications of null mouse models, the subtle differences between conventional and conditional approaches will be outlined (Figures 3-2 and 3-3). Generating conditional null mouse models is essentially the same process as for conventional mice with the exception of vector design and the application of Cre/*Flp* recombinase for creating the conditionally mutant allele. The Cre/*LoxP* system from bacteriophage P1 and the Flp/*FRT*system from yeast are recombinases that provide invaluable tools for selectively silencing gene products (or enhancing them when used with transgenics). For the purposes of this chapter, the Cre recombinase and associated *LoxP* sites will be used for illustration of conditional systems, but Flp and *FRT* sites could be substituted for interpretation. Targeting-vector design is similar to conventional methods, although insertion of the positive selection cassette is typically in an intronic section of DNA flanked by *LoxP* sites, in order to facilitate removal of the cassette and prevent possible interference with phenotypical characterization. *LoxP* sites are also subcloned to flank the section of genomic DNA one wishes to delete, which can range from an exon or two to a large section of chromosome. After the targeting vector has been constructed, electroporation of ES cells is performed and heterozygous ES cells with the recombinant allele are obtained. The recombinant ES cells are then typically treated with a viral Cre expression vector, and ES clones are screened for recombinant events mediated by Cre, which will include three possible outcomes (Figure 3-2): (1) a deleted allele-this can be used to generate a "conventional" null mouse; (2) a deleted allele with a *neo* insert that is flanked by *LoxP* site ("floxed")-this orientation is not desired because the *neo* insert could influence phenotype; and (3) a floxed target gene with *neo* excised-this is the conditional construct. These ES

Since a "conventional" deletion can be obtained with the Cre/LoxP (Flp/FRT) system, it is becoming more common to utilize this approach to construct null mouse models, which eliminates the influence of the positive selection cassette (NEO), and allows for selective deletion.

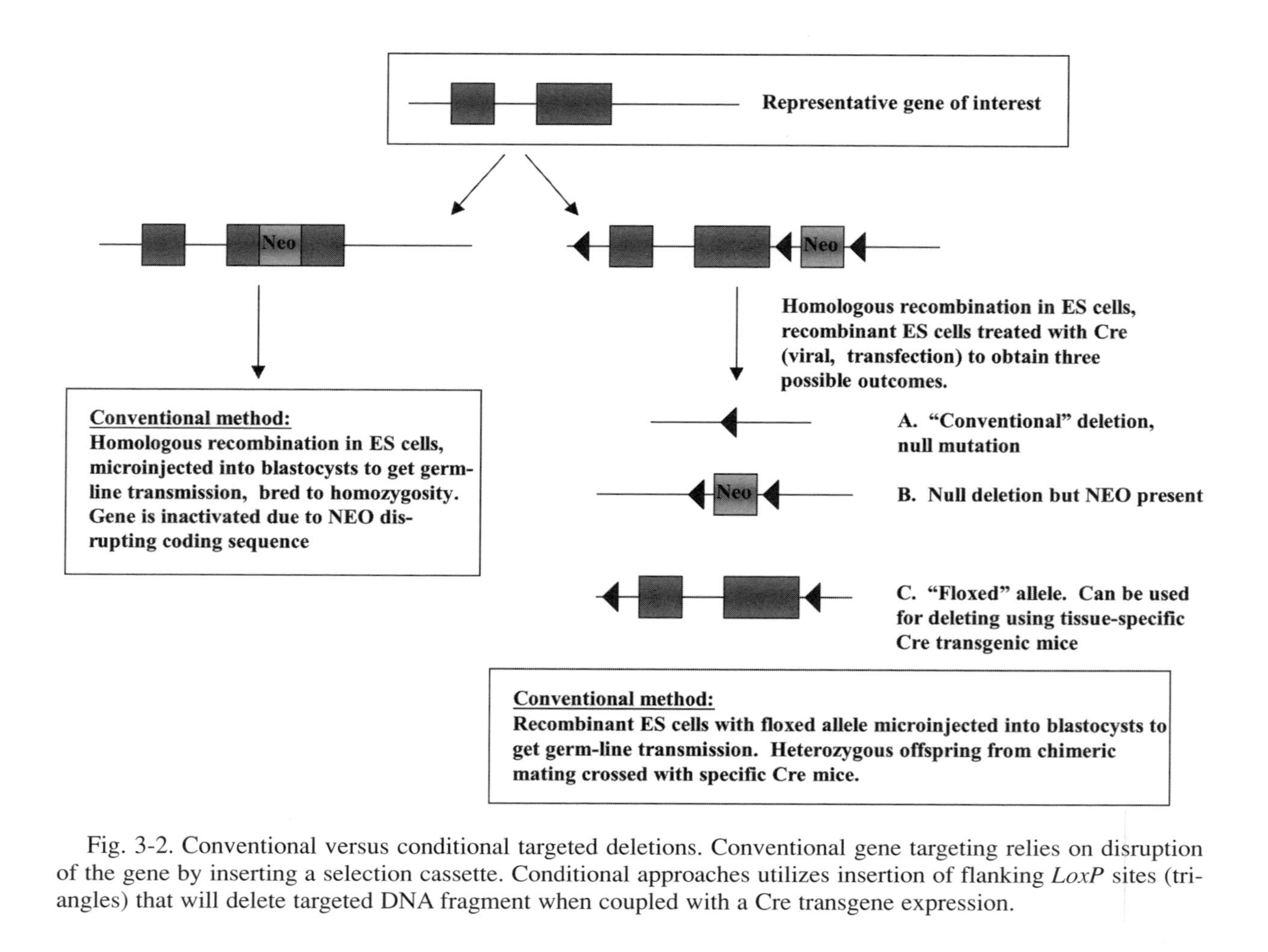

Fig. 3-2. Conventional versus conditional targeted deletions. Conventional gene targeting relies on disruption of the gene by inserting a selection cassette. Conditional approaches utilizes insertion of flanking *LoxP* sites (triangles) that will delete targeted DNA fragment when coupled with a Cre transgene expression.

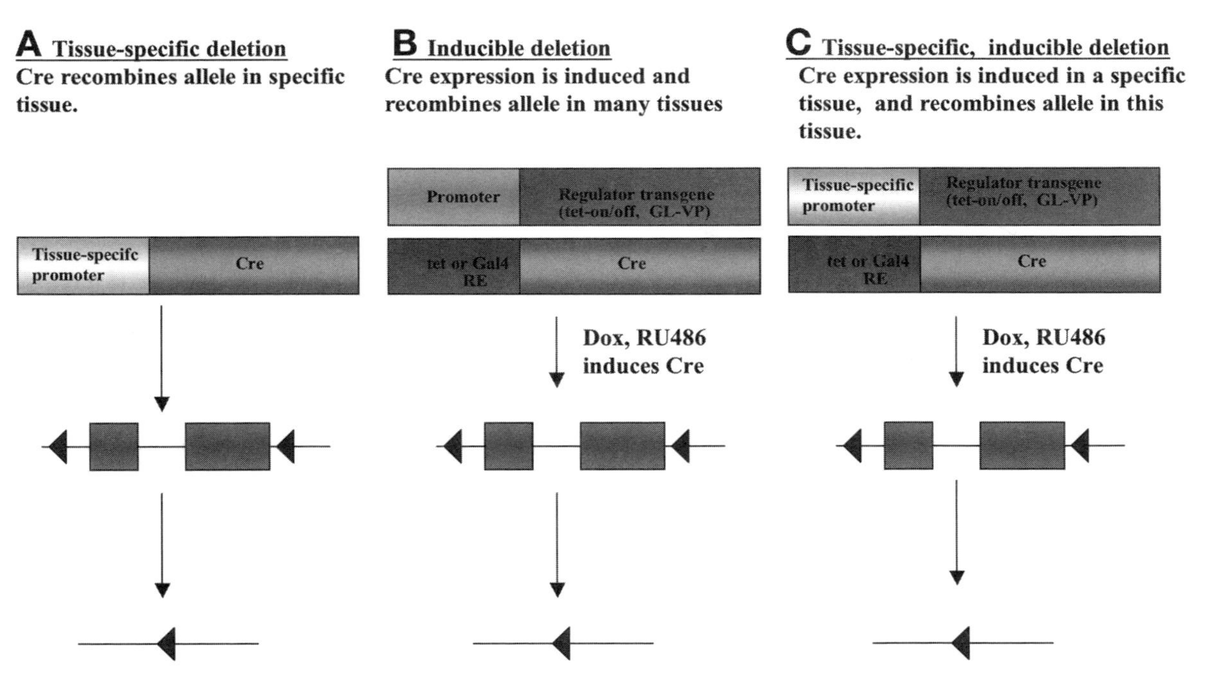

Fig. 3-3. Conditional deletion of floxed target gene. Conditional gene targeting can be accomplished by a number of methods. When Cre expression is limited to a specific tisue, deletion occurs in this tissue only **(A).** When Cre expression is driven by a regulator protein whose expression is driven by a nonspecific promoter, targeted deletion will occur in response to ligand activation, which drives Cre expression and subsequent recombination in more than one tissue **(B).** When Cre expression is driven by a regulator protein whose expression is driven by a tissue-specific promoter, targeted deletion will occur in response to ligand activation, which drives Cre expression and subsequent recombination in this specific tissue **(C).** RE, response element; Dox, doxycyline. For other abbreviations, see Table 2-1, p. 275.

cells are then used for microinjection into blastocysts (Figure 3-1) and germ-line transmission obtained. Heterozygous offspring with the floxed allele can then be crossed with transgenic (or knock-in) Cre (or *Flp*) mice to conditionally delete the target gene of interest.

Conditional deletion of target genes to produce null mouse models has evolved significantly in the last 10 yr and requires the use of transgenic (or knock-in) mice overexpressing Cre recombinase. For transgenic Cre mice, there are three methods that are used most frequently for this purpose. The first is tissue-specific deletion that utilizes a transgenic Cre mouse line in which Cre expression is driven by a tissue-specific promoter (e.g., albumin for liver expression). Using this approach (Figure 3-3A), the floxed target gene can be deleted in liver, and typically this does not occur in other tissues because Cre expression is limited to liver. The disadvantage of this system is that target gene deletion occurs in the tissue at all times and one cannot control for the timing of such deletion. To improve this method, bi-transgenic Cre mice with inducible promoters have been developed. Using this approach (Figure 3-3B), Cre expression can be turned on in response to an inducing agent, such as doxycycline or RU486, depending on the system, and targeted deletion occurs in a defined timeframe. The disadvantage of this system is that targeted deletion will occur in more than one tissue, where the Cre transgene is induced in response to global regulator expression. This system has been improved for a bi-transgenic Cre mouse line, whereby the regulator transgene is driven by a tissue-specific promoter, such that Cre expression is limited to a specific tissue, thereby limiting targeted deletion to this tissue (Figure 3-3C). With this system, the regulator is expressed in a specific tissue, and in response to doxycycline, RU486, or appropriate ligand, Cre is expressed to effectively delete the floxed gene of interest. Thus, this system is the most advantageous since the investigator can not only target a specific tissue for deletion but also determine the timeframe at which the deletion occurs. Cre-expressing mice can also be generated using knock-in approaches, but for the purposes of this chapter this will not be discussed. The reader is encouraged to consult other reference books to learn more about the many other details that are important in generating various null mouse models, including vector design, various methods for producing null mutations or subtle mutations in ES cells, combining Cre/*LoxP* and Flp/*FRT* systems, determining efficiency of Cre expression patterns for conditional deletions, and many other important concepts that are too detailed to be covered within the scope of this chapter.

2. Applications of Null Mouse Models

Transgenic, "gain of function" models described in the previous section provide a useful model system in which to determine how overexpression of specific proteins modulates changes in gene expression that ultimately produce a physiological response in the whole organism. Similarly, in the last twenty years, "loss of function" models using targeted disruption or deletion of mouse genomic DNA have become common in science. A large database of these resources is available (http://www.jax.org/). By deleting, or disrupting, a specific gene product, a powerful genetic model is created that can be utilized in approaches similar to those described for transgenic mouse lines. Overexpressing a specific protein can provide observable evidence of protein function at end points that are regulated or modulated by the given protein. In contrast, by deleting a particular gene product, one can delineate how certain pathways and processes are altered by the absence of a specific protein. Similar to transgenic mice, specific null mouse models can be used for a number of applications to delineate mechanisms underlying regulation of gene expression, including (but not limited to) (1) identifying roles and target genes of transcription factors, (2) determining novel roles of coeffector proteins in physiology, disease, and cancer, (3) determining biological functions of growth factors and cytokines, (4) elucidating functions of specific enzymes and proteins, (5) "humanizing" mouse models, (6) creating multiple mutant mice to delineate biological pathways, (7) modeling disease states, (8) conditionally deleting a gene product, and (9) identifying target genes using microarrays. As with transgenic mouse models, there are numerous examples of null mouse models that have been developed for these purposes, but a select number have been chosen for illustration purposes. The application of null mouse models has advantages over using chemical inhibitors as outlined in Figure 3-4. For example, a receptor antagonist can be used to effectively inhibit activation of both soluble and membrane-bound receptors, thus preventing downstream events, including changes in transcription mediated by receptor transcription factors, as well as alterations in signal transduction pathways mediated by both soluble and membrane-bound receptors (Figure 3-4A). However, since chemical antagonists can have nonspecific effects on other molecules, null mouse models provide an alternative method for delineating the molecular regulation of these pathways. Similarly, chemical inhibitors that have been used to inhibit an enzyme/protein will ultimately also affect downstream events associated with this reaction. Since chemical inhibitors can also nonspecifically influence other molecules, null mouse models can be designed to overcome this disadvantage (Figure 3-4B).

2.1. Transcription Factor Null Mice

Nuclear hormone receptors regulate transcription of target genes primarily in response to ligand activation. Through this mechanism numerous biological processes are modulated to accommodate a wide variety of stimuli including

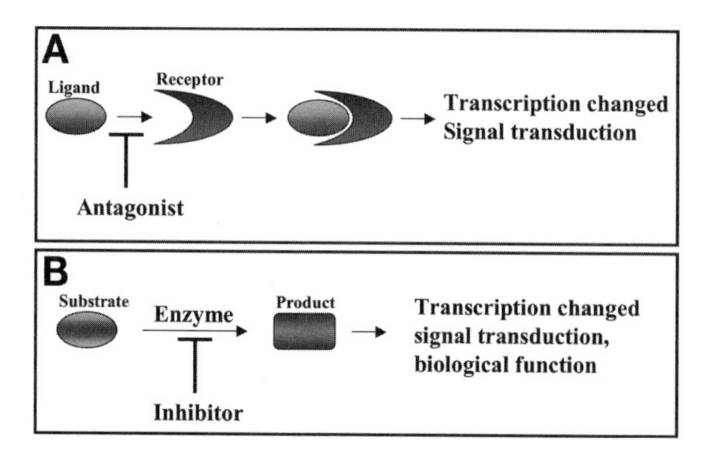

Fig. 3-4. Contrasting classic inhibition studies and null mouse applications. **(A)** Inhibition of ligand binding to either soluble or membrane-bound receptor can be accomplished with an antagonist that will prevent subsequent downstream events, including altered transcription and signal-transduction pathways. The disadvantage of this system is that the antagonist can have a nonspecific effect on other effector molecules. Null mouse models can be developed to eliminate ligand production, receptor presence/expression, and specific downstream molecules in the pathway under study. **(B)** Inhibitors of proteins can be used to prevent an enzyme/protein from functioning. Increased presence of substrate, decreased presence of product, and reduced function imply specificity of inhibition. Similar to receptor antagonists, chemical inhibitors can have nonspecific effects that can be overcome by specifically deleting the enzyme/protein of interest, substrate presence, or downstream effector molecules in a null mouse model.

dietary factors (e.g., fatty acids), endogenous compounds (e.g., bile acids, eicosanoids), drugs, toxicants, and chemical carcinogens. A great deal of information delineating these fundamental processes have been obtained using null mouse models. Several examples of this include the peroxisome proliferator-activated receptor α (PPARα), farnesoid X receptor- (FXR) null-, and aryl hydrocarbon receptor-(AhR) null mice.

By deleting a nuclear receptor, one can determine putative target genes/functions by examining the phenotype of mice with and without ligand activation.

Based on data from transient transactivation assays, site-directed mutagenesis studies, nuclear run-on assays, and analysis of RNA after ligand treatment, a number of putative PPARα target genes were identified [49]. These included peroxisomal acyl-CoA oxidase (ACO), cytochrome P450 4A (CYP4A), and fatty acid binding protein (FABP). However, the former methods were not conclusive for identification of target genes, because secondary events not directly linked to receptor activation could not be controlled for. Using the PPARα-null mouse, definitive evidence that ACO, CYP4A, and FABP represent PPARα target genes was provided, since induction of these mRNAs is absent in ligand-treated null mice. These data show that PPARα is essential in order for these changes in gene expression to occur in response to ligand activation. Ligand activation of PPARα also modulates numerous changes in gene expression that lead to increased transport, release, and oxidation of fatty acids, some of which may be related to the hepatocarcinogenic effects of these ligands. Similar to demonstrating specificity of target genes, the PPARα-null mice have also been used to show that they are refractory to both modulation of lipid metabolism and liver cancer induced by PPARα ligands, demonstrating the absolute requirement of this receptor to mediate these biological processes. It is also of interest to note that since constitutive expression of mitochondrial fatty acid metabolizing enzymes is reduced in PPARα-null mice, this mouse has also been used to suggest the presence of an endogenous natural ligand.

The role of FXR in physiology has only recently begun to be understood, in large part owing to analysis of a null mouse. In the late 1990s, experiments were reported demonstrating that bile acids could bind to FXR and modulate activity of reporter constructs and putative FXR target genes. Analysis of FXR-null mice confirmed a significant role for this receptor in bile-acid homeostasis because they exhibited elevated levels of serum bile acids, cholesterol, and triglycerides, as well as simαhydroxylase and ileal bile-acid binding protein represent FXR target genes, since alterations in expression in response to the bile-acid-rich diet were absent in FXR-null mice [50]. An interaction between FXR and PPARα activity has also been shown using both lines of null mice. Bile acids repress basal expression of PPARα target genes independently of PPARα, as shown by the occurrence of these effects in the PPARα-null mouse. In contrast, bile acids

49. Lee, S.S. et al. Mol. Cell. Biol. 15 (1995) 3012–3022.

Demonstrating a lack of change in gene expression in response to ligand activation is a powerful and definitive in vivo approach to demonstrate that a transcription factor mediates this and other downstream event(s).

50. Sinal, C.J. et al. Cell 102 (2000) 731–744.

inhibit Wy-14,643-induced expression of PPARα target genes, which is independent of FXR, as this effect is observed in FXR-null mice, and is likely attributable to antagonism of PPARα receptor activation by bile acids. Thus, comparisons between the two null mouse lines has helped to delineate potential pathways of interaction between receptor activation of FXR and PPARα both of which have critical roles in the regulation of lipid homeostasis.

Another classic example of a null mouse that has been used to elucidate the specific molecular changes induced by exogenous chemicals is the aryl hydrocarbon receptor (AhR)-null mouse. Similar to observations made for PPARα, a number of data sets suggested that AhR target genes included CYP1A1, CYP1A2, CYP1B1, and the phase II enzymes UDP-glucuronidyl transferases. By treating wild-type and AhR-null mice with the potent AhR ligand tetrachlorodibenzo[*p*]dioxin (TCDD) and showing that mRNAs encoding these target genes are induced in wild-type but absent in AhR-null mice [51], definitive evidence for the requirement of AhR to mediate these changes in gene expression was provided. In addition, the AhR-null mouse has been used to show that the teratogenicity of TCDD requires the AhR, most likely through alterations in tissue-specific gene expression. Interestingly, unexpected roles for the AhR in immune cells and liver metabolism of vitamin A have also been discovered through analysis of AhR-null mice, as significantly depressed accumulation of lymphocytes in spleen and lymph nodes and significantly increased levels of hepatic retinoic acid have been observed in this line. For the latter phenotype, this difference is attributable to reduced catabolism of retinoic acid, likely through reduced expression of an unidentified AhR target gene.

51. Fernandez-Salguero, P. et al. Science 268 (1995) 722–726.

2.2. Transcriptional Coeffector Null Mice

The use of null mice to delineate the roles of transcriptional coeffector molecules represents a rather novel application that cannot be equally compared to other inhibition studies depicted in Figure 3-3, since inhibitors of this type of protein–protein interaction are not well established. Thus, by creating a null mouse for a coeffector, one can determine functional differences in vivo by specific inhibition of coeffector interactions with nuclear transcription factors.

Nuclear receptor coactivators and co-repressors represent a growing list of proteins known to function in the initiation of transcription, or in repression of receptor activation, respectively. Much of the work describing the roles of these effector molecules has been derived from studies using yeast homologs, or from in vitro studies. Null mouse models for a number of these proteins have been described. The steroid receptor coactivator-1 (SRC-1)-null mouse was one of the first coactivator-null mouse models to be described [52]. It is surprising that, despite the fact that this protein is likely required to modulate transcriptional changes that occur after ligand activation of PPARs and other nuclear hormone receptors, the null mice do not exhibit a lethal phenotype. Further, alterations in gene expression in response to the PPARα activator Wy-14,643 is similar between SRC-1-null and wild-type mice. This suggests that there may be some redundancy in nuclear receptor coactivators. Alternatively, it is also possible that transcriptional regulation mediated by other nuclear hormone receptors will be dependent on SRC-1 and could be determined through analysis of SRC-1-null mice treated with various ligands.

52. Qi, C. et al. PNAS 96 (1999) 1585–1590.

In contrast to SRC-1-null mice, peroxisome proliferator-activated receptor binding protein (PBP)-null mice exhibit an embryo lethal-phenotype, demonstrating the essential nature of this coactivator during development [53]. Obviously, in order to completely characterize the function of this coactivator in transcriptional regulation, a conditional mutant model will be required. However, analysis of early embryos (and embryonic fibroblasts

53. Crawford, S.E. et al. J. Biol. Chem. 277 (2002) 3585–3592.

described later in this chapter) has provided additional clues about the functional roles of PBP. While it is clear from the null mouse study, that PBP facilitates transcriptional regulation mediated by PPARγ similarities in embryonic phenotypes between this and the GATA-3-null mice, including abnormalities in cardiac development, suggested a possible interaction between PBP and GATA-3. Indeed, further analysis based on this comparison led to the discovery that, in addition to PPARγ, PBP also interacts with the GATA family of zinc finger transcription factors. This illustrates how careful comparisons between null mouse models can provide further clues into the biological function of various proteins.

Interestingly, analysis of heterozygous coactivator-null mice has also shown that some of these proteins have critical roles in preventing age-associated malignancies. For example, while CREB-binding protein (CBP)-null mice exhibit an embryo-lethal phenotype, mice heterozygous for this allele are characterized by an increased incidence of hematological malignancies [54]. Loss of heterozygosity of the CBP locus is associated with these conversions, and illustrates how heterozygous mutations can also be used to demonstrate specific functions of a protein of interest. It is worth noting that not all heterozygous mice exhibit a phenotype dissimilar from wild-type, although this conclusion should be carefully drawn only after thorough analysis. There are a number of examples of gene deletions where the absence of one allele does lead to a significant phenotype, such as APC*min*-null or p53-null mice, but this is not always true.

54. Kung, A.L. et al. Genes Dev. 14 (2000) 272–277.

The transducer of ErbB2 (tob) protein is thought to be a co-repressor, and null mouse studies support this idea. In contrast to some of the coactivator-null mice, tob-null mice are all viable, suggesting that tob is not required for embryo development [55]. Enhanced cell proliferation and increased expression of cyclin D1 in null-mouse liver both provide

55. Yoshida, Y. et al. Genes Dev. 17 (2003) 1201–1206.

good evidence that tob helps to repress cyclin D1 expression. Consistent with this, tob-null mice exhibit an increased incidence of tumors in liver, lymph nodes, lung, and pancreas, ranging from adenomas to malignant carcinomas, demonstrating that this co-repressor appears to function as a tumor suppressor by preventing changes in gene expression that contribute to tumor formation. The specific mechanisms of these events are not known, but the tob-null mouse will be a useful tool for determining how this co-repressor mediates this tumor suppressive influence.

2.3. Growth Factors and Cytokine-Null Mice

Similar to soluble nuclear hormone receptor-null mice, targeted deletion of membrane-bound receptors that mediate changes in response to growth factors or cytokines can also facilitate determination of functional roles for this class of receptors. For example, epidermal growth factor (EGF) mediates changes in signal transduction pathways leading to changes in gene expression after binding to its receptor, epidermal growth factor receptor (EGFR). Deletion of EGFR results in embryo lethality, but in some cases, neonates do survive shortly after birth [56,57]. These null mice exhibit significant abnormalities in epithelial cells, including those of skin and gastrointestinal tract, supporting a critical role of EGFR in signaling required for proper function and development of epidermal cells. Conditional deletion of this receptor will provide a more valuable tool for examining EGF and EGFR function. Cells derived from null embryos have also been used for similar purposes. The EGFR-null mouse is also an excellent example of how differences in genetic background can lead to differences in phenotype, since significant differences in embryo lethality are observed depending on the genetic background of the mouse.

As illustrated in Figure 3-4, another approach to delineating a protein function is to inhibit ligand

56. Sibilia, M. and Wagner, E.F. Science 269 (1995) 234–238.

57. Threadgill, D.W. et al. Science 269 (1995) 230–234.

activation that can be accomplished by antagonists, or by deleting the production of the ligand via targeted mutations in the gene encoding this gene product. A good example of this is the tumor necrosis factor-α (TNFα)-null mouse. Deleting the gene encoding TNFα results in mice that do not express TNFα, and are resistant to lipopolysaccharide-induced lethality and TNFα-induced toxicity [58]. These and other results using TNFα-null mice demonstrate how preventing ligand presence can allow examination of ligand-specific effects, and provide strong genetic evidence that TNFα is essential in eliciting responses that contribute to host defense against invading pathogens, regulation of inflammation, and generation of adaptive B-cell immune responses. The TNFα-null mouse has also been used to determine the role of this cytokine in modulating downstream events associated with PPARα activation. It had been postulated that the hepatocarcinogenic effects of PPARα activators was attributable in part to increased cell-proliferation signaling mediated by induction of TNFα in the liver of PPARα-ligand treated rodents. To examine this hypothesis, investigators treated TNFα-null (and TNF-α-receptor-null) mice with Wy-14,643 and determined relative differences in markers of cell proliferation between wild-type and TNFα-null mice. Surprisingly, relative cell proliferation in the livers of mice from both genotypes was similar [59], indicating that TNFα signaling is not essential for increased cell proliferative signaling in response to Wy-14,643, while PPARα activation is. This type of application to examine the role of downstream effectors is an excellent approach that can be applied to numerous systems. Another related application is to construct multiple mutant mice, which will be described in Section 2.6.

58. Marino, M.W. et al. PNAS, 94 (1997) 8093–8098.

59. Lawrence, J.W. et al. Carcinogenesis 22 (2001) 381–386.

2.4. Enzyme/Protein-Null Mice

Classic enzyme inhibition studies using chemicals capable of interfering with substrate binding and product formation have shown how specific enzymes metabolize endogenous and exogenous chemicals including kinetic differences between various substrates (Figure 3-3B). When more than one form of an enzyme is present in an organism, or when the chemical inhibitor is not specific for a given enzyme, the results from this approach can be less than convincing. A good example of enzymes with more than one isoform is the cytochrome P450s (CYP), which include a large family of proteins that can metabolize chemicals of similar structure. Thus, null mouse models for CYPs can lead to a better understanding of substrate specificity and subsequent product function. For example, deletion of CYP1B1 significantly reduces bioactivation of dimethylbenzanthracene (DMBA) compared to wild-type mice, which leads to significantly reduced incidence of lymphomas [60]. This is surprising, given that in vitro evidence had suggested that CYP1A1 preferentially bioactivated DMBA. Since CYP1B1-null mice express CYP1A1 at normal levels, this result provides strong in vivo evidence that the CYP1B1 isoform is likely more important in the bioactivation of DMBA and its subsequent carcinogenic effects. Similar to CYP1B1-null mice, CYP2E1-null mice are also resistant to acetaminophen-induced liver toxicity [61], demonstrating that this CYP contributes, at least in part (discussed later), to the toxicity of this compound.

Knockout mice provide in vivo tools to examine the effect of enzyme inhibition, with an advantage over a chemical—gene deletion can be more specific than a chemical.

Another good example of how deleting a xenobiotic metabolizing enzyme can lead to a better understanding of how these enzymes protect against DNA damage is the microsomal epoxide hydrolase (MEH)-null mouse. MEH-null mice do not metabolize DMBA to an active carcinogenic metabolite, and they are also protected against tumor formation induced from this chemical [62]. This shows that MEH metabolism in mammalian organisms leads to the production of carcinogenic metabolites from various substrates and implies that individuals with mutations in this gene could be protected against

60. Buters, J.T. et al. PNAS 96 (1999) 1977–1982.

61. Lee, S.S. et al. J. Biol. Chem. 271 (1996) 12,063–12,067.

62. Miyata, M. et al. J. Biol. Chem. 274 (1999) 23,963–23,968.

carcinogenic chemicals. In addition to its known role of metabolizing xenobiotics, soluble epoxide hydrolase (SEH) is also known to metabolize endogenous substrates. CYPs can metabolize arachidonic acid to epoxyeicosatrienoic acids, which are further metabolized by SEH to form dihydroxyeicosatrienoic acids that in turn regulate blood pressure. Indeed, SEH-null mice have significantly lower levels of renal dihydroxyeicosatrienoic acids, and blood pressure is also significantly lower in male SEH-null mice. This suggests that SEH could be a novel target protein that could be used to regulate blood pressure associated with hypertension.

Although not an enzymatically active protein, metallothionein (MT) is an important protein that regulates zinc, copper, and iron functions. Similar to the example of coeffector-null mice, no specific inhibitors are available to help determine physiological roles of MT, thus a null mouse model provides a unique tool for this purpose. Deletion of MT in mutant mice has shown that this protein is important in protecting against heavy metal toxicity, since MT-null mice are very sensitive to cadmium-induced liver and developmental toxicity [63]. Additionally, there is also evidence from MT-null cells suggesting that MT functions as an antioxidant since these cells exhibit increased measures of oxidative damage compared with wild-type cells. Further, analysis of MT-null mice suggests that, in the absence of MT, expression of other proteins is altered, including that of transketolase, vanin-3, and contrapsin. The precise mechanism for these differences is not known, but could be attributable to indirect regulation resulting from compensatory changes, or to differences in the availability of zinc, copper, or iron.

63. Masters, B.A. et al. PNAS 91 (1994) 584–588.

2.5. "Humanizing" Mice/"Knock-In" Mice

A novel approach—to replace murine genes with the human homolog—has been described and represents an invaluable application of null mouse models. Through this method one can replace a murine gene with the human isoform and study how a human gene product functions. This could be applied when there may be subtle differences between protein function, allowing for evaluation in an in vivo system. This application has been used extensively to introduce mutations found in human diseases into the mouse genome, to determine if these changes led to similar diseases in mice. A mouse whose murine genes are replaced with human homologs or a construct containing subtle mutations is referred to as "knock-in." A similar approach has been described that introduces human transgenes into a null mouse model, but this system has the disadvantage of not being able to control for the number of gene copies inserted into the mouse genome; in addition, the site of integration is random when using a transgenic approach. Using double replacement gene targeting, the mouse α-lactalbumin gene has been successfully replaced with the human isoform [64,65]. Interestingly, mice expressing the human homolog exhibit much greater mRNA and protein abundance, suggesting that there are major determinants of human α-lactalbumin expression close to, or within, the human gene, and that the mouse gene does not exert a negative influence on expression. Mice that display signs of Huntington's disease have been produced by replacing the short CAG repeat of the mouse gene associated with this disease [66], with the length range found to cause Huntington's disease in humans. These knock-in mice exhibit late-onset behavioral and neuroanatomic abnormalities consistent with this disease, in addition to increased glial fibrillary acidic protein in the striatum that is not found in standard transgenic models, illustrating how knock-in mouse models can improve on the former method to overexpress a human gene. Lastly, by introducing a point mutation found in a human cancer patient with immunodeficiency and intracellular accumulation of DNA replication inter-

Creating mice that express human gene products provides an in vivo tool that may model humans very closely.

64. Stacey, A. et al. PNAS 92 (1995) 2835–2839.

65. Stacey A. et al. Mol. Cell. Biol. 14 (1994) 1009–1016.

66. Lin, C.H. et al. Hum. Mol. Genet. 10 (2001) 137–144.

mediates into the mouse DNA ligase gene, mutant mice were generated exhibiting reduced proliferation of cells associated with immune function, increased genome instability, and increased epithelial-cell tumor formation [67]. Thus, by introducing a point mutation found in human cancer into the mouse genome, an invaluable model for studying human cancer was created.

67. Harrison, C. et al. Cancer Res. 62 (2002) 4065–4074.

By introducing a known human mutation or polymorphism into a mouse model, the functional significance of these alterations in DNA structure can be examined in vivo.

The knock-in approach has also been used to evaluate gene expression patterns during development and in adult mice. However, this usually involves inserting a reporter construct such as a LacZ or GFP, under the control of the promoter for the gene of interest. Thus, this method also effectively deletes expression by disrupting the coding sequence of the gene target. Considerable care must be taken when generating the targeting vector to ensure that the reporter construct will be regulated by the endogenous promoter and that the cDNA encoding the reporter will be efficiently expressed. One example of this application is an AhR-null mouse line. By inserting a LacZ reporter into an exon encoding the AhR, the AhR allele was disrupted [68]. Subsequent analysis of tissue sections from developing mice using β-galactosidase staining reveals patterns of AhR expression that can be used to identify specific tissue and time points for further examination of the biological roles of this receptor. Treating mice with TCDD causes cleft palate, and examination of these AhR-null/reporter mice confirms that expression of the AhR is temporally associated with fusion of the palate. Similar applications of mice containing this type of reporter system allows for identification of target tissues and the ability to correlate protein expression with a specific function.

68. Mimura, J. et al. Genes Cells 2 (1997) 645–654.

2.6. Multiple Null Mice

The technology of null mouse production, coupled with the large number of mouse models generated in the last 10 yr, has helped scientists construct complicated models to carefully distinguish distinct and overlapping roles for related proteins that participate in the same pathways. Multiple mutant mice containing targeted deletions for more than one gene product are becoming increasingly common and provide more powerful genetic models. There is more than one way to generate multiple mutant mice and a few of these models will be described.

Multiple mutant mice can be used to elucidate complex gene regulation (e.g., RXR), or complex metabolic pathways (e.g., CYP).

One of the more obvious ways to generate a mouse with more than one mutation is to cross two null mice and screen the offspring until homozygous mutations for both alleles are achieved. For example, double mutant mice that are null for both CYP1A2 and CYP2E1 have been produced by this approach and used to determine if acetaminophen toxicity can be ameliorated since both CYP isoforms function in the metabolism of this analgesic [69]. Production of the electrophilic *N*-acetyl-*p*-benzoquinone imine (NAPQI) is significantly reduced in double CYP1A2/CYP2E1 mice compared with controls, and liver damage induced by a very high dose of acetaminophen is not found in the double mutant mice. Interestingly, the double mutant mice also tolerate a higher dose of acetaminophen than single CYP2E1-null mice, suggesting that both enzymes are required to effectively bioactivate acetaminophen and that metabolism by CYP1A1 also leads to some toxicity. Comparing metabolism between single and double CYP-mutant mice is a good tool to dissect out the metabolic fate of various substrates.

69. Zaher, H. et al. Toxicol. Appl. Pharmacol. 152 (1998) 193–199.

TNFα and lymphotoxins are structurally related cytokines that are all encoded within an approx 11-kb fragment of the mouse genome. Single mutant mice for all of these gene products have been made and used to delineate unique yet related functions for each protein, including roles in the regulation of immune function. To determine the possibility of a common redundancy between these gene products, and to examine possible interactions in receptor activation, a mutant mouse was created that effectively deleted TNFα, lymphotoxin (LT)α, and LTβ [70]. This approach utilized Cre-mediated deletion of all three gene products via flanking *LoxP* sites that were inserted into the targeting vector. Interestingly, these triple mutant mice still exhibited specific functional roles for each cytokine, as evidenced by unique activation of target genes that were absent in single mutant mice, and completely absent in the triple mutants. This type of analysis clearly shows that there is no redundancy in gene product function and illustrates how multiple mutations can be introduced for a number of important receptor ligands when gene products are conveniently located fairly close together within the genome.

70. Kuprash D.V. et al. Mol. Cell. Biol. 22 (2002) 8626–8634.

The phenotype of homozygous RXRα mutant mice is embryo lethal, with ocular malformations being observed consistent with the established role of this receptor during eye development. In contrast, the RARγ-null mouse exhibits no gross developmental defects. Crossing RXRα mutant mice with RARγ mutant mice revealed a synergistic phenotype as RXRα/RARγ double mutants exhibit several malformations not seen in single mutants [71]. Results from this approach suggest that RXRα/RARγ heterodimers modulate retinoid signaling during development and are dependent on functional roles for both receptors. Similar approaches have been used to generate other double mutant mice for RAR/RXR and RAR/RAR isoforms and have demonstrated unique interactions for the heterodimeric receptor partners that were not apparent from analysis of single mutant mice. By creating multiple mutant mice, related roles for both nuclear receptors were elucidated and this illustrates the use multiple mutations to determine how these receptors mediate important developmental signaling.

71. Mark, M. et al. Proc. Nutr. Soc. 58 (1999) 609–613.

2.7. Null Mouse Models of Disease States

Many human diseases have been linked to mutations in specific genes; thus it is not surprising that null mouse models have given supportive evidence that these mutations are functionally related to human disease. Additionally, when a disease can be modeled in a mouse, scientists can carefully delineate pathways that lead to disease progression, and in some cases they can test approaches that may prevent these diseases. Some excellent examples of targeted deletions in mice that have modeled human diseases include the apolipoprotein E (ApoE)-null mouse [72], which is a good model for studying events associated with atherosclerosis; hepatocyte nuclear factor 1α (HNF1α)-null mice, which exhibit elevated serum glucose for studying diabetes [73]; adenomatous polyposis coli (APC)-null mouse that develops intestinal tumors for evaluating gastrointestinal tract cancers [74]; and leptin-null mice, which have been used to study obesity and diabetes. In some cases, deletion of one gene can lead to the formation of a disease state, while deletion of a different gene can prevent disease occurrence. For example, targeted deletion of Notch 1 leads to the spontaneous development of skin tumors [75]. In contrast, deletion of Rac activator Tiam1 results in mice that are resistant to chemically induced skin tumors [76]. It is also worth noting that null mouse models that mimic human disease are not always discovered, due to "expected" phenotypes (e.g., delete an apolipoprotein and observe differences in serum lipid metabolism similar to hyperlipidemias). In many cases, target genes are deleted and unexpected phenotypes are found to resemble human disease. For example, targeted deletion of adenosine deaminase results in a phenotype that resembles asthma and was discovered by careful evaluation of neonates that died shortly after birth [77]. While transgenic mouse models can also be used for this application, null mice may represent

72. Tordjman, K. et al. J. Clin. Invest. 107 (2001) 1025–1034.

73. Lee, Y.H. et al. Mol. Cell. Biol. 18 (1998) 3059–3068.

74. Su, L-K. et al. Science 256 (1992) 668–670.

75. Nicolas, M. et al. Nat. Genet. 33 (2003) 416–421.

76. Malliri, A. et al. Nature 417 (2002) 867–871.

77. Blackburn, M. R. et al. J. Exp. Med. 192 (2000) 159–170.

a more suitable model given that disease etiology is attributable to mutations (deletions) in specific gene products. Molecular analysis of these mouse models can lead to identification of targets for therapeutic intervention and/or prevention.

2.8. Conditional Deletions

Targeted deletion of some gene products results in embryo lethality, preventing further analysis of protein function beyond that allowed for prior to embryo death. Applications utilizing cells derived from embryos will be described later and serve as an alternative application to mice exhibiting lethal phenotypes. To overcome the problem of embryo lethality associated with deletion of target genes with significant roles during development, conditional gene targeting strategies have been developed as described in Part I. Conditional gene targeting can be applied to the study of all the events previously described, including analysis of receptors, ligands, enzymes, protein–protein interactions, and signal transduction pathways.

Tissue-specific, inducible, conditional null mouse models are state of the art for studying gene regulation in vivo.

For example, using a tissue-specific promoter Cre (or *Flp*)-expressing mouse crossed with a mouse containing homozygous floxed alleles of the gene target of interest, deletion of the gene product will occur specifically in the tissue of interest. This allows for analysis of the gene product in the tissue of interest but is limited because the timing of deletion cannot be controlled, and the efficiency of gene deletion mediated by Cre is limited by the strength of the promoter driving its expression. For example, using mutant mice expressing a muscle-specific Cre mouse (muscle creatine kinase promoter) and containing floxed insulin receptor alleles, muscle-specific deletion of the insulin receptor was accomplished [78]. In addition to approx 95% reduction in insulin-receptor expression, downstream events associated with insulin-receptor activation were also significantly reduced, including glucose uptake. While this system provides a useful

Flanking LoxP *sites around a particular gene locus are typically referred to as "floxed" alleles.*

78. Bruning, J.C. et al. Mol. Cell 2 (1998) 559–569.

model for studying the role of insulin receptor in muscle, it is limited in the sense that the timing of gene product deletion is regulated by, or dependent on, the tissue-specific promoter.

The ability to selectively choose the exact timing of gene deletion has been facilitated with sophisticated Cre mice that respond to various inducing agents. For example, Cre expression driven by the alpha/beta interferon-inducible promoter (MX-Cre) can be induced in response to interferon or pIpC. By crossing MX-Cre mice with floxed PPARγ mice and treating with pIpC, conditional deletion of PPARγ was accomplished in macrophages, and PPARγ target gene expression can be significantly reduced, demonstrating functional roles for this receptor in this tissue [79]. Thus, this approach allows for deleting a gene product at specific time points. However, MX-Cre expression is also induced in other tissues, which can limit the application of this system.

One of the most sophisticated approaches to conditionally delete a gene product involves mice that have inducible and tissue-specific Cre expression systems in place. For example, using a transgene that expresses Cre solely in adipose tissue and is only inducible in response to tamoxifen, Cre expression can be induced in adipose at specific time points to delete a target gene. This mouse line was used and crossed with mice containing floxed RXRα alleles, and selective deletion of RXRα in adipose tissue was accomplished by injecting mice with tamoxifen [80]. Using this approach, RXRα expression can be effectively diminished in adipocytes, and target genes known to be regulated by RXRα and one of its heterodimeric binding partners (PPARγ) are essentially not expressed. In addition to these changes in gene expression, these mice do not develop obesity in response to high fat diets and exhibit significantly impaired adipocyte differentiation. Thus, using tissue-specific and inducible Cre mice coupled with conditional floxed alleles for

79. Akiyama, T.E. et al. Mol. Cell. Biol. 22 (2002) 2607–2619.

80. Imai, T. et al. PNAS 98 (2001) 224–228.

target genes, one can selectively delete genes in tissues of interest, with the additional obvious advantage of allowing for analysis of gene deletions without the embryo lethality attendant on conventional methods.

2.9. Identification of Regulatory Pathways Using Microarrays

Null mouse models are ideally suited for applications utilizing microarray technology (described in Chapter 2, Section 2.8.) to tentatively identify putative target genes or genes that are dependent on the protein of interest. This section will briefly summarize how microarray analysis of RNA from null mice can be used for this purpose. By examining differential gene regulation between wild-type and null mouse tissue, one can putatively determine logical and novel roles for ligands, receptors, enzymes, or other functional proteins that have been deleted. However, simply determining increased or decreased expression of a particular gene product in wild-type tissue as compared with null mouse tissue does not demonstrate the identity of a target gene. Confirmation by Northern blotting, Real-time PCR, RT-PCR, or RPAs is necessary to confirm that changes detected in the microarray are reproducible, and subsequent analysis is also required to clearly link these functional changes in gene expression to a particular protein (e.g., identifying functional response elements within the promoter of a novel target gene for a transcription factor). Examples of this are provided in Part I, Chapters 1–4. In addition to allowing for identification of target genes and regulatory pathways that can be constitutively regulated, microarray analysis using RNA samples from wild-type and null mice treated with ligands or chemicals can also be applied to elucidating inducible target genes and regulatory pathways through similar analysis.

4

Isolation of Cells and Cell Lines From Transgenic and Knockout Animals

1. Standard Production

There are countless examples how cells derived from transgenic and null mouse models can be applied to study gene regulation.

Cells and cell lines can be created from transgenic and null mice providing a way to differentiate between direct effects and those that are dependent on secondary changes in neighboring cells. Isolation of various cell types can be accomplished using a variety of methods that can be obtained from standard textbooks. Additionally, cell lines can be derived from cells obtained for this purpose. The application of cells and cell lines from transgenic and null mice is particularly well suited when embryo lethality occurs, assuming midgestation-stage embryos can be obtained for collection of the cell type of interest. A number of cells have been used for these purposes, but for illustration the remaining portion of this section will describe the use of embryonic stem (ES) cells, embryonic fibroblasts (EF), and hepatocytes.

1.1. Embryonic Stem Cells

ES cells are pluripotent cells that maintain their ability to form many different cell types. Differentiation into specific cell types can be facilitated using various culture conditions. Homozygous null ES cells can be obtained by performing electroporation of ES cells containing a heterozygous mutation of interest, introduced by previous homologous recombination, and screening for homozygous mutant ES cells. Homozygous null ES cells can then be cultured and forced to differentiate into

Use of Transgenic and Knockout Mice to Study Gene Regulation by J. M. Peters
From: *Regulation of Gene Expression*
By: G. H. Perdew et al. © Humana Press Inc., Totowa, NJ

a specific cell type to allow for studying events related to the protein of interest. For example, homozygous PPARβ-null ES cells have been differentiated into macrophages, and the role of this receptor in the regulation of lipoprotein metabolism in these cells is currently under investigation [81]. This and many other applications can be facilitated using ES cells containing homozygous mutations.

81. Chawla, A. et al. PNAS 100 (2003) 1268–1273.

1.2. Embryonic Fibroblast Cells

Murine EF cells have also been used in ways similar to ES cells. Often when an embryo-lethal phenotype is encountered, EF cells can be obtained from the embryos, allowing for further analysis not available with adult cells or tissues. Numerous examples of EF cell applications are available, and only a few will be highlighted here. Studies with EF cells from null mice have demonstrated that PPARγ-dependent transcription is reduced in the absence of PBP-null EF cells [82], and to demonstrate reduced metabolism of carcinogenic substrates by microsomal epoxide hydrolase (MEH). Further, EF cells have provided evidence for a role of AhR in regulation of cell cycle since MEFs proliferate much faster compared to MEFs from wild-type mice [83].

82. Qi, C. et al. J. Biol. Chem. 278 (2003) 25,281–25,284.

83. Tohkin, M. et al. Mol. Pharmacol. 58 (2000) 845–851.

1.3. Hepatocytes

Hepatocytes are one of the main cell types in the liver, and the culture of these cells is widely applied since the liver is important for so many different tissues. However, while primary culture of hepatocytes can be applied to cells from null mice, proliferation of these cells is limited in vitro. Thus, construction of transformed cell lines can facilitate in vitro analysis. For example, viral transformation of PPAR-null mouse hepatocytes results in a cell line that responds to the mitogenic stimulus of peroxisome proliferators [84], an event that does not occur in primary cultures of mouse hepatocytes. Similar applications from other null mice will likely provide in vitro models that can be effectively manipulated to determine direct effects on cells from various conditions.

84. Tien, E.S. et al. Cancer Res. 63 (2003) 5767–5780.

Study Questions

1. What led to the ability to insert into mice foreign DNA that can be expressed?
2. What led to the ability to perform homologous recombination in ES cells to construct null mouse models?
3. What are the differences between mono- and bi-transgenic mice?
4. How do inducible transgenic systems work?
5. How can inducible transgenic mice be used to conditionally delete genes?
6. Contrast the differences between transgenic and null mice.
7. Contrast the similarities in applications of transgenic and null mice.
8. How can double or triple mutant mice be used to delineate molecular pathways?
9. Why/how are both transgenic and null mice useful to study molecular regulation underlying human diseases?
10. What are the advantages/disadvantages of using null mice versus classic enzyme inhibition to study gene regulation?
11. How could one effectively delete two target genes of interest using conditional methods? Could this be accomplished at different timepoints? In different tissues?

References

1. Jaenisch R and Mintz B. Simian virus 40 DNA sequences in DNA of healthy adult mice derived from preimplantation blastocysts injected with viral DNA. Proc Natl Acad Sci USA 71(4): 1250–1254, 1974.
2. Jaenisch R. Germ line integration and Mendelian transmission of the exogenous Moloney leukemia virus. Proc Natl Acad Sci USA 73(4): 1260–1264, 1976.
3. Gordon JW, Scangos GA, Plotkin DJ, Barbosa JA and Ruddle FH. Genetic transformation of mouse embryos by microinjection of purified DNA. Proc Natl Acad Sci USA 77(12): 7380–7384, 1980.
4. Costantini F and Lacy E. Introduction of a rabbit beta-globin gene into the mouse germ line. Nature 294(5836): 92–94, 1981.
5. Brinster RL, Chen HY, Trumbauer M, Senear AW, Warren R and Palmiter RD. Somatic expression of herpes thymidine kinase in mice following injection of a fusion gene into eggs. Cell 27(1 Pt 2): 223–231, 1981.
6. Gordon JW and Ruddle FH. Integration and stable germ line transmission of genes injected into mouse pronuclei. Science 214(4526): 1244–1246, 1981.
7. Harbers K, Jahner D and Jaenisch R. Microinjection of cloned retroviral genomes into mouse zygotes: integration and expression in the animal. Nature 293(5833): 540–542, 1981.
8. Evans MJ and Kaufman MH. Establishment in culture of pluripotential cells from mouse embryos. Nature 292(5819): 154–156, 1981.
9. Martin GR. Isolation of a pluripotent cell line from early mouse embryos cultured in medium conditioned by teratocarcinoma stem cells. Proc Natl Acad Sci USA 78(12): 7634–7638, 1981.
10. Robertson EJ. Using embryonic stem cells to introduce mutations into the mouse germ line. Biol Reprod 44(2): 238–245, 1991.
11. Nagy A, Gertsenstein M, Vintersten K and Behringer R. Manipulating the Mouse Embryo: A Laboratory Manual. Cold Spring Harbor Laboratory, Cold Spring Harbor, NY, 2003.
12. Pinkert CA. Transgenic Animal Technology: A Laboratory Handbook. Academic Press, San Diego, CA, 2002.
13. DeMayo FJ and Tsai S. Targeted gene regulation and gene ablation. Trends Endocrinol Metab 12(8): 348–353, 2001.
14. Lewandoski M. Conditional control of gene expression in the mouse. Nat Rev Genet 2(10): 743–755, 2001.
15. Woitach JT, Conner EA, Wirth PJ and Thorgeirsson SS. Aberrant expression and regulation of hepatic epidermal growth factor receptor in a c-myc transgenic mouse model. J Cell Biochem 64(4): 651–660, 1997.
16. Murakami H, Sanderson ND, Nagy P, Marino PA, Merlino G and Thorgeirsson SS. Transgenic mouse model for synergistic effects of nuclear oncogenes and growth factors in tumorigenesis: interaction of c-myc and transforming growth factor alpha in hepatic oncogenesis. Cancer Res 53(8): 1719–1723, 1993.

17. Sanders S and Thorgeirsson SS. Promotion of hepatocarcinogenesis by phenobarbital in c-myc/TGF-alpha transgenic mice. Mol Carcinog 28(3): 168–173, 2000.
18. Riu E, Ferre T, Mas A, Hidalgo A, Franckhauser S and Bosch F. Overexpression of c-myc in diabetic mice restores altered expression of the transcription factor genes that regulate liver metabolism. Biochem J 368(Pt 3): 931–937, 2002.
19. Gardiner EM, Baldock PA, Thomas GP, et al. Increased formation and decreased resorption of bone in mice with elevated vitamin D receptor in mature cells of the osteoblastic lineage. FASEB J 14(13): 1908–1916, 2000.
20. Davis VL, Couse JF, Goulding EH, Power SG, Eddy EM and Korach KS. Aberrant reproductive phenotypes evident in transgenic mice expressing the wild-type mouse estrogen receptor. Endocrinology 135(1): 379–386, 1994.
21. Couse JF, Davis VL, Hanson RB, et al. Accelerated onset of uterine tumors in transgenic mice with aberrant expression of the estrogen receptor after neonatal exposure to diethylstilbestrol. Mol Carcinog 19(4): 236–242, 1997.
22. Lin J, Wu H, Tarr PT, et al. Transcriptional co-activator PGC-1 alpha drives the formation of slow-twitch muscle fibres. Nature 418(6899): 797–801, 2002.
23. Yu CT, Feng MH, Shih HM and Lai MZ. Increased p300 expression inhibits glucocorticoid receptor-T-cell receptor antagonism but does not affect thymocyte positive selection. Mol Cell Biol 22(13): 4556–4566, 2002.
24. Sueoka N, Sueoka E, Miyazaki Y, et al. Molecular pathogenesis of interstitial pneumonitis with TNF-alpha transgenic mice. Cytokine 10(2): 124–131, 1998.
25. Kimura S, Umeno M, Skoda RC, Meyer UA and Gonzalez FJ. The human debrisoquine 4-hydroxylase (CYP2D) locus: sequence and identification of the polymorphic CYP2D6 gene, a related gene, and a pseudogene. Am J Hum Genet 45(6): 889–904, 1989.
26. Li Y, Yokoi T, Kitamura R, et al. Establishment of transgenic mice carrying human fetus-specific CYP3A7. Arch Biochem Biophys 329(2): 235–240, 1996.
27. Granvil CP, Yu AM, Elizondo G, et al. Expression of the human CYP3A4 gene in the small intestine of transgenic mice: in vitro metabolism and pharmacokinetics of midazolam. Drug Metab Disp. 31(5): 548–558, 2003.
28. Imaoka S, Hayashi K, Hiroi T, Yabusaki Y, Kamataki T and Funae Y. A transgenic mouse expressing human CYP4B1 in the liver. Biochem Biophys Res Commun 284(3): 757–762, 2001.
29. Hwang DY, Chae KR, Shin DH, et al. Xenobiotic response in humanized double transgenic mice expressing tetracycline-controlled transactivator and human CYP1B1. Arch Biochem Biophys 395(1): 32–40, 2001.
30. Hinshelwood MM and Mendelson CR. Tissue-specific expression of the human CYP19 (aromatase) gene in ovary and adipose tissue of transgenic mice. J Steroid Biochem Mol Biol 79(1-5): 193–201, 2001.
31. Xie W, Barwick JL, Downes M, et al. Humanized xenobiotic response in mice expressing nuclear receptor SXR. Nature 406(6794): 435–439, 2000.
32. Greenberg NM, DeMayo F, Finegold MJ, et al. Prostate cancer in a transgenic mouse. Proc Natl Acad Sci USA 92(8): 3439–3443, 1995.

33. Guy CT, Webster MA, Schaller M, Parsons TJ, Cardiff RD and Muller WJ. Expression of the neu protooncogene in the mammary epithelium of transgenic mice induces metastatic disease. Proc Natl Acad Sci USA 89(22): 10,578–10,582, 1992.
34. Moechars D, Dewachter I, Lorent K, et al. Early phenotypic changes in transgenic mice that overexpress different mutants of amyloid precursor protein in brain. J Biol Chem 274(10): 6483–6492, 1999.
35. Finck BN, Lehman JJ, Leone TC, et al. The cardiac phenotype induced by PPARalpha overexpression mimics that caused by diabetes mellitus. J Clin Invest 109(1): 121–130, 2002.
36. Walter CA, Zhou ZQ, Manguino D, et al. Health span and life span in transgenic mice with modulated DNA repair. Ann N Y Acad Sci 928: 132–140, 2001.
37. Carvajal JJ, Cox D, Summerbell D and Rigby PW. A BAC transgenic analysis of the Mrf4/Myf5 locus reveals interdigitated elements that control activation and maintenance of gene expression during muscle development. Development 128(10): 1857–1868, 2001.
38. Rincon M and Flavell RA. Regulation of AP-1 and NFAT transcription factors during thymic selection of T cells. Mol Cell Biol 16(3): 1074–1084, 1996.
39. Millet I, Phillips RJ, Sherwin RS, et al. Inhibition of NF-kappaB activity and enhancement of apoptosis by the neuropeptide calcitonin gene-related peptide. J Biol Chem 275(20): 15,114–15,121, 2000.
40. Swan KA, Alberola-Ila J, Gross JA, et al. Involvement of p21ras distinguishes positive and negative selection in thymocytes. EMBO J 14(2): 276–285, 1995.
41. Fentzke RC, Korcarz CE, Lang RM, Lin H and Leiden JM. Dilated cardiomyopathy in transgenic mice expressing a dominant-negative CREB transcription factor in the heart. J Clin Invest 101(11): 2415–2426, 1998.
42. Eckhart AD, Fentzke RC, Lepore J, et al. Inhibition of betaARK1 restores impaired biochemical beta-adrenergic receptor responsiveness but does not rescue CREB(A133) induced cardiomyopathy. J Mol Cell Cardiol 34(6): 669–677, 2002.
43. Pepin MC, Pothier F and Barden N. Impaired type II glucocorticoid-receptor function in mice bearing antisense RNA transgene. Nature 355(6362): 725–728, 1992.
44. Beggah AT, Escoubet B, Puttini S, et al. Reversible cardiac fibrosis and heart failure induced by conditional expression of an antisense mRNA of the mineralocorticoid receptor in cardiomyocytes. Proc Natl Acad Sci USA 99(10): 7160–7165, 2002.
45. Lottmann H, Vanselow J, Hessabi B and Walther R. The Tet-On system in transgenic mice: inhibition of the mouse pdx-1 gene activity by antisense RNA expression in pancreatic beta-cells. J Mol Med 79(5-6): 321–328, 2001.
46. Hasuwa H, Kaseda K, Einarsdottir T and Okabe M. Small interfering RNA and gene silencing in transgenic mice and rats. FEBS Lett 532(1-2): 227–230, 2002.
47. Joyner AL. Gene Targeting: A Practical Approach. Oxford University Press, Oxford; New York, 2000.
48. Copeland NG, Jenkins NA and Court DL. Recombineering: a powerful new tool for mouse functional genomics. Nat Rev Genet 2(10): 769–779, 2001.

49. Lee SS, Pineau T, Drago J, et al. Targeted disruption of the alpha isoform of the peroxisome proliferator-activated receptor gene in mice results in abolishment of the pleiotropic effects of peroxisome proliferators. Mol Cell Biol 15(6): 3012–3022, 1995.
50. Sinal CJ, Tohkin M, Miyata M, Ward JM, Lambert G and Gonzalez FJ. Targeted disruption of the nuclear receptor FXR/BAR impairs bile acid and lipid homeostasis. Cell 102(6): 731–744, 2000.
51. Fernandez-Salguero P, Pineau T, Hilbert DM, et al. Immune system impairment and hepatic fibrosis in mice lacking the dioxin-binding Ah receptor. Science 268(5211): 722–726, 1995.
52. Qi C, Zhu Y, Pan J, Yeldandi AV, et al. Mouse steroid receptor coactivator-1 is not essential for peroxisome proliferator-activated receptor alpha-regulated gene expression. Proc Natl Acad Sci USA 96(4): 1585–1590, 1999.
53. Crawford SE, Qi C, Misra P, et al. Defects of the heart, eye, and megakaryocytes in peroxisome proliferator activator receptor-binding protein (PBP) null embryos implicate GATA family of transcription factors. J Biol Chem 277(5): 3585–3592, 2002.
54. Kung AL, Rebel VI, Bronson RT, et al. Gene dose-dependent control of hematopoiesis and hematologic tumor suppression by CBP. Genes Dev 14(3): 272–277, 2000.
55. Yoshida Y, Nakamura T, Komoda M, et al. Mice lacking a transcriptional corepressor Tob are predisposed to cancer. Genes Dev 17(10): 1201–1206, 2003.
56. Sibilia M and Wagner EF. Strain-dependent epithelial defects in mice lacking the EGF receptor. Science 269(5221): 234–238, 1995.
57. Threadgill DW, Dlugosz AA, Hansen LA, et al. Targeted disruption of mouse EGF receptor: effect of genetic background on mutant phenotype. Science 269(5221): 230–234, 1995.
58. Marino MW, Dunn A, Grail D, et al. Characterization of tumor necrosis factor-deficient mice. Proc Natl Acad Sci USA 94(15): 8093–8098, 1997.
59. Lawrence JW, Wollenberg GK and DeLuca JG. Tumor necrosis factor alpha is not required for WY14,643-induced cell proliferation. Carcinogenesis 22(3): 381–386, 2001.
60. Buters JT, Sakai S, Richter T, Pineau T, Alexander DL, Savas U, Doehmer J, Ward JM, Jefcoate CR and Gonzalez FJ. Cytochrome P450 CYP1B1 determines susceptibility to 7, 12-dimethylbenz[a]anthracene-induced lymphomas. Proc Natl Acad Sci USA 96(5): 1977–1982, 1999.
61. Lee SS, Buters JT, Pineau T, Fernandez-Salguero P and Gonzalez FJ, Role of CYP2E1 in the hepatotoxicity of acetaminophen. J Biol Chem 271(20): 12,063–12,067, 1996.
62. Miyata M, Kudo G, Lee YH, et al. Targeted disruption of the microsomal epoxide hydrolase gene. Microsomal epoxide hydrolase is required for the carcinogenic activity of 7,12-dimethylbenz[a]anthracene. J Biol Chem 274(34): 23,963–23,968, 1999.
63. Masters BA, Kelly EJ, Quaife CJ, Brinster RL and Palmiter RD. Targeted disruption of metallothionein I and II genes increases sensitivity to cadmium. Proc Natl Acad Sci USA 91(2): 584–588, 1994.

64. Stacey A, Schnieke A, Kerr M, et al. Lactation is disrupted by alpha-lactalbumin deficiency and can be restored by human alpha-lactalbumin gene replacement in mice. Proc Natl Acad Sci USA 92(7): 2835–2839, 1995.
65. Stacey A, Schnieke A, McWhir J, Cooper J, Colman A and Melton DW. Use of double-replacement gene targeting to replace the murine alpha-lactalbumin gene with its human counterpart in embryonic stem cells and mice. Mol Cell Biol 14(2): 1009–1016, 1994.
66. Lin CH, Tallaksen-Greene S, Chien WM, et al. Neurological abnormalities in a knock-in mouse model of Huntington's disease. Hum Mol Genet 10(2): 137–144, 2001.
67. Harrison C, Ketchen AM, Redhead NJ, O'Sullivan MJ and Melton DW. Replication failure, genome instability, and increased cancer susceptibility in mice with a point mutation in the DNA ligase I gene. Cancer Res 62(14): 4065–4074, 2002.
68. Mimura J, Yamashita K, Nakamura K, et al. Loss of teratogenic response to 2,3,7,8-tetrachlorodibenzo-*p*-dioxin (TCDD) in mice lacking the Ah (dioxin) receptor. Genes Cells 2(10): 645–654, 1997.
69. Zaher H, Buters JT, Ward JM, et al. Protection against acetaminophen toxicity in CYP1A2 and CYP2E1 double-null mice. Toxicol Appl Pharmacol 152(1): 193–199, 1998.
70. Kuprash DV, Alimzhanov MB, Tumanov AV, et al. Redundancy in tumor necrosis factor (TNF) and lymphotoxin (LT) signaling in vivo: mice with inactivation of the entire TNF/LT locus versus single-knockout mice. Mol Cell Biol 22(24): 8626–8634, 2002.
71. Mark M, Ghyselinck NB, Wendling O, et al. A genetic dissection of the retinoid signalling pathway in the mouse. Proc Nutr Soc 58(3): 609–613, 1999.
72. Tordjman K, Bernal-Mizrachi C, Zemany L, et al. PPARalpha deficiency reduces insulin resistance and atherosclerosis in apoE-null mice. J Clin Invest 107(8): 1025–1034, 2001.
73. Lee YH, Sauer B and Gonzalez FJ. Laron dwarfism and non-insulin-dependent diabetes mellitus in the Hnf-1alpha knockout mouse. Mol Cell Biol 18(5): 3059–3068, 1998.
74. Su LK, Kinzler KW, Vogelstein B, et al. Multiple intestinal neoplasia caused by a mutation in the murine homolog of the APC gene. Science 256(5057): 668–670, 1992.
75. Nicolas M, Wolfer A, Raj K, et al. Notch1 functions as a tumor suppressor in mouse skin. Nat Genet 33(3): 416–421, 2003.
76. Malliri A, van der Kammen RA, Clark K, van der Valk M, Michiels F and Collard JG. Mice deficient in the Rac activator Tiam1 are resistant to Ras-induced skin tumours. Nature 417(6891): 867–871, 2002.
77. Blackburn MR, Volmer JB, Thrasher JL, et al. Metabolic consequences of adenosine deaminase deficiency in mice are associated with defects in alveogenesis, pulmonary inflammation, and airway obstruction. J Exp Med 192(2): 159–170, 2000.
78. Bruning JC, Michael MD, Winnay JN, et al. A muscle-specific insulin receptor knockout exhibits features of the metabolic syndrome of NIDDM without altering glucose tolerance. Mol Cell 2(5): 559–569, 1998.

79. Akiyama TE, Sakai S, Lambert G, et al. Conditional Disruption of the Peroxisome Proliferator-Activated Receptor gamma gene in mice results in lowered expression of ABCA1, ABCG1, and apoE in macrophages and reduced cholesterol efflux. Mol Cell Biol 22(8): 2607–2619, 2002.
80. Imai T, Jiang M, Chambon P and Metzger D. Impaired adipogenesis and lipolysis in the mouse upon selective ablation of the retinoid X receptor alpha mediated by a tamoxifen-inducible chimeric Cre recombinase (Cre-ERT2) in adipocytes. Proc Natl Acad Sci USA 98(1): 224–228, 2001.
81. Chawla A, Lee CH, Barak Y, et al. PPARδ is a very low-density lipoprotein sensor in macrophages. Proc Natl Acad Sci USA 100(3): 1268–1273, 2003.
82. Qi C, Surapureddi S, Zhu YJ, et al. Transcriptional coactivator PRIP, the peroxisome proliferator-activated receptor gamma (PPARγ)-interacting protein, is required for PPARγ-mediated adipogenesis. J Biol Chem 278(28): 25,281–25,284, 2003.
83. Tohkin M, Fukuhara M, Elizondo G, Tomita S and Gonzalez FJ. Aryl hydrocarbon receptor is required for p300-mediated induction of DNA synthesis by adenovirus E1A. Mol Pharmacol 58(4): 845–851, 2000.
84. Tien ES, Gray JP, Peters JM and Vanden Heuvel JP. Comprehensive gene expression analysis of peroxisome proliferator-treated immortalized hepatocytes: identification of peroxisome proliferator-activated receptor alpha-dependent growth regulatory genes. Cancer Res 63(18): 5767–5780, 2003.

Index

K

L

M

N

O

P

T

U

V

W

X

Y